Geometry

Dr. Noah

International Books Publishing House

www.al-Qarni.com

The universe is pure geometry - basically, a beautiful shape twisting around and dancing over space-time.

What Is Geometry?

Geometry is the most elementary of the sciences that enable student to make predictions (based on observation) about our physical world.

It is a branch of mathematics that is concerned with the properties of configurations of geometric objects - points, lines, and circles being the most basic of these.

The power of mathematics is often to change one thing into another, to change geometry into language.

* Plane Geometry

is about flat shapes such as lines, circles and triangles ... shapes that can be drawn on a piece of paper.

* Solid Geometry

is about three dimensional objects like cubes, prisms, cylinders and spheres

Einstein's gravitational theory, which is said to be the greatest single achievement of theoretical physics, resulted in beautiful relations connecting gravitational phenomena with the geometry of space; this was an exciting idea.

Point, Line, Plane and Solid

* A Point has no dimensions, only position
* A Line is one-dimensional
* A Plane is two dimensional (2D)
* A Solid is three-dimensional (3D)

Why?

We do Geometry to discover and find patterns, areas, volumes, lengths and angles, and to better understand the world around us.

You can't criticize geometry. It's never wrong.

What physics tells us is that everything comes down to geometry and the interactions of elementary particles. And things can happen only if these interactions are perfectly balanced.

Contents

Introduction

Geometry
Points, Lines & Planes

Item	Illustration	Notation	Definition
Point	•	A	A location in space.
Segment	——	$\overline{AB}$	A straight path that has two endpoints.
Ray	⟶	$\overrightarrow{AB}$	A straight path that has one endpoint and extends infinitely in one direction.
Line	⟷	ℓ or $\overleftrightarrow{AB}$	A straight path that extends infinitely in both directions.
Plane	▱	m or ABD (points A, B, D not linear)	A flat surface that extends infinitely in two dimensions.

Collinear points are points that lie on the same line.

Coplanar points are points that lie on the same plane.

In the figure at right:

- A, B, C, D, E and F are points.
- ℓ is a line
- m and n are planes.

In addition, note that:

- C, D, E and F are **collinear points.**
- A, B and E are **coplanar points.**
- A, B and D are **coplanar points.**
- Ray $\overrightarrow{EF}$ goes off in a southeast direction.
- Ray $\overrightarrow{EC}$ goes off in a northwest direction.
- Together, rays $\overrightarrow{EF}$ and $\overrightarrow{EC}$ make up line ℓ.
- Line ℓ intersects both planes m and n.

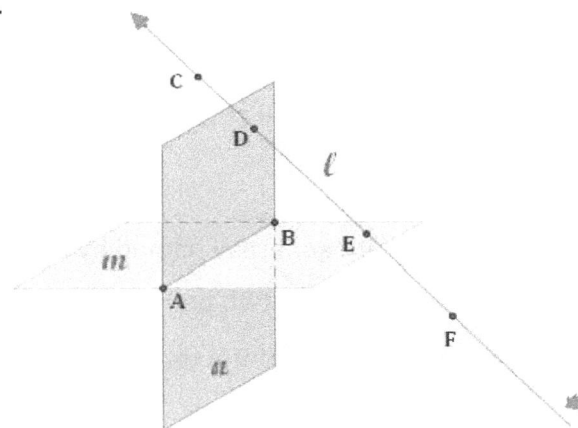

Note: In geometric figures such as the one above, it is important to remember that, even though planes are drawn with edges, they extend infinitely in the 2 dimensions shown.

An **intersection** of geometric shapes is the set of points they share in common.

ℓ and m intersect at point E.

ℓ and n intersect at point D.

m and n intersect in line $\overleftrightarrow{AB}$.

Geometry
Segments, Rays & Lines

Some Thoughts About ...

Line Segments

- Line segments are generally named by their endpoints, so the segment at right could be named either $\overline{AB}$ or $\overline{BA}$.

- Segment $\overline{AB}$ contains the two endpoints (A and B) and all points on line $\overleftrightarrow{AB}$ that are between them.

Rays

- Rays are generally named by their single endpoint, called an **initial point**, and another point on the ray.

- Ray $\overrightarrow{AB}$ contains its initial point A and all points on line $\overleftrightarrow{AB}$ in the direction of the arrow.

- Rays $\overrightarrow{AB}$ and $\overrightarrow{BA}$ are not the same ray.

- If point O is on line $\overleftrightarrow{AB}$ and is between points A and B, then rays $\overrightarrow{OA}$ and $\overrightarrow{OB}$ are called **opposite rays**. They have only point O in common, and together they make up line $\overleftrightarrow{AB}$.

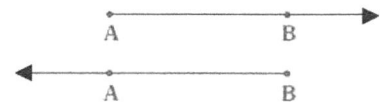

Lines

- Lines are generally named by either a single script letter (e.g., ℓ) or by two points on the line (e.g., $\overleftrightarrow{AB}$).

- A line extends infinitely in the directions shown by its arrows.

- Lines are **parallel** if they are in the same plane and they never intersect. Lines f and g, at right, are parallel.

- Lines are **perpendicular** if they intersect at a 90° angle. A pair of perpendicular lines is always in the same plane. Lines f and e, at right, are perpendicular. Lines g and e are also perpendicular.

- Lines are **skew** if they are not in the same plane and they never intersect. Lines k and ℓ, at right, are skew. (Remember this figure is 3-dimensional.)

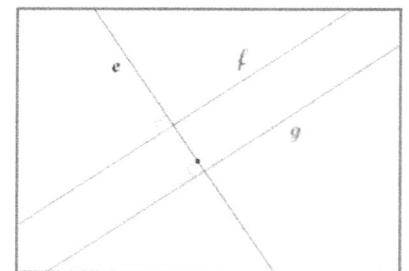

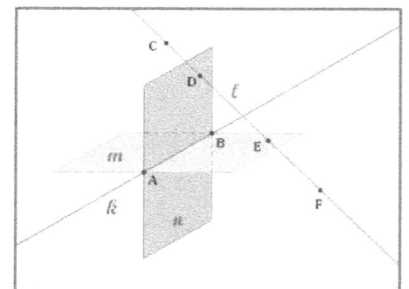

Geometry
Distance Between Points

Distance measures how far apart two things are. The distance between two points can be measured in any number of dimensions, and is defined as the length of the line connecting the two points. Distance is always a positive number.

1-Dimensional Distance

In one dimension the distance between two points is determined simply by subtracting the coordinates of the points.

Example: In this segment, the distance between -2 and 5 is calculated as: $5 - (-2) = 7$.

```
          _____
        -2                        5
```

2-Dimensional Distance

In two dimensions, the distance between two points can be calculated by considering the line between them to be the hypotenuse of a right triangle. To determine the length of this line:

- Calculate the difference in the x-coordinates of the points
- Calculate the difference in the y-coordinates of the points
- Use the Pythagorean Theorem.

This process is illustrated below, using the variable *"d"* for distance.

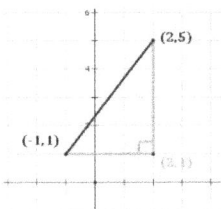

Example: Find the distance between (-1,1) and (2,5). Based on the illustration to the left:

x-coordinate difference: $2 - (-1) = 3$.
y-coordinate difference: $5 - 1 = 4$.

Then, the distance is calculated using the formula: $d^2 = (3^2 + 4^2) = (9 + 16) = 25$

So, $$d = 5$$

If we define two points generally as (x_1, y_1) and (x_2, y_2), then a 2-dimensional distance formula would be:

$$distance = \sqrt{(x_2 - x_1)^2 + (y_2 - y_1)^2}$$

Geometry

Distance Formula in "n" Dimensions

The distance between two points can be generalized to "n" dimensions by successive use of the Pythagorean Theorem in multiple dimensions. To move from two dimensions to three dimensions, we start with the two-dimensional formula and apply the Pythagorean Theorem to add the third dimension.

3 Dimensions

Consider two 3-dimensional points (x_1, y_1, z_1) and (x_2, y_2, z_2). Consider first the situation where the two z-coordinates are the same. Then, the distance between the points is 2-dimensional, i.e., $d = \sqrt{(x_2 - x_1)^2 + (y_2 - y_1)^2}$.

We then add a third dimension using the Pythagorean Theorem:

$$distance^2 = d^2 + (z_2 - z_1)^2$$
$$distance^2 = \left(\sqrt{(x_2 - x_1)^2 + (y_2 - y_1)^2}\right)^2 + (z_2 - z_1)^2$$
$$distance^2 = (x_2 - x_1)^2 + (y_2 - y_1)^2 + (z_2 - z_1)^2$$

And, finally the 3-dimensional difference formula:

$$distance = \sqrt{(x_2 - x_1)^2 + (y_2 - y_1)^2 + (z_2 - z_1)^2}$$

n Dimensions

Using the same methodology in "n" dimensions, we get the generalized n-dimensional difference formula (where there are n terms beneath the radical, one for each dimension):

$$distance = \sqrt{(x_2 - x_1)^2 + (y_2 - y_1)^2 + (z_2 - z_1)^2 + \cdots + (w_2 - w_1)^2}$$

Or, in higher level mathematical notation:
The distance between 2 points $A=(a_1, a_2, \ldots, a_n)$ and $B=(b_1, b_2, \ldots, b_n)$ is

$$d(A,B) = |A - B| = \sqrt{\sum_{i=1}^{n} (a_i - b_i)^2}$$

Geometry
Angles

Parts of an Angle

An angle consists of two rays with a common endpoint (or, initial point).

- Each ray is a side of the angle.
- The common endpoint is called the vertex of the angle.

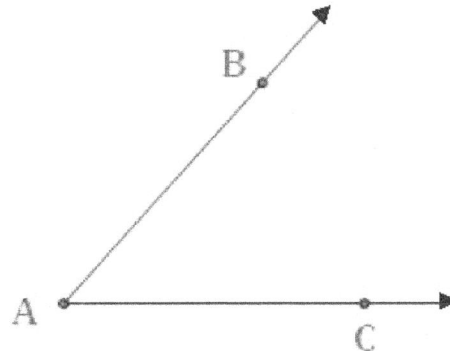

Naming Angles

Angles can be named in one of two ways:

- Point-vertex-point method. In this method, the angle is named from a point on one ray, the vertex, and a point on the other ray. This is the most unambiguous method of naming an angle, and is useful in diagrams with multiple angles sharing the same vertex. In the above figure, the angle shown could be named $\angle BAC$ or $\angle CAB$.

- Vertex method. In cases where it is not ambiguous, an angle can be named based solely on its vertex. In the above figure, the angle could be named $\angle A$.

Measure of an Angle

There are two conventions for measuring the size of an angle:

- In degrees. The symbol for degrees is $^\circ$. There are 360° in a full circle. The angle above measures approximately 45° (one-eighth of a circle).

- In radians. There are 2π radians in a complete circle. The angle above measures approximately $\frac{1}{4}\pi$ radians.

Some Terms Relating to Angles

Angle interior is the area between the rays.

Angle exterior is the area not between the rays.

Adjacent angles are angles that share a ray for a side. $\angle BAD$ and $\angle DAC$ in the figure at right are adjacent angles.

Congruent angles area angles with the same measure.

Angle bisector is a ray that divides the angle into two congruent angles. Ray $\overrightarrow{AD}$ bisects $\angle BAC$ in the figure at right.

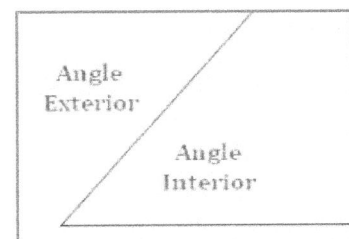

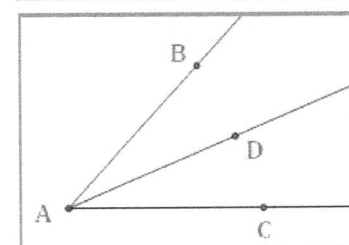

Geometry
Types of Angles

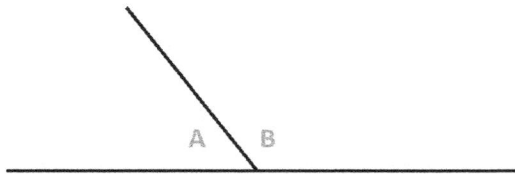

Supplementary Angles

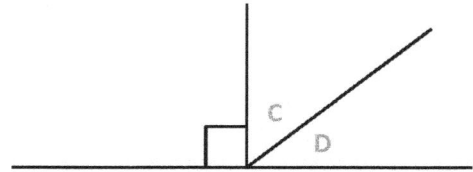

Complementary Angles

Angles A and B are supplementary.

Angles A and B form a linear pair.

$$m\angle A + m\angle B = 180^0$$

Angles C and D are complementary.

$$m\angle C + m\angle D = 90^0$$

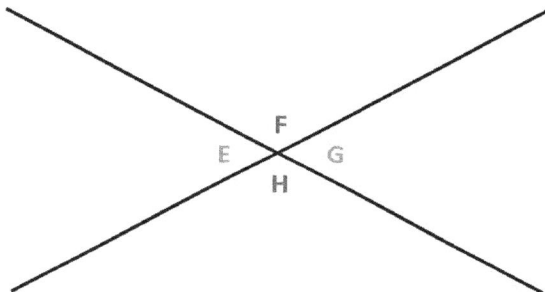

Vertical Angles

Angles which are opposite each other when two lines cross are vertical angles.

Angles E and G are vertical angles.
Angles F and H are vertical angles.

$$m\angle E = m\angle G \quad and \quad m\angle F = m\angle H$$

In addition, each angle is supplementary to the two angles adjacent to it. For example:

Angle E is supplementary to Angles F and H.

An acute angle is one that is less than 90°. In the illustration above, angles E and G are acute angles.

A right angle is one that is exactly 90°.

An obtuse angle is one that is greater than 90°. In the illustration above, angles F and H are obtuse angles.

A straight angle is one that is exactly 180°.

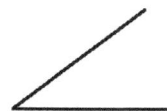

Acute

Obtuse

Right

Straight

Geometry
Conditional Statements

A conditional statement contains both a hypothesis and a conclusion in the following form:

If hypothesis, then conclusion.

For any conditional statement, it is possible to create three related conditional statements, as shown below. In the table, p is the hypothesis of the original statement and q is the conclusion of the original statement.

> Statements linked below by red arrows must be either both true or both false.

Type of Conditional Statement	Example Statement is:
Original Statement: If p, then q. $(p \rightarrow q)$ • Example: If a number is divisible by 6, then it is divisible by 3. • The original statement may be either true or false.	TRUE
Converse Statement: If q, then p. $(q \rightarrow p)$ • Example: If a number is divisible by 3, then it is divisible by 6. • The converse statement may be either true or false, and this does not depend on whether the original statement is true or false.	FALSE
Inverse Statement: If not p, then not q. $(\sim p \rightarrow \sim q)$ • Example: If a number is not divisible by 6, then it is not divisible by 3. • The inverse statement is always true when the converse is true and false when the converse is false.	FALSE
Contrapositive Statement: If not q, then not p. $(\sim q \rightarrow \sim p)$ • Example: If a number is not divisible by 3, then it is not divisible by 6. • The Contrapositive statement is always true when the original statement is true and false when the original statement is false.	TRUE

Note also that:
- When two statements must be either both true or both false, they are called equivalent statements.
  - The original statement and the contrapositive are equivalent statements.
  - The converse and the inverse are equivalent statements.
- If both the original statement and the converse are true, the phrase "if and only if" (abbreviated "iff") may be used. For example, "A number is divisible by 3 iff the sum of its digits is divisible by 3."

Geometry
Basic Properties of Algebra

Properties of Equality and Congruence.

Property	Definition for Equality For any real numbers **a**, **b**, and **c**:	Definition for Congruence For any geometric elements **a**, **b** and **c**. (e.g., segment, angle, triangle)
Reflexive Property	$a = a$	$a \cong a$
Symmetric Property	$If\ a = b, then\ b = a$	$If\ a \cong b, then\ b \cong a$
Transitive Property	$If\ a = b\ and\ b = c, then\ a = c$	$If\ a \cong b\ and\ b \cong c, then\ a \cong c$
Substitution Property	If $a = b$, then either can be substituted for the other in any equation (or inequality).	If $a \cong b$, then either can be substituted for the other in any congruence expression.

More Properties of Equality. For any real numbers **a**, **b**, and **c**:

Property	Definition for Equality
Addition Property	$If\ a = b, then\ a + c = b + c$
Subtraction Property	$If\ a = b, then\ a - c = b - c$
Multiplication Property	$If\ a = b, then\ a \cdot c = b \cdot c$
Division Property	$If\ a = b\ and\ c \neq 0, then\ a \div c = b \div c$

Properties of Addition and Multiplication. For any real numbers **a**, **b**, and **c**:

Property	Definition for Addition	Definition for Multiplication
Commutative Property	$a + b = b + a$	$a \cdot b = b \cdot a$
Associative Property	$(a + b) + c = a + (b + c)$	$(a \cdot b) \cdot c = a \cdot (b \cdot c)$
Distributive Property	$a \cdot (b + c) = (a \cdot b) + (a \cdot c)$	

Geometry
Inductive vs. Deductive Reasoning

Inductive Reasoning

Inductive reasoning uses observation to form a hypothesis or conjecture. The hypothesis can then be tested to see if it is true. The test must be performed in order to confirm the hypothesis.

Example: Observe that the sum of the numbers 1 to 4 is $(4 \cdot 5/2)$ and that the sum of the numbers 1 to 5 is $(5 \cdot 6/2)$. Hypothesis: the sum of the first n numbers is $(n * (n + 1)/2)$. Testing this hypothesis confirms that it is true.

Deductive Reasoning

Deductive reasoning argues that if something is true about a broad category of things, it is true of an item in the category.

> **Example:** All birds have beaks. A pigeon is a bird; therefore, it has a beak.

There are two key types of deductive reasoning of which the student should be aware:

- **Law of Detachment.** Given that $p \to q$, if p is true then q is true. In words, if one thing implies another, then whenever the first thing is true, the second must also be true.

    Example: Start with the statement: "If a living creature is human, then it has a brain." Then because you are human, we can conclude that you have a brain.

- **Syllogism.** Given that $p \to q$ and $q \to r$, we can conclude that $p \to r$. This is a kind of transitive property of logic. In words, if one thing implies a second and that second thing implies a third, then the first thing implies the third.

    Example: Start with the statements: "If my pencil breaks, I will not be able to write," and "if I am not able to write, I will not pass my test." Then I can conclude that "If my pencil breaks, I will not pass my test."

Geometry
An Approach to Proofs

Learning to develop a successful proof is one of the key skills students develop in geometry. The process is different from anything students have encountered in previous math classes, and may seem difficult at first. Diligence and practice in solving proofs will help students develop reasoning skills that will serve them well for the rest of their lives.

Requirements in Performing Proofs

- Each proof starts with a set of "givens," statements that you are supplied and from which you must derive a "conclusion." Your mission is to start with the givens and to proceed logically to the conclusion, providing reasons for each step along the way.

- Each step in a proof builds on what has been developed before. Initially, you look at what you can conclude from the" givens." Then as you proceed through the steps in the proof, you are able to use additional things you have concluded based on earlier steps.

- Each step in a proof must have a valid reason associated with it. So, each statement in the proof must be furnished with an answer to the question: "Why is this step valid?"

Tips for Successful Proof Development

- At each step, think about what you know and what you can conclude from that information. Do this initially without regard to what you are being asked to prove. Then look at each thing you can conclude and see which ones move you closer to what you are trying to prove.

- Go as far as you can into the proof from the beginning. If you get stuck, …

- Work backwards from the end of the proof. Ask yourself what the last step in the proof is likely to be. For example, if you are asked to prove that two triangles are congruent, try to see which of the several theorems about this is most likely to be useful based on what you were given and what you have been able to prove so far.

- Continue working backwards until you see steps that can be added to the front end of the proof. You may find yourself alternating between the front end and the back end until you finally bridge the gap between the two sections of the proof.

- Don't skip any steps. Some things appear obvious, but actually have a mathematical reason for being true. For example, $a = a$ might seem obvious, but "obvious" is not a valid reason in a geometry proof. The reason for $a = a$ is a property of algebra called the "reflexive property of equality." Use mathematical reasons for all your steps.

Geometry
Parallel Lines and Transversals

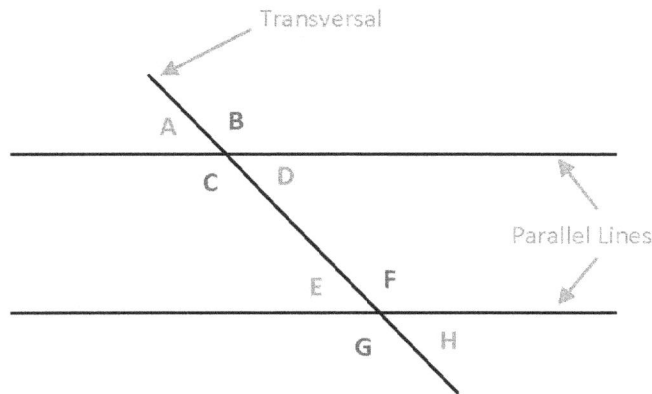

Alternate: refers to angles that are on opposite sides of the transversal.

Consecutive: refers to angles that are on the same side of the transversal.

Interior: refers to angles that are between the parallel lines.

Exterior: refers to angles that are outside the parallel lines.

Corresponding Angles

Corresponding Angles are angles in the same location relative to the parallel lines and the transversal. For example, the angles on top of the parallel lines and left of the transversal (i.e., top left) are corresponding angles.

Angles A and E (top left) are Corresponding Angles. So are angle pairs B and F (top right), C and G (bottom left), and D and H (bottom right). Corresponding angles are congruent.

Alternate Interior Angles

Angles D and E are Alternate Interior Angles. Angles C and F are also alternate interior angles. Alternate interior angles are congruent.

Alternate Exterior Angles

Angles A and H are Alternate Exterior Angles. Angles B and G are also alternate exterior angles. Alternate exterior angles are congruent.

Consecutive Interior Angles

Angles C and E are Consecutive Interior Angles. Angles D and F are also consecutive interior angles. Consecutive interior angles are supplementary.

Note that angles A, D, E, and H are congruent, and angles B, C, F, and G are congruent. In addition, each of the angles in the first group are supplementary to each of the angles in the second group.

Geometry
Multiple Sets of Parallel Lines

Two Transversals

Sometimes, the student is presented two sets of intersecting parallel lines, as shown above. Note that each pair of parallel lines is a set of transversals to the other set of parallel lines.

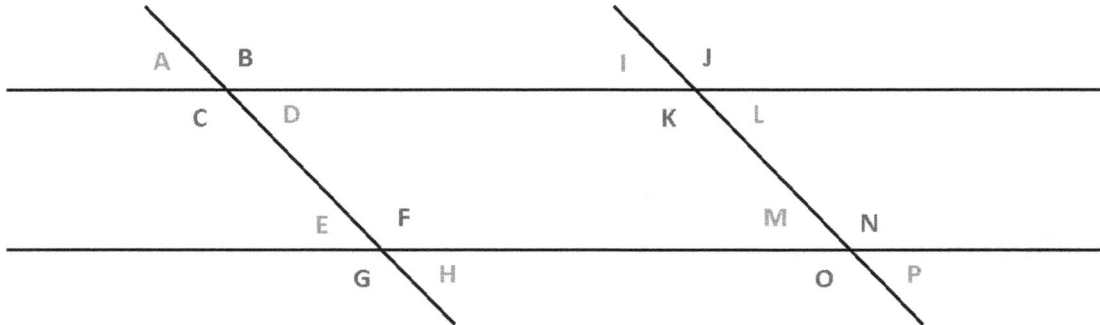

In this case, the following groups of angles are congruent:

- Group 1: Angles A, D, E, H, I, L, M and P are all congruent.
- Group 2: Angles B, C, F, G, J, K, N, and O are all congruent.
- Each angle in the Group 1 is supplementary to each angle in Group 2.

Geometry
Proving Lines are Parallel

The properties of parallel lines cut by a transversal can be used to prove two lines are parallel.

Corresponding Angles

If two lines cut by a transversal have congruent corresponding angles, then the lines are parallel. Note that there are 4 sets of corresponding angles.

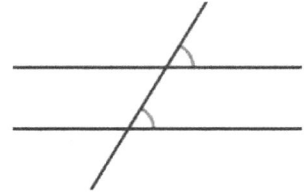

Alternate Interior Angles

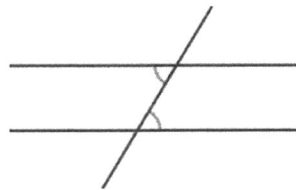

If two lines cut by a transversal have congruent alternate interior angles congruent, then the lines are parallel. Note that there are 2 sets of alternate interior angles.

Alternate Exterior Angles

If two lines cut by a transversal have congruent alternate exterior angles, then the lines are parallel. Note that there are 2 sets of alternate exterior angles.

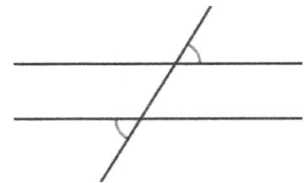

Consecutive Interior Angles

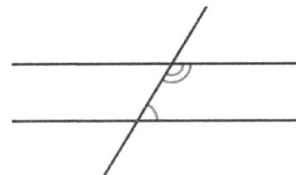

If two lines cut by a transversal have supplementary consecutive interior angles, then the lines are parallel. Note that there are 2 sets of consecutive interior angles.

Geometry
Parallel and Perpendicular Lines in the Coordinate Plane

Parallel Lines

Two lines are parallel if their slopes are equal.

- In $y = mx + b$ form, if the values of m are the same.

  Example: $y = 2x - 3$ and
  $$y = 2x + 1$$

- In Standard Form, if the coefficients of x and y are proportional between the equations.

  Example: $3x - 2y = 5$ and
  $$6x - 4y = -7$$

- Also, if the lines are both vertical (i.e., their slopes are undefined).

  Example: $x = -3$ and
  $$x = 2$$

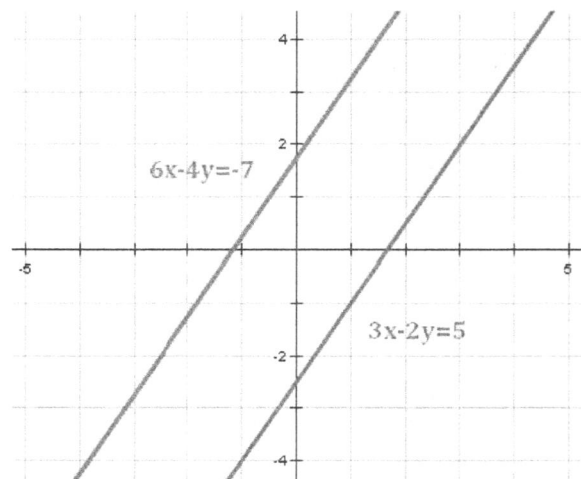

Perpendicular Lines

Two lines are perpendicular if the product of their slopes is -1. That is, if the slopes have different signs and are multiplicative inverses.

- In $y = mx + b$ form, the values of m multiply to get -1..

  Example: $y = 6x + 5$ and
  $$y = -\frac{1}{6}x - 3$$

- In Standard Form, if you add the product of the x-coefficients to the product of the y-coefficients and get zero.

  Example: $4x + 6y = 4$ and
  $$3x - 2y = 5 \quad \text{because} \quad (4 \cdot 3) + (6 \cdot (-2)) = 0$$

- Also, if one line is vertical (i.e., m is undefined) and one line is horizontal (i.e., $m = 0$).

  Example: $x = 6$ and
  $$y = 3$$

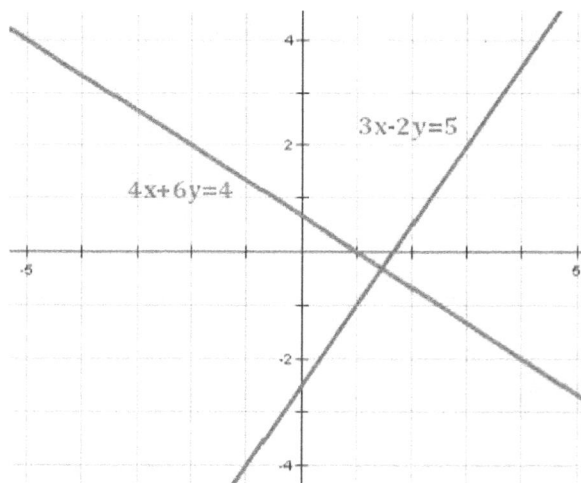

Geometry
Types of Triangles

Scalene

A Scalene Triangle has 3 sides of different lengths. Because the sides are of different lengths, the angles must also be of different measures.

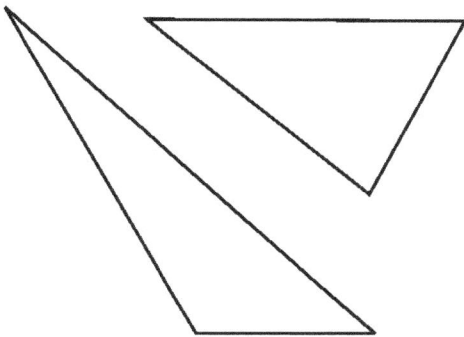

Isosceles

An Isosceles Triangle has 2 sides the same length (i.e., congruent). Because two sides are congruent, two angles must also be congruent.

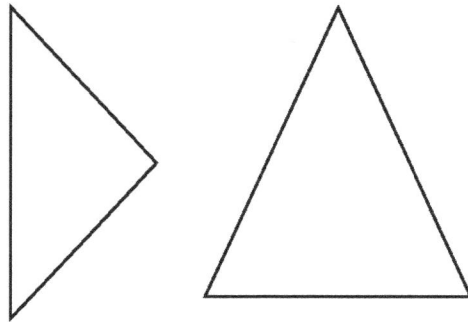

Equilateral

An Equilateral Triangle has all 3 sides the same length (i.e., congruent). Because all 3 sides are congruent, all 3 angles must also be congruent. This requires each angle to be 60°.

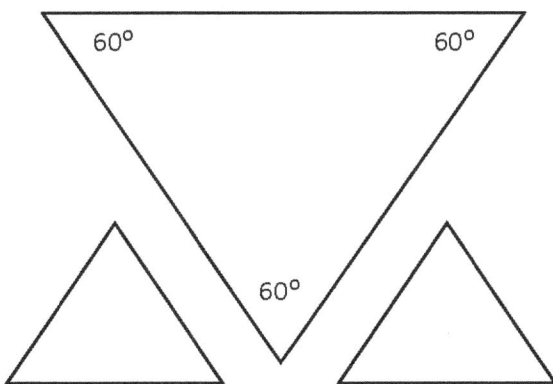

60°   60°

60°

Right

A Right Triangle is one that contains a 90° angle. It may be scalene or isosceles, but cannot be equilateral. Right triangles have sides that meet the requirements of the Pythagorean Theorem.

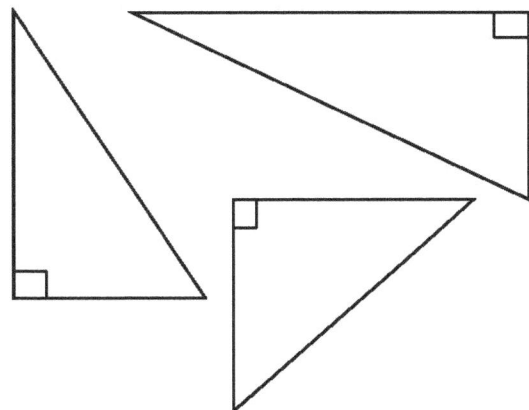

Geometry
Congruent Triangles

The following theorems present conditions under which triangles are congruent.

Side-Angle-Side (SAS) Congruence

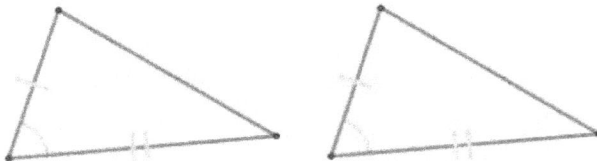

SAS congruence requires the congruence of two sides and the angle between those sides. Note that there is no such thing as SSA congruence; the congruent angle must be between the two congruent sides.

Side-Side-Side (SSS) Congruence

SSS congruence requires the congruence of all three sides. If all of the sides are congruent then all of the angles must be congruent. The converse is not true; there is no such thing as AAA congruence.

Angle-Side-Angle (ASA) Congruence

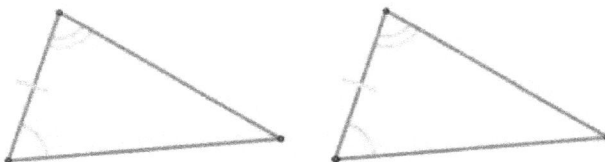

ASA congruence requires the congruence of two angles and the side between those angles.

Note: ASA and AAS combine to provide congruence of two triangles whenever any two angles and any one side of the triangles are congruent.

Angle-Angle-Side (AAS) Congruence

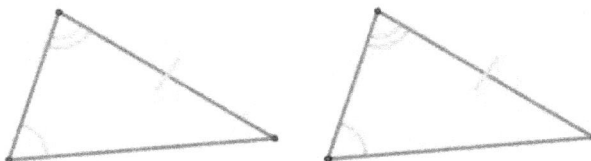

AAS congruence requires the congruence of two angles and a side which is not between those angles.

CPCTC

CPCTC means "corresponding parts of congruent triangles are congruent." It is a very powerful tool in geometry proofs and is often used shortly after a step in the proof where a pair of triangles is proved to be congruent.

Geometry
Centers of Triangles

The following are all points which can be considered the center of a triangle.

Centroid (Medians)

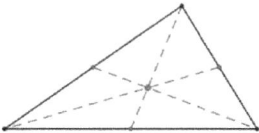

The centroid is the intersection of the three medians of a triangle. A median is a line segment drawn from a vertex to the midpoint of the line opposite the vertex.

- The centroid is located 2/3 of the way from a vertex to the opposite side. That is, the distance from a vertex to the centroid is double the length from the centroid to the midpoint of the opposite line.
- The medians of a triangle create 6 inner triangles of equal area.

Orthocenter (Altitudes)

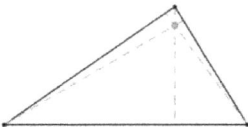

The orthocenter is the intersection of the three altitudes of a triangle. An altitude is a line segment drawn from a vertex to a point on the opposite side (extended, if necessary) that is perpendicular to that side.

- In an acute triangle, the orthocenter is inside the triangle.
- In a right triangle, the orthocenter is the right angle vertex.
- In an obtuse triangle, the orthocenter is outside the triangle.

Circumcenter (Perpendicular Bisectors)

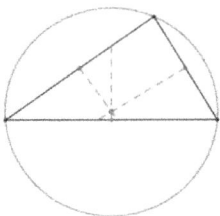

The circumcenter is the intersection of the perpendicular bisectors of the three sides of the triangle. A perpendicular bisector is a line which both bisects the side and is perpendicular to the side. The circumcenter is also the center of the circle circumscribed about the triangle.

Euler Line: Interestingly, the centroid, orthocenter and circumcenter of a triangle are collinear (i.e., lie on the same line, which is called the Euler Line).

- In an acute triangle, the circumcenter is inside the triangle.
- In a right triangle, the circumcenter is the midpoint of the hypotenuse.
- In an obtuse triangle, the circumcenter is outside the triangle.

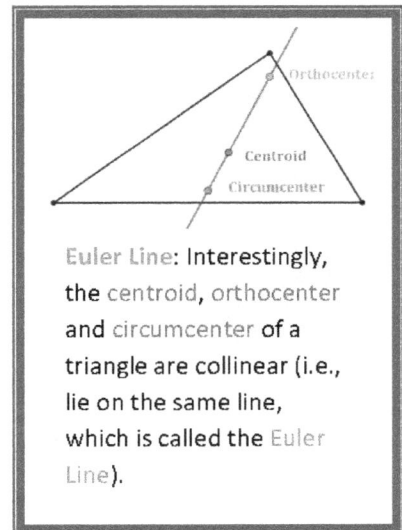

Incenter (Angle Bisectors)

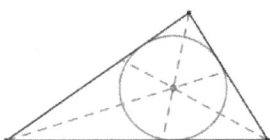

The incenter is the intersection of the angle bisectors of the three angles of the triangle. An angle bisector cuts an angle into two congruent angles, each of which is half the measure of the original angle. The incenter is also the center of the circle inscribed in the triangle.

Geometry
Length of Height, Median and Angle Bisector

Height

The formula for the length of a height of a triangle is derived
from Heron's formula for the area of a triangle:

$$h = \frac{2\sqrt{s\,(s-a)\,(s-b)\,(s-c)}}{c}$$

where, $s = \frac{1}{2}(a+b+c)$, and

a, b, c are the lengths of the sides of the triangle.

Height

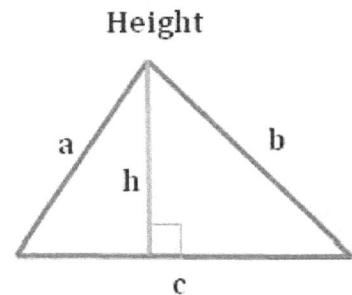

Median

The formula for the length of a median of a triangle is:

$$m = \frac{1}{2}\sqrt{2a^2 + 2b^2 - c}$$

where, a, b, c are the lengths of the sides of the triangle.

Median

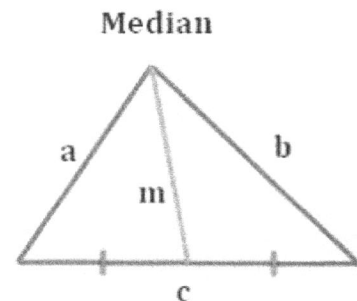

Angle Bisector

The formula for the length of an angle bisector of a triangle is:

$$t = \sqrt{ab\left(1 - \frac{c^2}{(a+b)^2}\right)}$$

where, a, b, c are the lengths of the sides of the triangle.

Angle Bisector

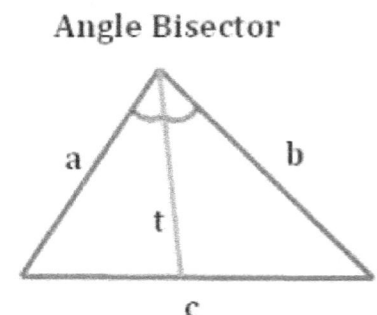

Geometry
Inequalities in Triangles

Angles and their opposite sides in triangles are related. In fact, this is often reflected in the labeling of angles and sides in triangle illustrations.

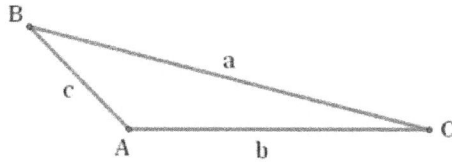

Angles and their opposite sides are often labeled with the same letter. An upper case letter is used for the angle and a lower case letter is used for the side.

The relationship between angles and their opposite sides translates into the following triangle inequalities:

$$\text{If } m\angle C < m\angle B < m\angle A, \text{ then } c < b < a$$

$$\text{If } m\angle C \leq m\angle B \leq m\angle A, \text{ then } c \leq b \leq a$$

That is, in any triangle,

- The largest side is opposite the largest angle.
- The medium side is opposite the medium angle.
- The smallest side is opposite the smallest angle.

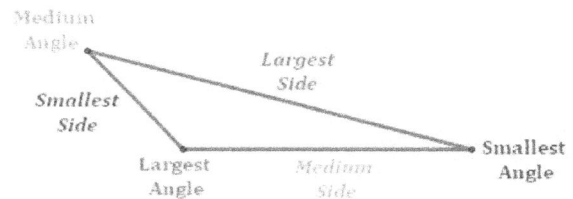

Other Inequalities in Triangles

Triangle Inequality: The sum of the lengths of any two sides of a triangle is greater than the length of the third side. This is a crucial element in deciding whether segments of any 3 lengths can form a triangle.

$$a + b > c \quad and \quad b + c > a \quad and \quad c + a > b$$

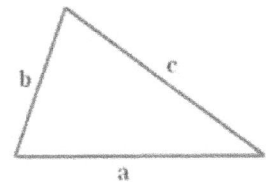

Exterior Angle Inequality: The measure of an external angle is greater than the measure of either of the two non-adjacent interior angles. That is, in the figure below:

$$m\angle DAB > m\angle B \quad and \quad m\angle DAB > m\angle C$$

Note: the Exterior Angle Inequality is much less relevant than the Exterior Angle Equality.

Exterior Angle Equality: The measure of an external angle is equal to the sum of the measures of the two non-adjacent interior angles. That is, in the figure below:

$$m\angle DAB = m\angle B + m\angle C$$

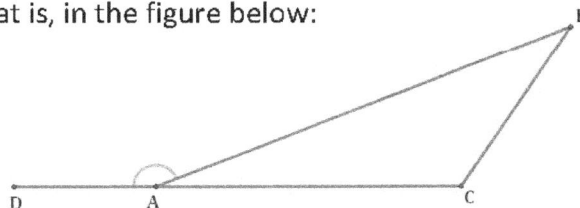

Geometry
Polygons - Basics

Basic Definitions

Polygon: a closed path of three or more line segments, where:
- no two sides with a common endpoint are collinear, and
- each segment is connected at its endpoints to exactly two other segments.

Side: a segment that is connected to other segments (which are also sides) to form a polygon.

Vertex: a point at the intersection of two sides of the polygon. (plural form: **vertices**)

Diagonal: a segment, from one vertex to another, which is not a side.

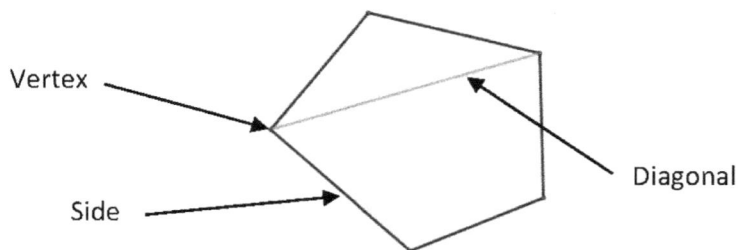

Vertex

Diagonal

Side

Concave: A polygon in which it is possible to draw a diagonal "outside" the polygon. (Notice the orange diagonal drawn outside the polygon at right.) Concave polygons actually look like they have a "cave" in them.

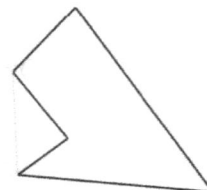

Convex: A polygon in which it is **not** possible to draw a diagonal "outside" the polygon. (Notice that all of the orange diagonals are inside the polygon at right.) Convex polygons appear more "rounded" and do not contain "caves."

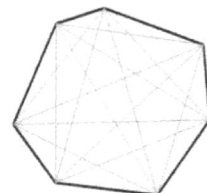

Names of Some Common Polygons

Number of Sides	Name of Polygon	Number of Sides	Name of Polygon
3	Triangle	9	Nonagon
4	Quadrilateral	10	Decagon
5	Pentagon	11	Undecagon
6	Hexagon	12	Dodecagon
7	Heptagon	20	Icosagon
8	Octagon	n	n-gon

Names of polygons are generally formed from the Greek language; however, some hybrid forms of Latin and Greek (e.g., undecagon) have crept into common usage.

Geometry
Polygons – More Definitions

Definitions

Equilateral: a polygon in which all of the sides are equal in length.

Equiangular: a polygon in which all of the angles have the same measure.

Regular: a polygon which is both equilateral and equiangular. That is, a *regular polygon* is one in which all of the sides have the same length and all of the angles have the same measure.

Interior Angle: An angle formed by two sides of a polygon. The angle is inside the polygon.

Exterior Angle: An angle formed by one side of a polygon and the line containing an adjacent side of the polygon. The angle is outside the polygon.

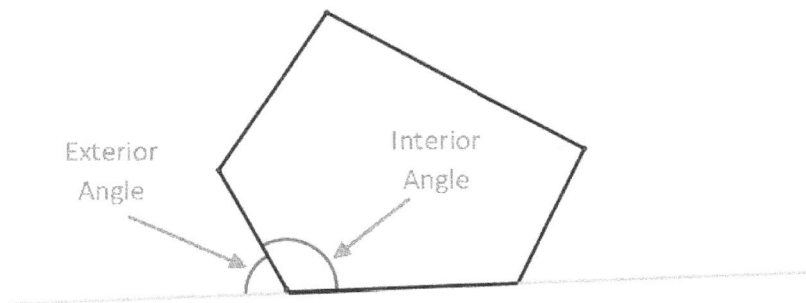

"Advanced" Definitions:

Simple Polygon: *a polygon whose sides do not intersect at any location other than its endpoints.* Simple polygons always divide a plane into two regions – one inside the polygon and one outside the polygon.

Complex Polygon: *a polygon with sides that intersect someplace other than their endpoints (i.e., not a simple polygon).* Complex polygons do not always have well-defined insides and outsides.

Skew Polygon: *a polygon for which not all of its vertices lie on the same plane.*

How Many Diagonals Does a Convex Polygon Have?

Believe it or not, this is a common question with a simple solution. Consider a polygon with n sides and, therefore, n vertices.

- Each of the n vertices of the polygon can be connected to $(n-3)$ other vertices with diagonals. That is, it can be connected to all other vertices except itself and the two to which it is connected by sides. So, there are $[\,n \cdot (n-3)]$ lines to be drawn as diagonals.

- However, when we do this, we draw each diagonal twice because we draw it once from each of its two endpoints. So, the number of diagonals is actually half of the number we calculated above.

- Therefore, the number of diagonals in an n-sided polygon is:

$$\frac{n \cdot (n-3)}{2}$$

Geometry
Interior and Exterior Angles of a Polygon

Interior Angles

The sum of the interior angles in an n-sided polygon is:

$$\Sigma = (n - 2) \cdot 180°$$

If the polygon is regular, you can calculate the measure of each interior angle as:

$$\frac{(n-2) \cdot 180°}{n}$$

Notation: The Greek letter **"Σ"** is equivalent to the English letter **"S"** and is math short-hand for a summation (i.e., addition) of things.

Interior Angles		
Sides	Sum of Interior Angles	Each Interior Angle
3	180°	60°
4	360°	90°
5	540°	108°
6	720°	120°
7	900°	129°
8	1,080°	135°
9	1,260°	140°
10	1,440°	144°

Exterior Angles

No matter how many sides there are in a polygon, the sum of the exterior angles is:

$$\Sigma = 360°$$

If the polygon is regular, you can calculate the measure of each exterior angle as:

$$\frac{360°}{n}$$

Exterior Angles		
Sides	Sum of Exterior Angles	Each Exterior Angle
3	360°	120°
4	360°	90°
5	360°	72°
6	360°	60°
7	360°	51°
8	360°	45°
9	360°	40°
10	360°	36°

Geometry
Definitions of Quadrilaterals

Name	Definition
Quadrilateral	A polygon with 4 sides.
Kite	A quadrilateral with two consecutive pairs of congruent sides, but with opposite sides not congruent.
Trapezoid	A quadrilateral with exactly one pair of parallel sides.
Isosceles Trapezoid	A trapezoid with congruent legs.
Parallelogram	A quadrilateral with both pairs of opposite sides parallel.
Rectangle	A parallelogram with all angles congruent (i.e., right angles).
Rhombus	A parallelogram with all sides congruent.
Square	A quadrilateral with all sides congruent and all angles congruent.

Quadrilateral Tree:

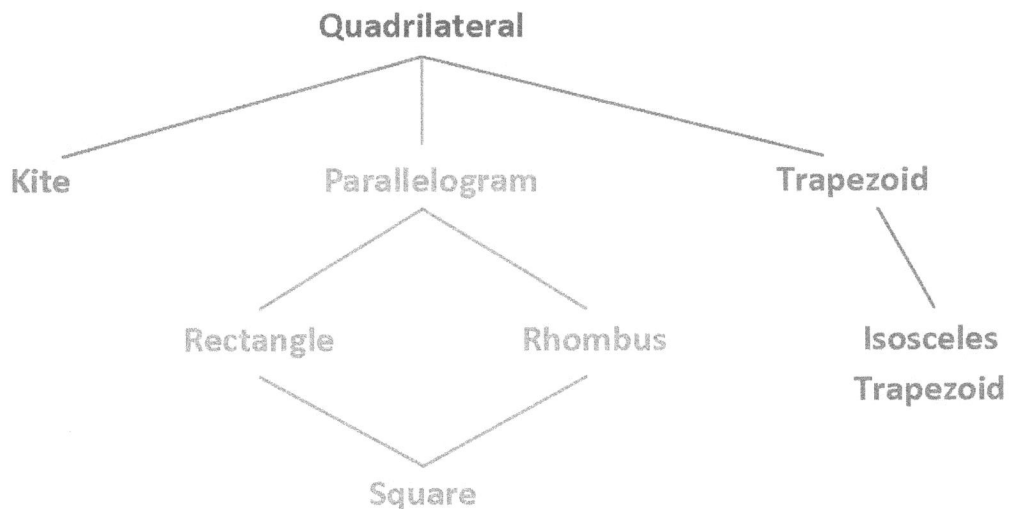

Geometry
Figures of Quadrilaterals

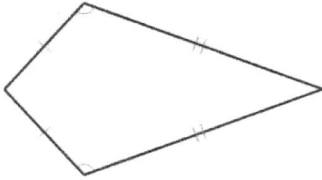

Kite
- 2 consecutive pairs of congruent sides
- 1 pair of congruent opposite angles
- Diagonals perpendicular

Trapezoid
- 1 pair of parallel sides (called "bases")
- Angles on the same "side" of the bases are supplementary

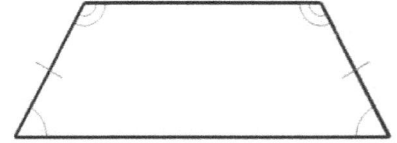

Isosceles Trapezoid
- 1 pair of parallel sides
- Congruent legs
- 2 pair of congruent base angles
- Diagonals congruent

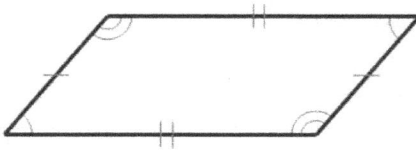

Parallelogram
- Both pairs of opposite sides parallel
- Both pairs of opposite sides congruent
- Both pairs of opposite angles congruent
- Consecutive angles supplementary
- Diagonals bisect each other

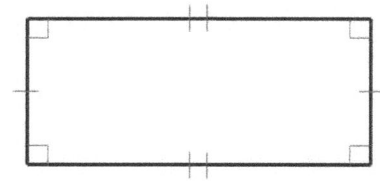

Rectangle
- Parallelogram with all angles congruent (i.e., right angles)
- Diagonals congruent

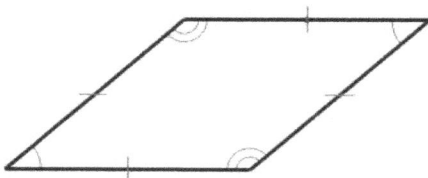

Rhombus
- Parallelogram with all sides congruent
- Diagonals perpendicular
- Each diagonal bisects a pair of opposite angles

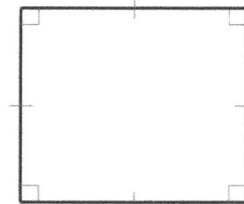

Square
- Both a Rhombus and a Rectangle
- All angles congruent (i.e., right angles)
- All sides congruent

Geometry
Parallelogram Proofs

Proving a Quadrilateral is a Parallelogram

To prove a quadrilateral is a parallelogram, prove any of the following conditions:

1. Both pairs of opposite sides are parallel. (note: this is the definition of a parallelogram)
2. Both pairs of opposite sides are congruent.
3. Both pairs of opposite angles are congruent.
4. An interior angle is supplementary to both of its consecutive angles.
5. Its diagonals bisect each other.
6. A pair of opposite sides is both parallel and congruent.

Proving a Quadrilateral is a Rectangle

To prove a quadrilateral is a rectangle, prove any of the following conditions:

1. All 4 angles are congruent.
2. It is a parallelogram and its diagonals are congruent.

Proving a Quadrilateral is a Rhombus

To prove a quadrilateral is a rhombus, prove any of the following conditions:

1. All 4 sides are congruent.
2. It is a parallelogram and Its diagonals are perpendicular.
3. It is a parallelogram and each diagonal bisects a pair of opposite angles.

Proving a Quadrilateral is a Square

To prove a quadrilateral is a square, prove:

1. It is both a Rhombus and a Rectangle.

Geometry
Kites and Trapezoids

Facts about a Kite

To prove a quadrilateral is a kite, prove:

- It has two pair of congruent sides.
- Opposite sides are not congruent.

Also, if a quadrilateral is a kite, then:

- Its diagonals are perpendicular
- It has exactly one pair of congruent opposite angles.

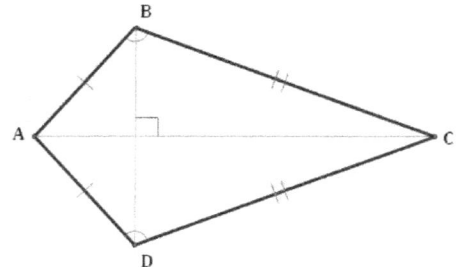

Parts of a Trapezoid

Trapezoid ABCD has the following parts:

- $\overline{AD}$ and $\overline{BC}$ are bases.
- $\overline{AB}$ and $\overline{CD}$ are legs.
- $\overline{EF}$ is the midsegment.
- $\overline{AC}$ and $\overline{BD}$ are diagonals.
- Angles A and D form a pair of base angles.
- Angles B and C form a pair of base angles.

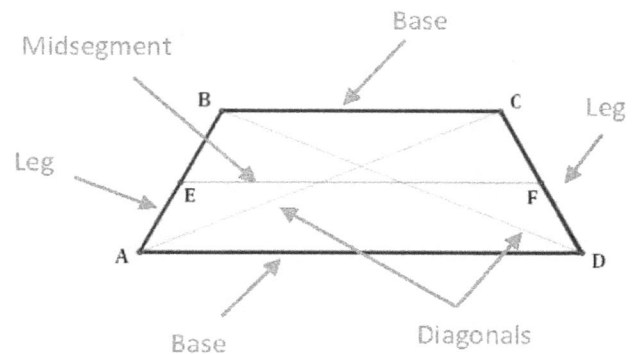

Trapezoid Midsegment Theorem

The **midsegment** of a trapezoid is parallel to each of its bases and: $EF = \frac{1}{2}(AD + BC)$.

Proving a Quadrilateral is an Isosceles Trapezoid

To prove a quadrilateral is an isosceles trapezoid, prove any of the following conditions:

1. It is a trapezoid and has a pair of congruent legs. **(definition of isosceles trapezoid)**
2. It is a trapezoid and has a pair of congruent base angles.
3. It is a trapezoid and its diagonals are congruent.

Geometry
Introduction to Transformation

A Transformation is a mapping of the pre-image of a geometric figure onto an image that retains key characteristics of the pre-image.

Definitions

The Pre-Image is the geometric figure before it has been transformed.

The Image is the geometric figure after it has been transformed.

A mapping is an association between objects. Transformations are types of mappings. In the figures below, we say $ABCD$ is mapped onto $A'B'C'D'$, or $ABCD \longrightarrow A'B'C'D'$. The order of the vertices is critical to a properly named mapping.

An Isometry is a one-to-one mapping that preserves lengths. Transformations that are isometries (i.e., preserve length) are called rigid transformations.

Isometric Transformations

Reflection is flipping a figure across a line called a "mirror." The figure retains its size and shape, but appears "backwards" after the reflection.	Rotation is turning a figure around a point. Rotated figures retain their size and shape, but not their orientation.	Translation is sliding a figure in the plane so that it changes location but retains its shape, size and orientation.

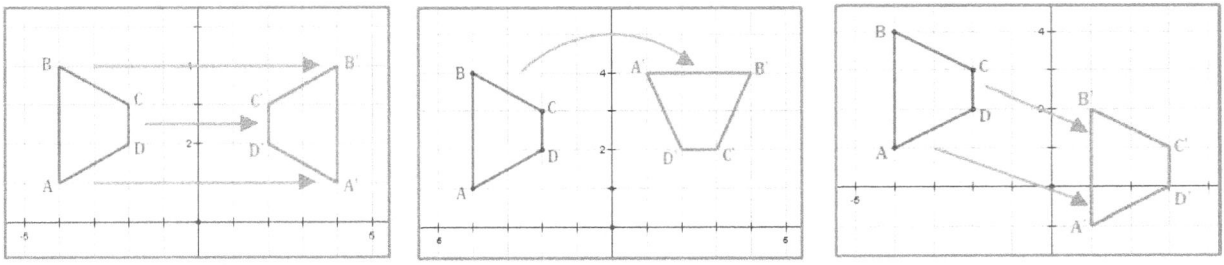

Table of Characteristics of Isometric Transformations

Transformation	Reflection	Rotation	Translation
Isometry (Retains Lengths)?	Yes	Yes	Yes
Retains Angles?	Yes	Yes	Yes
Retains Orientation to Axes?	No	No	Yes

Geometry
Introduction to Transformation (cont'd)

Transformation of a Point

A point is the easiest object to transform. Simply reflect, rotate or translate it following the rules for the transformation selected. By transforming key points first, any transformation becomes much easier.

Transformation of a Geometric Figure

To transform any geometric figure, it is only necessary to transform the items that define the figure, and then re-form it. For example:

- To transform a line segment, transform its two endpoints, and then connect the resulting images with a line segment.
- To transform a ray, transform the initial point and any other point on the ray, and then construct a ray using the resulting images.
- To transform a line, transform any two points on the line, and then fit a line through the resulting images.
- To transform a polygon, transform each of its vertices, and then connect the resulting images with line segments.
- To transform a circle, transform its center and, if necessary, its radius. From the resulting images, construct the image circle.
- To transform other conic sections (parabolas, ellipses and hyperbolas), transform the foci, vertices and/or directrix. From the resulting images, construct the image conic section.

Example: Reflect Quadrilateral ABCD

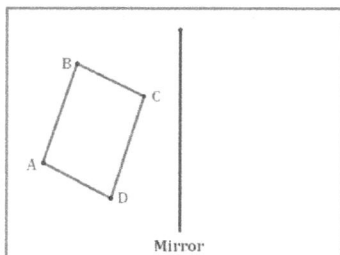

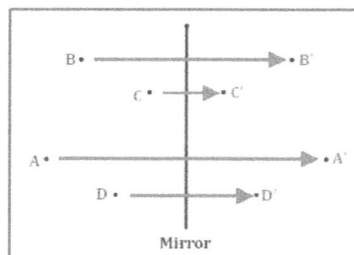

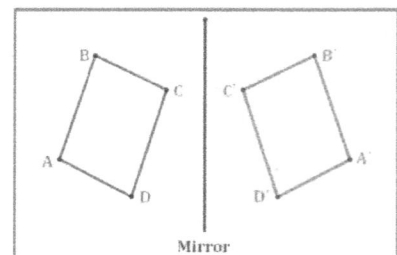

Geometry
Reflection

Definitions

Reflection is flipping a figure across a mirror.

The Line of Reflection is the mirror through which the reflection takes place.

Note that:

- The line segment connecting corresponding points in the image and pre-image is bisected by the mirror.

- The line segment connecting corresponding points in the image and pre-image is perpendicular to the mirror.

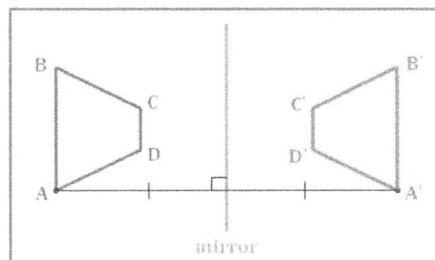

Reflection through an Axis or the Line $y = x$

Reflection of the point (a, b) through the x- or y-axis or the line $y = x$ gives the following results:

Pre-Image Point	Mirror Line	Image Point
(a, b)	x-axis	(a, -b)
(a, b)	y-axis	(-a, b)
(a, b)	the line: $y = x$	(b, a)

If you forget the above table, start with the point $(3, 2)$ on a set of coordinate axes. Reflect the point through the selected line and see which set of "a, b" coordinates works.

Line of Symmetry

A Line of Symmetry is any line through which a figure can be mapped onto itself. The thin black lines in the following figures show their axes of symmetry:

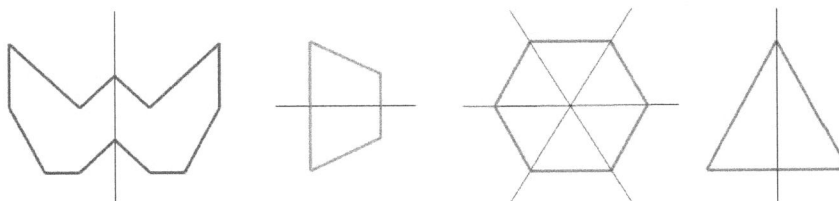

Geometry
Rotation

Definitions

Rotation is turning a figure by an angle about a fixed point.

The Center of Rotation is the point about which the figure is rotated. Point P, at right, is the center of rotation.

The Angle of Rotation determines the extent of the rotation. The angle is formed by the rays that connect the center of rotation to the pre-image and the image of the rotation. Angle P, at right, is the angle of rotation. Though shown only for Point A, the angle is the same for any of the figure's 4 vertices.

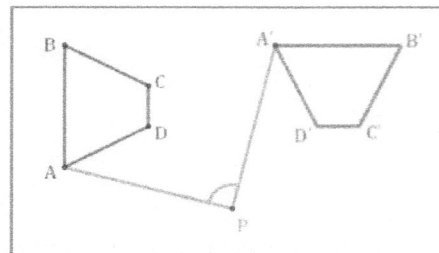

Note: In performing rotations, it is important to indicate the direction of the rotation – clockwise or counterclockwise.

Rotation about the Origin

Rotation of the point (a, b) about the origin (0, 0) gives the following results:

Pre-Image Point	Clockwise Rotation	Counterclockwise Rotation	Image Point
(a, b)	90^0	270^0	(b, -a)
(a, b)	180^0	180^0	(-a, -b)
(a, b)	270^0	90^0	(-b, a)
(a, b)	360^0	360^0	(a, b)

If you forget the above table, start with the point $(3, 2)$ on a set of coordinate axes. Rotate the point by the selected angle and see which set of "a, b" coordinates works.

Rotational Symmetry

A figure in a plane has Rotational Symmetry if it can be mapped onto itself by a rotation of 180^0 or less. Any regular polygon has rotational symmetry, as does a circle. Here are some examples of figures with rotational symmetry:

Geometry
Rotation by 90° about a Point (x_0, y_0)

Rotating an object by 90° about a point involves rotating each point of the object by 90° about that point. For a polygon, this is accomplished by rotating each vertex and then connecting them to each other, so you mainly have to worry about the vertices, which are points. The mathematics behind the process of rotating a point by 90° is described below:

Let's define the following points:
- The point about which the rotation will take place: (x_0, y_0)
- The initial point (before rotation): (x_1, y_1)
- The final point (after rotation): (x_2, y_2)

The problem is to determine (x_2, y_2) if we are given (x_0, y_0) and (x_1, y_1). It involves 3 steps:
1. Convert the problem to one of rotating a point about the origin (a much easier problem).
2. Perform the rotation.
3. Convert the result back to the original set of axes.

We'll consider each step separately and provide an example:

Problem: Rotate a point by 90° about another point.

Step 1: Convert the problem to one of rotating a point about the origin:
First, we ask how the point (x_1, y_1) relates to the point about which it will be rotated (x_0, y_0) and create a new ("translated") point. This is essentially an "axis-translation," which we will reverse in Step 3.

General Situation	Example
Points in the Problem • Rotation Center: (x_0, y_0) • Initial point: (x_1, y_1) • Final point: (x_2, y_2)	Points in the Problem • Rotation Center: (2, 3) • Initial point: (-2, 1) • Final point: to be determined
Calculate a new point that represents how (x_1, y_1) relates to (x_0, y_0). That point is: (x_1-x_0, y_1-y_0)	Calculate a new point that represents how (-2, 1) relates to (2, 3). That point is: (-4, -2)

The next steps depend on whether we are making a clockwise or counter clockwise rotation.

Geometry
Rotation by 90° about a Point (cont'd)

Clockwise Rotation:

Step 2: Perform the rotation about the origin:

Rotating by 90° clockwise about the origin (0, 0) is simply a process of switching the x- and y-values of a point and negating *the new y-term*. That is (x, y) becomes (y, -x) after rotation by 90°.

General Situation	Example
Pre-rotated point (from Step 1): (x_1-x_0, y_1-y_0) Point after rotation: $(y_1-y_0, -x_1+x_0)$	Pre-rotated point (from Step 1): (-4, -2) Point after rotation: (-2, 4)

Step 3: Convert the result back to the original set of axes.

To do this, simply add back the point of rotation (which was subtracted out in Step 1.

General Situation	Example
Point after rotation: $(y_1-y_0, -x_1+x_0)$ Add back the point of rotation (x_0, y_0): $(y_1-y_0+x_0 , -x_1+x_0+y_0)$ which gives us the values of (x_2, y_2)	Point after rotation: (-2, 4) Add back the point of rotation (2, 3): (0, 7)

Finally, look at the formulas for x_2 and y_2:

Clockwise Rotation

$$x_2 = y_1 - y_0 + x_0$$

$$y_2 = -x_1 + x_0 + y_0$$

Notice that the formulas for clockwise and counter-clockwise rotation by 90° are the same except the terms in blue are negated between the formulas.

Interesting note: *If you are asked to find the point about which the rotation occurred, you simply substitute in the values for the starting point (x₁, y₁) and the ending point (x₂, y₂) and solve the resulting pair of simultaneous equations for x₀ and y₀.*

Geometry
Rotation by 90° about a Point (cont'd)

Counter-Clockwise Rotation:

Step 2: Perform the rotation about the origin:

Rotating by 90° counter-clockwise about the origin (0, 0) is simply a process of switching the x- and y-values of a point and negating *the new x-term*. That is (x, y) becomes (-y, x) after rotation by 90°.

General Situation	Example
Pre-rotated point (from Step 1): (x_1-x_0, y_1-y_0) Point after rotation: $(-y_1+y_0, x_1-x_0)$	Pre-rotated point (from Step 1): (-4, -2) Point after rotation: (2, -4)

Step 3: Convert the result back to the original set of axes.

To do this, simply add back the point of rotation (which was subtracted out in Step 1.

General Situation	Example
Point after rotation: $(-y_1+y_0, x_1-x_0)$ Add back the point of rotation (x_0, y_0): $(-y_1+y_0+x_0 , x_1-x_0+y_0)$ which gives us the values of (x_2, y_2)	Point after rotation: (2,-4) Add back the point of rotation (2, 3): (4, -1)

Finally, look at the formulas for x_2 and y_2:

Counter-Clockwise Rotation

$$x_2 = -y_1 + y_0 + x_0$$

$$y_2 = x_1 - x_0 + y_0$$

Notice that the formulas for clockwise and counter-clockwise rotation by 90° are the same except the terms in blue are negated between the formulas.

Interesting note: *The point half-way between the clockwise and counter-clockwise rotations of 90° is the center of rotation itself, **(x₀, y₀)**. In the example, **(2, 3)** is halfway between **(0, 7)** and **(4, -1)**.*

Geometry
Translation

Definitions

Translation is sliding a figure in the plane. Each point in the figure is moved the same distance in the same direction. The result is an image that looks the same as the pre-image in every way, except it has been moved to a different location in the plane.

Each of the four orange line segments in the figure at right has the same length and direction.

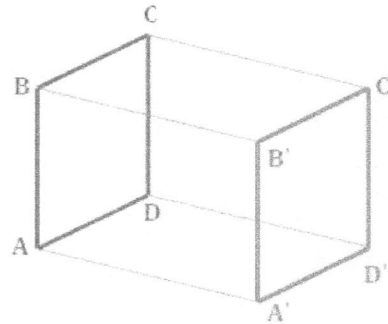

When Two Reflections = One Translation

If two mirrors are parallel, then reflection through one of them, followed by a reflection through the second is a translation.

In the figure at right, the black lines show the paths of the two reflections; this is also the path of the resulting translation. Note the following:

- The distance of the resulting translation (e.g., from A to A'') is double the distance between the mirrors.

- The black lines of movement are perpendicular to both mirrors.

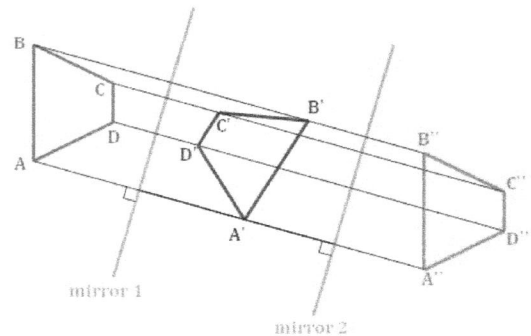

Defining Translations in the Coordinate Plane (Using Vectors)

A translation moves each point by the same distance in the same direction. In the coordinate plane, this is equivalent to moving each point the same amount in the *x-direction* and the same amount in the *y-direction*. This combination of *x*- and *y-direction* movement is described by a mathematical concept called a vector.

In the above figure, translation from A to A'' moves *10* in the *x-direction* and the *-3* in the *y-direction*. In vector notation, this is: $\overrightarrow{AA''} = \langle 10, -3 \rangle$. Notice the "half-ray" symbol over the two points and the funny-looking brackets around the movement values.

So, the translation resulting from the two reflections in the above figure moves each point of the pre-image by the vector $\overrightarrow{AA''}$. Every translation can be defined by the vector representing its movement in the coordinate plane.

Geometry
Compositions

When multiple transformations are combined, the result is called a Composition of the Transformations. Two examples of this are:

- Combining two reflections through parallel mirrors to generate a translation (see the previous page).

- Combining a translation and a reflection to generate what is called a glide reflection. The glide part of the name refers to translation, which is a kind of gliding of a figure on the plane.

> **Note:** In a **glide reflection**, if the line of reflection is parallel to the direction of the translation, it does not matter whether the reflection or the translation is performed first.

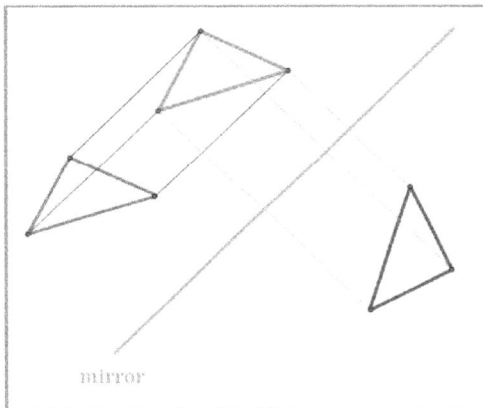

Figure 1: Translation followed by Reflection.   Figure 2: Reflection followed by Translation.

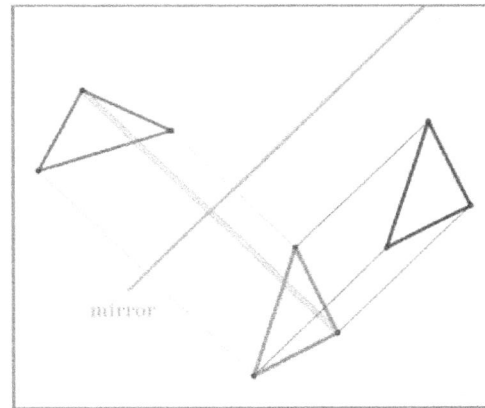

Composition Theorem

The composition of multiple isometries is as Isometry. Put more simply, if transformations that preserve length are combined, the composition will preserve length. This is also true of compositions of transformations that preserve angle measure.

Order of Composition

Order matters in most compositions that involve more than one class of transformation. If you apply multiple transformations of the same kind (e.g., reflection, rotation, or translation), order generally does not matter; however, applying transformations in more than one class may produce different final images if the order is switched.

Geometry
Ratios Involving Units

Ratios Involving Units

When simplifying ratios containing the same units:

- Simplify the fraction.

- Notice that the units disappear. They behave just like factors; if the units exist in the numerator and denominator, the cancel and are not in the answer.

Example:

$$\frac{3 \ inches}{12 \ inches} = \frac{1}{4}$$

Note: the unit "inches cancel out, so the answer is $\frac{1}{4}$, not $\frac{1}{4} \ inch$.

When simplifying ratios containing different units:

- Adjust the ratio so that the numerator and denominator have the same units.

- Simplify the fraction.

- Notice that the units disappear.

Example:

$$\frac{3 \ inches}{2 \ feet} = \frac{3 \ inches}{(2 \ feet) \cdot (12 \ inches/foot)} = \frac{3 \ inches}{24 \ inches} = \frac{1}{8}$$

Dealing with Units

Notice in the above example that units can be treated the same as factors; they can be used in fractions and they cancel when they divide. This fact can be used to figure out whether multiplication or division is needed in a problem. Consider the following:

Example: How long did it take for a car traveling at 48 miles per hour to go 32 miles?

Consider the units of each item: $32 \ miles$ $48 \ \frac{miles}{hour}$

- If you multiply, you get: $(32 \ miles) \cdot \left(48 \frac{miles}{hour}\right) = 1{,}536 \frac{miles^2}{hour}$. This is clearly wrong!

- If you divide, you get: $(32 \ miles) \div \left(48 \frac{miles}{hour}\right) = \frac{32}{48} \ miles \cdot \left(\frac{hour}{miles}\right) = \frac{2}{3} hour$. Now, this looks reasonable. Notice how the "$miles$" unit cancel out in the final answer.

Now you could have solved this problem by remembering that $distance = rate \cdot time$, or $d = rt$. However, paying close attention to the units also generates the correct answer. In addition, the "units" technique always works, no matter what the problem!

Geometry
Similar Polygons

In similar polygons,

- Corresponding angles are congruent, and
- Corresponding sides are proportional.

Both of these conditions are necessary for two polygons to be similar. Conversely, when two polygons are similar, all of the corresponding angles are congruent and all of the sides are proportional.

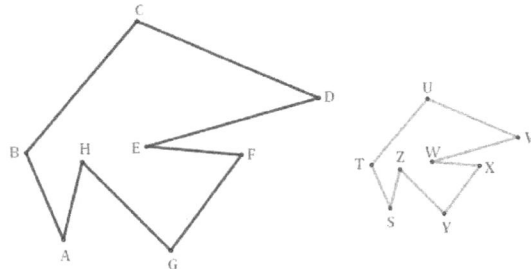

Naming Similar Polygons

Similar polygons should be named such that corresponding angles are in the same location in the name, and the order of the points in the name should "follow the polygon around."

Example: The polygons above could be shown similar with the following names:

$$ABCDEFGHI \sim STUVWXYZ$$

It would also be acceptable to show the similarity as:

$$DEFGHIABC \sim VWXYZSTU$$

Any names that preserve the order of the points and keeps corresponding angles in corresponding locations in the names would be acceptable.

Proportions

One common problem relating to similar polygons is to present three side lengths, where two of the sides correspond, and to ask for the length of the side corresponding to the third length.

Example: In the above similar polygons, if $BC = 20, EF = 12, and WX = 6, what is TU?$

This problem is solvable with proportions. To do so properly, it is important to relate corresponding items in the proportion:

$$\frac{BC}{TU} = \frac{EF}{WX} \quad \longrightarrow \quad \frac{20}{TU} = \frac{12}{6} \quad \longrightarrow \quad TU = 10$$

Notice that the left polygon is represented on the top of both proportions and that the left-most segments of the two polygons are in the left fraction.

Geometry
Scale Factors of Similar Polygons

From the similar polygons below, the following is known about the lengths of the sides:

$$\frac{AB}{ST} = \frac{BC}{TU} = \frac{CD}{UV} = \frac{DE}{VW} = \frac{EF}{WX} = \frac{FG}{XY} = \frac{GH}{YZ} = \frac{HA}{ZA} = k$$

That is, the ratios of corresponding sides in the two polygons are the same and they equal some constant k, called the scale factor of the two polygons. The value of k, then, is all you need to know to relate corresponding sides in the two polygons.

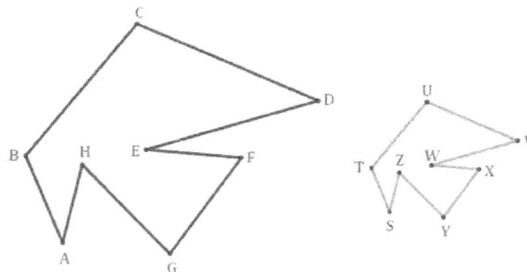

Finding the Missing Length

Any time the student is asked to find the missing length in similar polygons:

- Look for two corresponding sides for which the values are known.
- Calculate the value of k.
- Use the value of k to solve for the missing length.

k is a measure of the relative size of the two polygons. Using this knowledge, it is possible to put into words an easily understandable relationship between the polygons.

- Let Polygon 1 be the one whose sides are in the numerators of the fractions.
- Let Polygon 2 be the one whose sides are in the denominators of the fractions.
- Then, it can be said that Polygon 1 is k times the size of the Polygon 2.

Example: In the above similar polygons, if $BC = 20, EF = 12, and\ WX = 6, what\ is\ TU$?

Seeing that EF and WX relate, calculate:

$$\frac{EF}{WX} = \frac{12}{6} = 2 = k$$

Then solve for TU based on the value of k:

$$\frac{BC}{TU} = k \quad \rightarrow \quad \frac{20}{TU} = 2 \quad \rightarrow \quad TU = 10$$

Also, since $k = 2$, the length of every side in the blue polygon is double the length of its corresponding side in the orange polygon.

Geometry
Dilation of Polygons

A dilation is a special case of transformation involving similar polygons. It can be thought of as a transformation that creates a polygon of the same shape but a different size from the original. Key elements of a dilation are:

- Scale Factor – The scale factor of similar polygons is the constant k which represents the relative sizes of the polygons.

- Center – The center is the point from which the dilation takes place.

Note that $k > 0$ and $k \neq 1$ in order to generate a second polygon. Then,

- If $k > 1$, the dilation is called an "enlargement."
- If $k < 1$, the dilation is called a "reduction."

Dilations with Center (0, 0)

In coordinate geometry, dilations are often performed with the center being the origin $(0, 0)$. In that case, to obtain the dilation of a polygon:

- Multiply the coordinates of each vertex by the scale factor k, and
- Connect the vertices of the dilation with line segments (i.e., connect the dots).

Examples:

In the following examples:

- The green polygon is the original.
- The blue polygon is the dilation.
- The dashed orange lines show the movement away from (enlargement) or toward (reduction) the center, which is the origin in all 3 examples.

Notice that, in each example:

$$\begin{pmatrix} distance\ from\ center \\ to\ a\ vertex\ of\ the \\ dilated\ polygon \end{pmatrix} = k \cdot \begin{pmatrix} distance\ from\ center \\ to\ a\ vertex\ of\ the \\ original\ polygon \end{pmatrix}$$

This fact can be used to construct dilations when coordinate axes are not available. Alternatively, the student could draw a set of coordinate axes as an aid to performing the dilation.

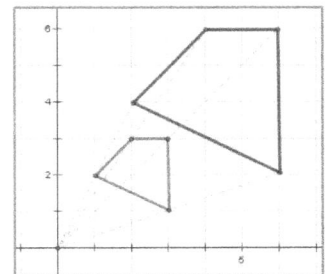

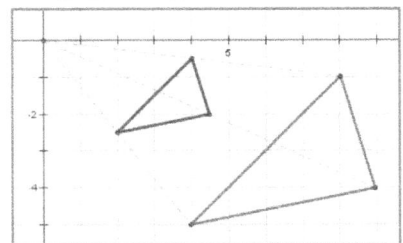

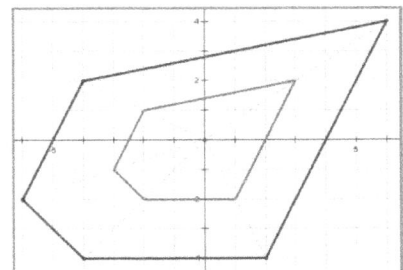

Geometry
More on Dilation

Dilations of Non-Polygons

Any geometric figure can be dilated. In the dilation of the green circle at right, notice that:

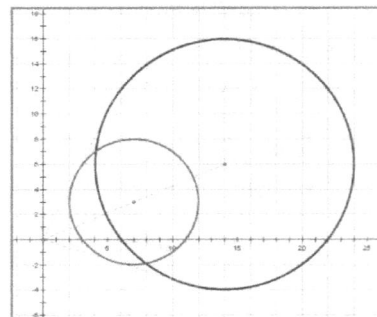

- The dilation factor is 2.
- The original circle has center $(7, 3)$ and radius $= 5$.
- The dilated circle has center $(14, 6)$ and radius $= 10$.

So, the center and radius are both increased by a factor of $k = 2$. This is true of any figure in a dilation with the center at the origin. All of the key elements that define the figure are increased by the scale factor k.

Dilations with Center (a, b)

In the figures below, the green quadrilaterals are dilated to the blue ones with a scale factor of $k = 2$. Notice the following:

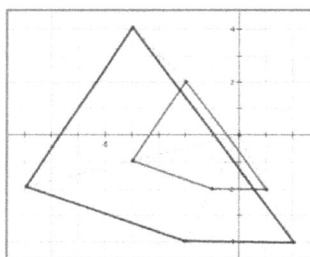

In the figure to the left, the dilation has center $(0, 0)$, whereas in the figure to the right, the dilation has center $(-4, -3)$. The size of the resulting figure is the same in both cases (because $k = 2$ in both figures), but the location is different.

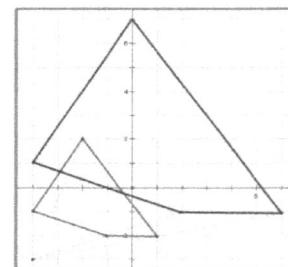

Graphically, the series of transformations that is equivalent to a dilation from a point (a, b) other than the origin is shown below. Compare the final result to the figure above (right).

- Step 1: Translate the original figure by $(-a, -b)$ to reset the center at the origin.
- Step 2: Perform the dilation.
- Step 3: Translate the dilated figure by (a, b). These steps are illustrated below.

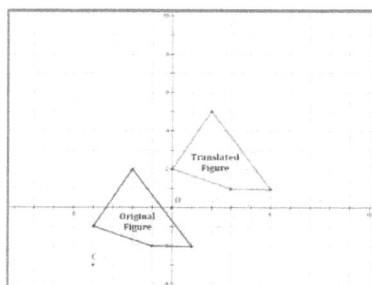

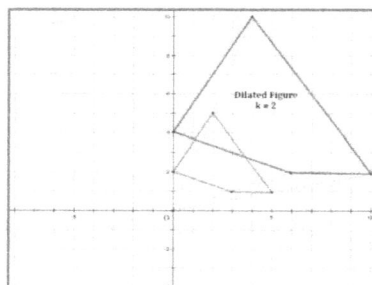

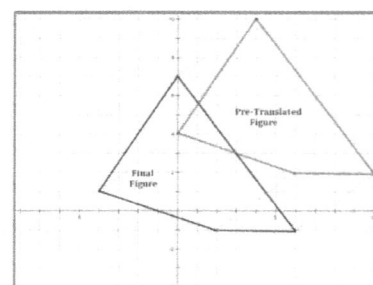

Step 1   Step 2   Step 3

Geometry
Similar Triangles

The following theorems present conditions under which triangles are similar.

Side-Angle-Side (SAS) Similarity

SAS similarity requires the proportionality of two sides and the congruence of the angle between those sides. Note that there is no such thing as SSA similarity; the congruent angle must be between the two proportional sides.

Side-Side-Side (SSS) Similarity

SSS similarity requires the proportionality of all three sides. If all of the sides are proportional, then all of the angles must be congruent.

Angle--Angle (AA) Similarity

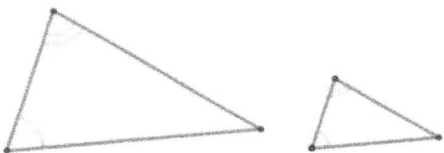

AA similarity requires the congruence of two angles and the side between those angles.

Similar Triangle Parts

In similar triangles,

- Corresponding sides are proportional.
- Corresponding angles are congruent.

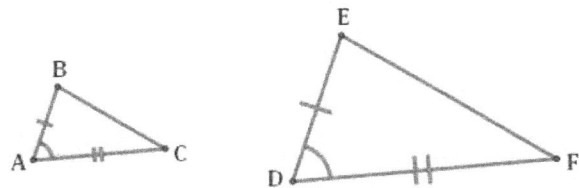

Establishing the proper names for similar triangles is crucial to line up corresponding vertices. In the picture above, we can say:

$$\triangle ABC \sim \triangle DEF \quad \text{or} \quad \triangle BCA \sim \triangle EFD \quad \text{or} \quad \triangle CAB \sim \triangle FDE \quad \text{or}$$
$$\triangle ACB \sim \triangle DFE \quad \text{or} \quad \triangle BAC \sim \triangle EDF \quad \text{or} \quad \triangle CBA \sim \triangle FED$$

All of these are correct because they match corresponding parts in the naming. Each of these similarities implies the following relationships between parts of the two triangles:

$$\angle A \cong \angle D \quad \text{and} \quad \angle B \cong \angle E \quad \text{and} \quad \angle C \cong \angle F$$

$$\frac{AB}{DE} = \frac{BC}{EF} = \frac{CA}{FD}$$

Geometry
Proportion Tables for Similar Triangles

Setting Up a Table of Proportions

It is often useful to set up a table to identify the proper proportions in a similarity. Consider the figure to the right. The table might look something like this:

Triangle	Left Side	Right Side	Bottom Side
Top Δ	AB	BC	CA
Bottom Δ	DE	EF	FD

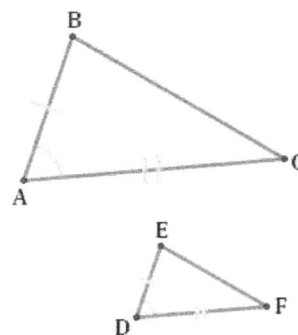

The purpose of a table like this is to organize the information you have about the similar triangles so that you can readily develop the proportions you need.

Developing the Proportions

To develop proportions from the table:

- Extract the columns needed from the table:

AB	BC
DE	EF

- Eliminate the table lines.

- Replace the horizontal lines with "division lines."

- Put an equal sign between the two resulting fractions:

$$\frac{AB}{DE} = \frac{BC}{EF}$$

Also from the above table,

$$\frac{AB}{DE} = \frac{CA}{FD}$$

$$\frac{BC}{EF} = \frac{CA}{FD}$$

Solving for the unknown length of a side:

You can extract any two columns you like from the table. Usually, you will have information on lengths of three of the sides and will be asked to calculate a fourth.

Look in the table for the columns that contain the 4 sides in question, and then set up your proportion. Substitute known values into the proportion, and solve for the remaining variable.

Geometry
Three Similar Triangles

A common problem in geometry is to find the missing value in proportions based on a set of three similar triangles, two of which are inside the third. The diagram often looks like this:

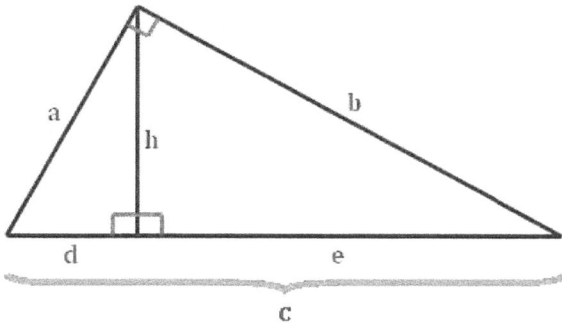

Pythagorean Relationships

Inside triangle on the left: $d^2 + h^2 = a^2$

Inside triangle on the right: $h^2 + e^2 = b^2$

Outside (large) triangle: $a^2 + b^2 = c^2$

Similar Triangle Relationships

Because all three triangles are similar, we have the relationships in the table below. These relationships are not obvious from the picture, but are very useful in solving problems based on the above diagram. Using similarities between the triangles, 2 at a time, we get:

From the two inside triangles	From the inside triangle on the left and the outside triangle	From the inside triangle on the right and the outside triangle
$$\frac{h}{d} = \frac{e}{h}$$	$$\frac{a}{d} = \frac{c}{a}$$	$$\frac{b}{e} = \frac{c}{b}$$
or	or	or
$$h^2 = d \cdot e$$	$$a^2 = d \cdot c$$	$$b^2 = e \cdot c$$
The height squared = the product of: the two parts of the base	The left side squared = the product of: the part of the base below it and the entire base	The right side squared = the product of: the part of the base below it and the entire base

Geometry
Pythagorean Theorem

In a right triangle, the Pythagorean Theorem says:

$$a^2 + b^2 = c^2$$

where,

- a and b are the lengths of the legs of a right triangle, and
- c is the length of the hypotenuse.

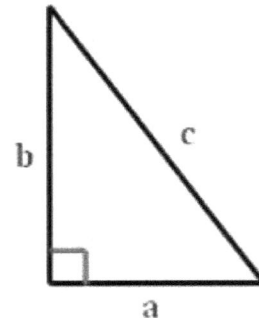

Right, Acute, or Obtuse Triangle?

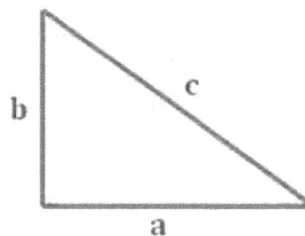

In addition to allowing the solution of right triangles, the Pythagorean Formula can be used to determine whether a triangle is a right triangle, an acute triangle, or an obtuse triangle.

To determine whether a triangle is obtuse, right, or acute:

- Arrange the lengths of the sides from low to high; call them a, b, and c, in increasing order
- Calculate: $a^2, b^2,$ and c^2.
- Compare: $a^2 + b^2$ vs. c^2
- Use the illustrations below to determine which type of triangle you have.

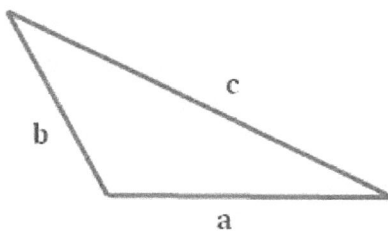

Obtuse Triangle	**Right Triangle**	**Acute Triangle**
$a^2 + b^2 < c^2$	$a^2 + b^2 = c^2$	$a^2 + b^2 > c^2$

Example:
Triangle with sides: 7, 9, 12

$7^2 + 9^2$ *vs.* 12^2

$49 + 81 < 144$

$\rightarrow$ *Obtuse Triangle*

Example:
Triangle with sides: 6, 8, 10

$6^2 + 8^2$ *vs.* 10^2

$36 + 64 = 100$

$\rightarrow$ *Right Triangle*

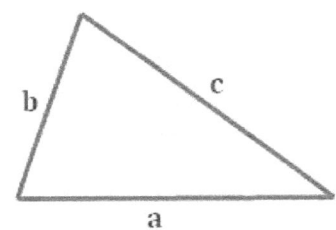

Example:
Triangle with sides: 5, 8, 9

$5^2 + 8^2$ *vs.* 9^2

$25 + 64 > 81$

$\rightarrow$ *Acute Triangle*

Geometry
Pythagorean Triples

Pythagorean Theorem: $\qquad a^2 + b^2 = c^2$

Pythagorean triples are sets of 3 positive integers that meet the requirements of the Pythagorean Theorem. Because these sets of integers provide "pretty" solutions to geometry problems, they are a favorite of geometry books and teachers. Knowing what triples exist can help the student quickly identify solutions to problems that might otherwise take considerable time to solve.

3-4-5 Triangle Family

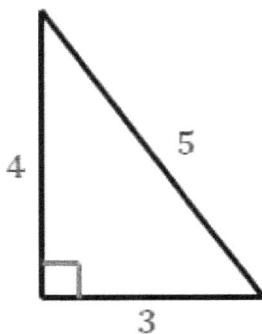

Sample Triples

3-4-5
6-8-10
9-12-15
12-16-20
30-40-50

$3^2 + 4^2 = 5^2$

$9 + 16 = 25$

7-24-25 Triangle Family

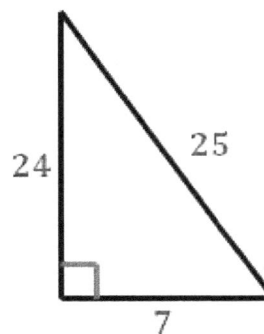

Sample Triples

7-24-25
14-48-50
21-72-75
. . .
70-240-250

$7^2 + 24^2 = 25^2$

$49 + 576 = 625$

5-12-13 Triangle Family

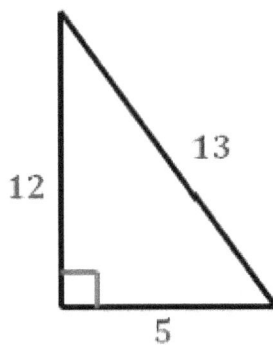

Sample Triples

5-12-13
10-24-26
15-36-39
. . .
50-120-130

$5^2 + 12^2 = 13^2$

$25 + 144 = 169$

8-15-17 Triangle Family

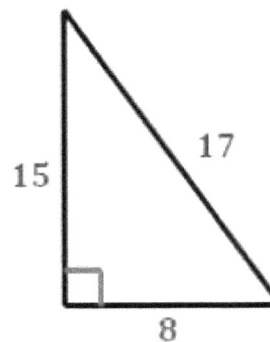

Sample Triples

8-15-17
16-30-34
24-45-51
. . .
80-150-170

$8^2 + 15^2 = 17^2$

$64 + 225 = 289$

Geometry
Special Triangles

The relationship among the lengths of the sides of a triangle is dependent on the measures of the angles in the triangle. For a right triangle (i.e., one that contains a 90^0 angle), two special cases are of particular interest. These are shown below:

45^0-45^0-90^0 Triangle

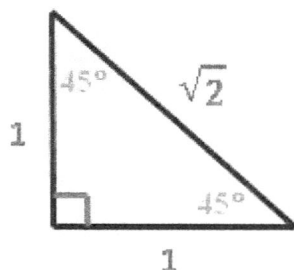

In a **45^0-45^0-90^0 triangle**, the congruence of two angles guarantees the congruence of the two legs of the triangle. The proportions of the three sides are: $1 : 1 : \sqrt{2}$. That is, the two legs have the same length and the hypotenuse is $\sqrt{2}$ times as long as either leg.

30^0-60^0-90^0 Triangle

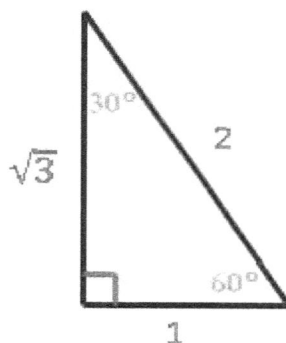

In a **30^0-60^0-90^0 triangle**, the proportions of the three sides are: $1 : \sqrt{3} : 2$. That is, the long leg is $\sqrt{3}$ times as long as the short leg, and the hypotenuse is 2 times as long as the short leg.

In a right triangle, we need to know the lengths of two sides to determine the length of the third. The power of the relationships in the special triangles lies in the fact that we need only know the length of one side of the triangle to determine the lengths of the other two sides.

Example Side Lengths

45^0-45^0-90^0 Triangle		30^0-60^0-90^0 Triangle	
$1 : 1 : \sqrt{2}$	$2 : 2 : 2\sqrt{2}$	$1 : \sqrt{3} : 2$	$2 : 2\sqrt{3} : 4$
$\sqrt{2} : \sqrt{2} : 2$	$\sqrt{3} : \sqrt{3} : \sqrt{6}$	$\sqrt{2} : \sqrt{6} : 2\sqrt{2}$	$\sqrt{3} : 3 : 2\sqrt{3}$
$3\sqrt{2} : 3\sqrt{2} : 6$	$25 : 25 : 25\sqrt{2}$	$3\sqrt{2} : 3\sqrt{6} : 6\sqrt{2}$	$25 : 25\sqrt{3} : 50$

Geometry
Trig Functions and Special Angles

Trigonometric Functions

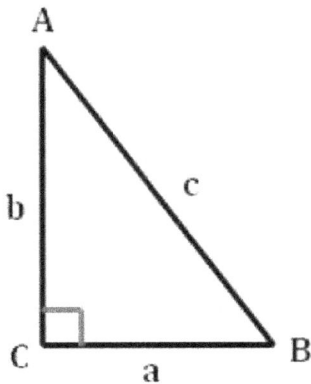

SOH-CAH-TOA

$$\sin = \frac{opposite}{hypoteneuse} \qquad \sin A = \frac{a}{c} \qquad \sin B = \frac{b}{c}$$

$$\cos = \frac{adjacent}{hypoteneuse} \qquad \cos A = \frac{b}{c} \qquad \cos B = \frac{a}{c}$$

$$\tan = \frac{opposite}{adjacent} \qquad \tan A = \frac{a}{b} \qquad \tan B = \frac{b}{a}$$

Special Angles

Trig Functions of Special Angles				
Radians	Degrees	$\sin \theta$	$\cos \theta$	$\tan \theta$
0	$0°$	$\frac{\sqrt{0}}{2} = 0$	$\frac{\sqrt{4}}{2} = 1$	$\frac{\sqrt{0}}{\sqrt{4}} = 0$
$\pi/6$	$30°$	$\frac{\sqrt{1}}{2} = \frac{1}{2}$	$\frac{\sqrt{3}}{2}$	$\frac{\sqrt{1}}{\sqrt{3}} = \frac{\sqrt{3}}{3}$
$\pi/4$	$45°$	$\frac{\sqrt{2}}{2}$	$\frac{\sqrt{2}}{2}$	$\frac{\sqrt{2}}{\sqrt{2}} = 1$
$\pi/3$	$60°$	$\frac{\sqrt{3}}{2}$	$\frac{\sqrt{1}}{2} = \frac{1}{2}$	$\frac{\sqrt{3}}{\sqrt{1}} = \sqrt{3}$
$\pi/2$	$90°$	$\frac{\sqrt{4}}{2} = 1$	$\frac{\sqrt{0}}{2} = 0$	undefined

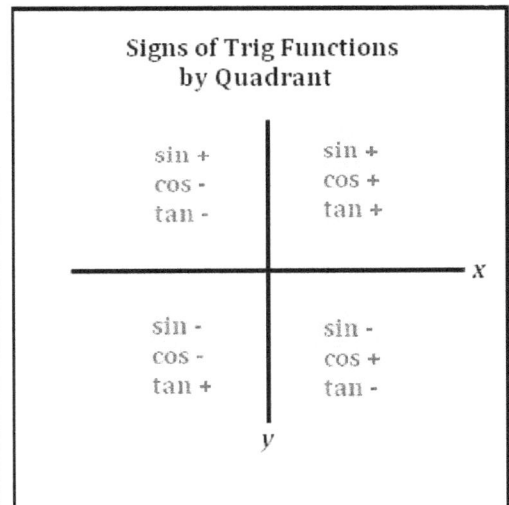

Signs of Trig Functions by Quadrant

sin + cos - tan -	sin + cos + tan +
sin - cos - tan +	sin - cos + tan -

Geometry
Trigonometric Function Values in Quadrants II, III, and IV

In quadrants other than Quadrant I, trigonometric values for angles are calculated in the following manner:

- Draw the angle θ on the Cartesian Plane.

- Calculate the measure of the angle from the x-axis to θ.

- Find the value of the trigonometric function of the angle in the previous step.

- Assign a "+" or "−" sign to the trigonometric value based on the function used and the quadrant θ is in.

Signs of Trig Functions by Quadrant

sin + cos - tan -	sin + cos + tan +
sin - cos - tan +	sin - cos + tan -

Examples:

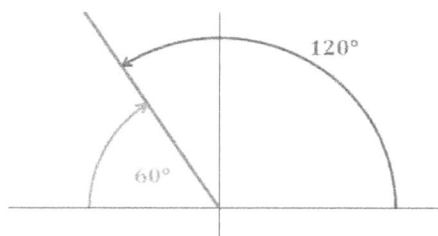

Θ in Quadrant II – Calculate: $(180^0 - m\angle\theta)$

For $\theta = 120^0$, base your work on $180° - 120° = 60°$

$\sin 60° = \dfrac{\sqrt{3}}{2}$, so: $\sin 120° = \dfrac{\sqrt{3}}{2}$

Θ in Quadrant III – Calculate: $(m\angle\theta - 180^0)$

For $\theta = 210^0$, base your work on $210° - 180° = 30°$

$\cos 30° = \dfrac{\sqrt{3}}{2}$, so: $\cos 210° = -\dfrac{\sqrt{3}}{2}$

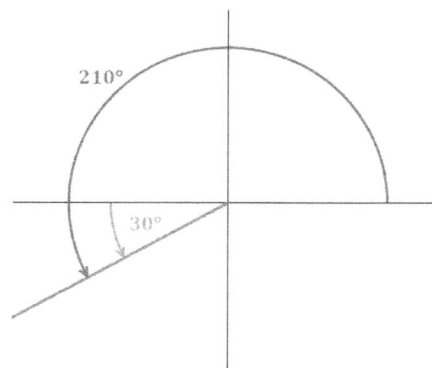

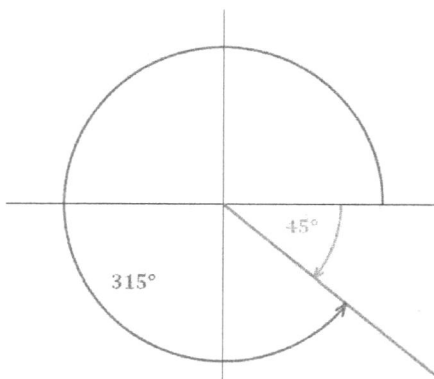

Θ in Quadrant IV – Calculate: $(360^0 - m\angle\theta)$

For $\theta = 315^0$, base your work on $360° - 315° = 45°$

$\tan 45° = 1$, so: $\tan 315° = -1$

Geometry
Graphs of Trigonometric Functions

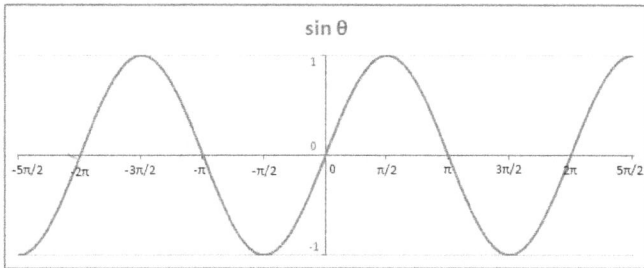

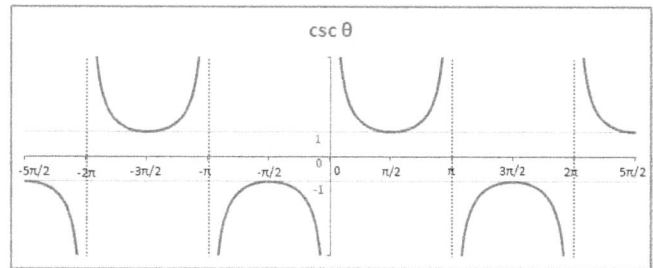

The sine and cosecant functions are inverses. So:

$$\sin \theta = \frac{1}{\csc \theta} \quad \text{and} \quad \csc \theta = \frac{1}{\sin \theta}$$

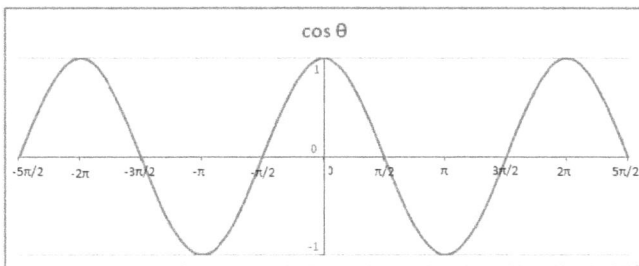

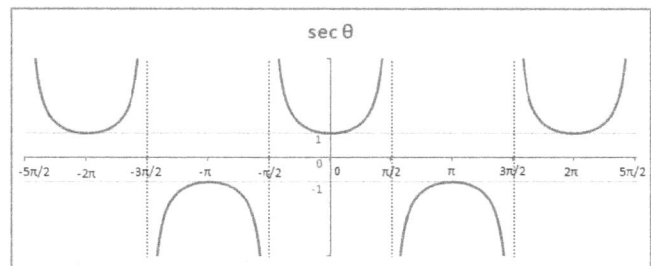

The cosine and secant functions are inverses. So:

$$\cos \theta = \frac{1}{\sec \theta} \quad \text{and} \quad \sec \theta = \frac{1}{\cos \theta}$$

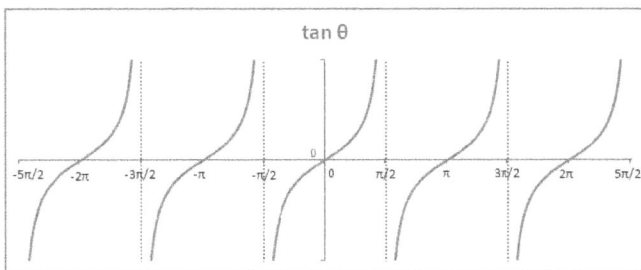

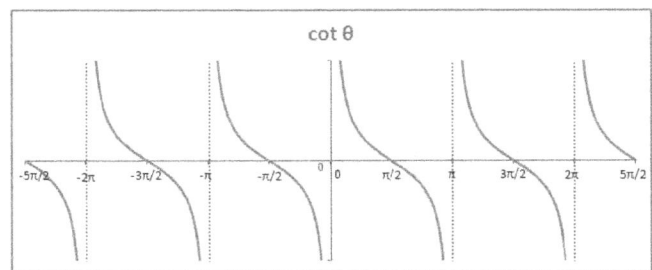

The tangent and cotangent functions are inverses. So:

$$\tan \theta = \frac{1}{\cot \theta} \quad \text{and} \quad \cot \theta = \frac{1}{\tan \theta}$$

Geometry
Vectors

Definitions

- A vector is a geometric object that has both magnitude (length) and direction.

- The Tail of the vector is the end opposite the arrow. It represents where the vector is moving from.

- The Head of the vector is the end with the arrow. It represents where the vector is moving to.

- The Zero Vector is denoted **0**. It has zero length and all the properties of zero.

- Two vectors are equal is they have both the same magnitude and the same direction.

- Two vectors are parallel if they have the same or opposite directions. That is, if the angles of the vectors are the same or $180°$ different.

- Two vectors are perpendicular if the difference of the angles of the vectors is $90°$ or $270°$.

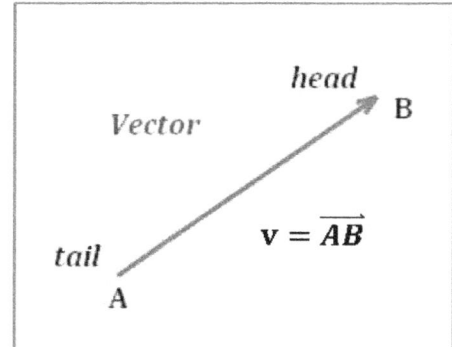

Magnitude of a Vector

The distance formula gives the magnitude of a vector. If the head and tail of vector **v** are the points $A = (x_1, y_1)$ and $B = (x_2, y_2)$, then the magnitude of **v** is:

$$|\mathbf{v}| = |\overrightarrow{AB}| = \sqrt{(x_2 - x_1)^2 + (y_2 - y_1)^2}$$

Note that $|\overrightarrow{AB}| = |\overrightarrow{BA}|$. The directions of the two vectors are opposite, but their magnitudes are the same.

Direction of a Vector

The direction of a vector is determined by the angle it makes with a horizontal line. In the figure at right, the direction is the angle θ. The value of θ can be calculated based on the lengths of the sides of the triangle the vector forms.

$$\tan \theta = \frac{3}{4} \quad \text{or} \quad \theta = \tan^{-1}\left(\frac{3}{4}\right)$$

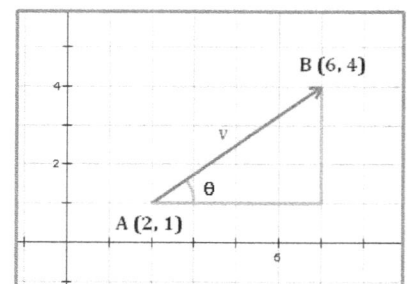

where the function **tan**$^{-1}$ is the inverse tangent function. The second equation in the line above reads "θ is the angle whose tangent is $\frac{3}{4}$."

Geometry
Operations with Vectors

It is possible to operate with vectors in some of the same ways we operate with numbers. In particular:

Adding Vectors

Vectors can be added in rectangular form by separately adding their *x*- and *y-components.* In general,

$$\mathbf{u} = \langle u_1, u_2 \rangle$$

$$\mathbf{v} = \langle v_1, v_2 \rangle$$

$$\mathbf{u} + \mathbf{v} = \langle u_1, u_2 \rangle + \langle v_1, v_2 \rangle = \langle u_1 + v_1, u_2 + v_2 \rangle$$

Example: In the figure at right,

$$\mathbf{u} = \langle 4, 3 \rangle$$

$$\mathbf{v} = \langle 2, -6 \rangle$$

$$\mathbf{w} = \mathbf{u} + \mathbf{v} = \langle 4, 3 \rangle + \langle 2, -6 \rangle = \langle 6, -3 \rangle$$

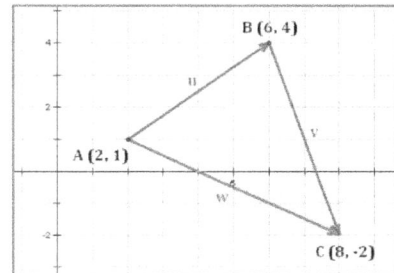

Vector Algebra

$\mathbf{u} + \mathbf{v} = \mathbf{v} + \mathbf{u}$	$\mathbf{u} + (-\mathbf{u}) = \mathbf{0}$	$a \cdot (\mathbf{u} + \mathbf{v}) = (a \cdot \mathbf{u}) + (a \cdot \mathbf{v})$
$(\mathbf{u} + \mathbf{v}) + \mathbf{w} = \mathbf{u} + (\mathbf{w} + \mathbf{v})$	$0 \cdot \mathbf{u} = \mathbf{0}$	$(a + b) \cdot \mathbf{u} = (a \cdot \mathbf{u}) + (b \cdot \mathbf{u})$
$\mathbf{u} + \mathbf{0} = \mathbf{u}$	$1 \cdot \mathbf{u} = \mathbf{u}$	$(ab) \cdot \mathbf{u} = a \cdot (b \cdot \mathbf{u}) = b \cdot (a \cdot \mathbf{u})$

Scalar Multiplication

Scalar multiplication changes the magnitude of a vector, but not the direction. In general,

$$\mathbf{u} = \langle u_1, u_2 \rangle$$

$$k \cdot \mathbf{u} = \langle k \cdot u_1, k \cdot u_2 \rangle$$

In the figure at right,

$$\mathbf{u} = \langle 4, 3 \rangle$$

$$2 \cdot \mathbf{u} = 2 \cdot \langle 4, 3 \rangle = \langle 8, 6 \rangle$$

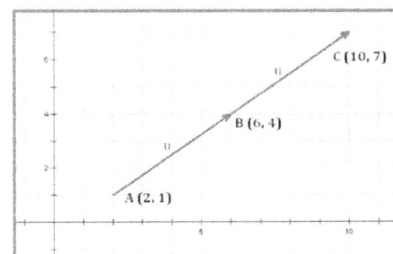

Geometry
Parts of Circles

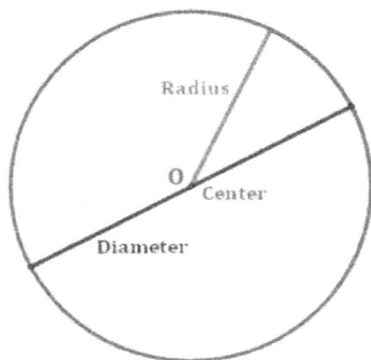

Center – the middle of the circle. All points on the circle are the same distance from the center.

Radius – a line segment with one endpoint at the center and the other endpoint on the circle. The term "radius" is also used to refer to the distance from the center to the points on the circle.

Diameter – a line segment with endpoints on the circle that passes through the center.

Arc – a path along a circle.

Minor Arc – a path along the circle that is less than 180°.

Major Arc – a path along the circle that is greater than 180°.

Semicircle – a path along a circle that equals 180°.

Sector – a region inside a circle that is bounded by two radii and an arc.

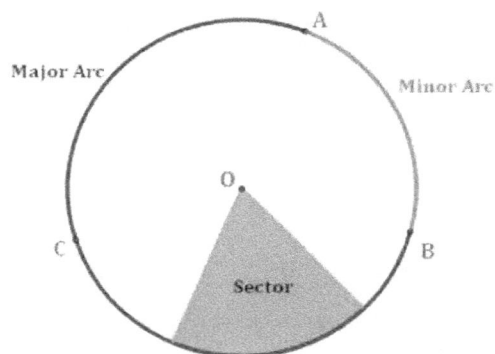

Secant Line – a line that intersects the circle in exactly two points.

Tangent Line– a line that intersects the circle in exactly one point.

Chord – a line segment with endpoints on the circle that does not pass through the center.

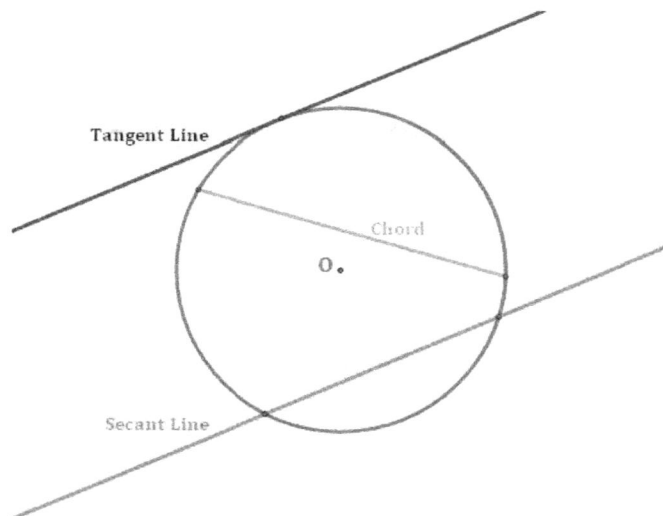

Geometry
Angles and Circles

Central Angle

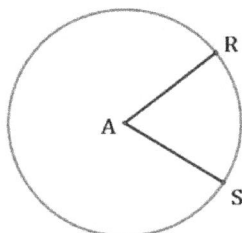

$$m\angle A = m\,\widehat{RS}$$

Inscribed Angle

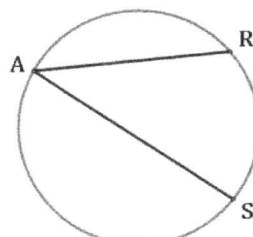

$$m\angle A = \frac{1}{2}\,m\,\widehat{RS}$$

Vertex inside the circle

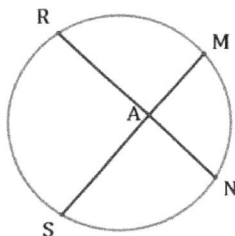

$$m\angle A = \frac{1}{2}\left(m\,\widehat{RS} + m\,\widehat{MN}\right)$$
$$RA \cdot AN = SA \cdot AM$$

Vertex outside the circle

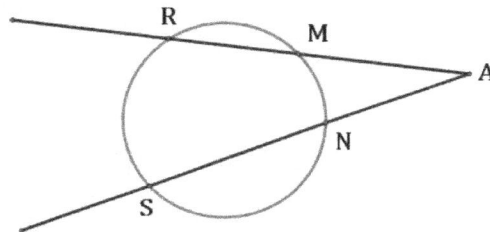

$$m\angle A = \frac{1}{2}\left(m\,\widehat{RS} - m\,\widehat{MN}\right)$$
$$AM \cdot AR = AN \cdot AS$$

Tangent on one side

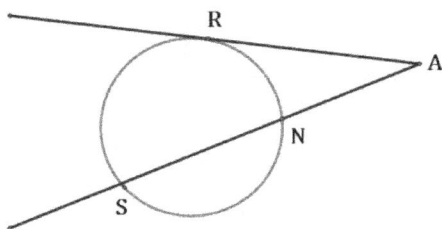

$$m\angle A = \frac{1}{2}\left(m\,\widehat{RS} - m\,\widehat{RN}\right)$$
$$AR^2 = AN \cdot AS$$

Tangents on two sides

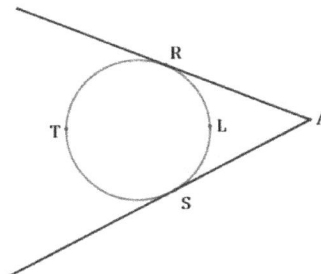

$$m\angle A = \frac{1}{2}\left(m\,\widehat{RTS} - m\,\widehat{RLS}\right)$$
$$AR = AS$$

Geometry
Perimeter and Area of a Triangle

Perimeter of a Triangle

The perimeter of a triangle is simply the sum of the measures of the three sides of the triangle.

$$P = a + b + c$$

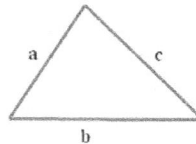

Area of a Triangle

There are two formulas for the area of a triangle, depending on what information about the triangle is available.

Formula 1: The formula most familiar to the student can be used when the base and height of the triangle are either known or can be determined.

$$A = \frac{1}{2}bh$$

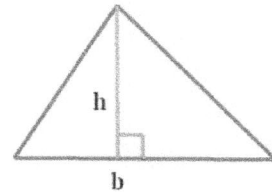

where, b is the length of the base of the triangle.

h is the height of the triangle.

Note: The base can be any side of the triangle. The height is the measure of the altitude of whichever side is selected as the base. So, you can use:

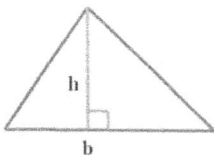 or 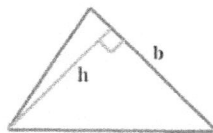 or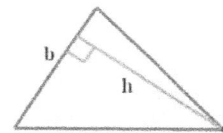

Formula 2: Heron's formula for the area of a triangle can be used when the lengths of all of the sides are known. Sometimes this formula, though less appealing, can be very useful.

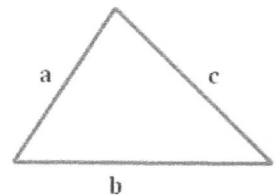

$$A = \sqrt{s(s - a)(s - b)(s - c)}$$

where, $s = \frac{1}{2}P = \frac{1}{2}(a + b + c)$.   **Note:** s is sometimes called the semi-perimeter of the triangle.

a, b, c are the lengths of the sides of the triangle.

Geometry
More on the Area of a Triangle

Trigonometric Formulas

The following formulas for the area of a triangle come from trigonometry. Which one is used depends on the information available:

Two angles and a side:

$$A = \frac{1}{2} \cdot \frac{a^2 \cdot \sin B \cdot \sin C}{\sin A} = \frac{1}{2} \cdot \frac{b^2 \cdot \sin A \cdot \sin C}{\sin B} = \frac{1}{2} \cdot \frac{c^2 \cdot \sin A \cdot \sin B}{\sin C}$$

Two sides and an angle:

$$A = \frac{1}{2} ab \sin C = \frac{1}{2} ac \sin B = \frac{1}{2} bc \sin A$$

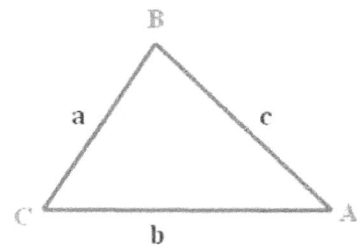

Coordinate Geometry

If the three vertices of a triangle are displayed in a coordinate plane, the formula below, using a determinant, will give the area of a triangle.

Let the three points in the coordinate plane be: (x_1, y_1), (x_2, y_2), (x_3, y_3). Then, the area of the triangle is one half of the absolute value of the determinant below:

$$A = \frac{1}{2} \cdot \left| \begin{vmatrix} x_1 & y_1 & 1 \\ x_2 & y_2 & 1 \\ x_3 & y_3 & 1 \end{vmatrix} \right|$$

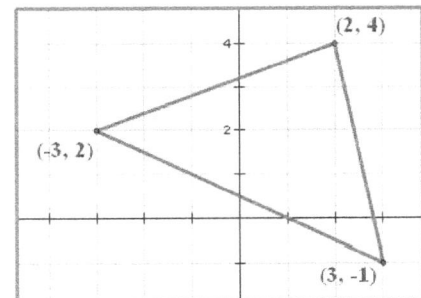

Example: For the triangle in the figure at right, the area is:

$$A = \frac{1}{2} \cdot \left| \begin{vmatrix} 2 & 4 & 1 \\ -3 & 2 & 1 \\ 3 & -1 & 1 \end{vmatrix} \right|$$

$$= \frac{1}{2} \cdot \left| \left(2 \begin{vmatrix} 2 & 1 \\ -1 & 1 \end{vmatrix} - 4 \begin{vmatrix} -3 & 1 \\ 3 & 1 \end{vmatrix} + \begin{vmatrix} -3 & 2 \\ 3 & -1 \end{vmatrix} \right) \right| = \frac{1}{2} \cdot 27 = \frac{27}{2}$$

Geometry
Perimeter and Area of Regular Polygons

Definitions – Regular Polygons

- The center of a polygon is the center of its circumscribed circle. Point O is the center of the hexagon at right.

- The radius of the polygon is the radius of its circumscribed circle. $\overline{OA}$ and $\overline{OB}$ are both radii of the hexagon at right.

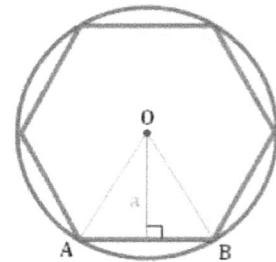

- The apothem of a polygon is the distance from the center to the midpoint of any of its sides. a is the apothem of the hexagon at right.

- The central angle of a polygon is an angle whose vertex is the center of the circle and whose sides pass through consecutive vertices of the polygon. In the figure above, $\angle AOB$ is a central angle of the hexagon.

Area of a Regular Polygon

$$A = \frac{1}{2}aP$$ where, a is the apothem of the polygon

P is the perimeter of the polygon

Perimeter and Area of Similar Figures

Let k be the scale factor relating two similar geometric figures F_1 and F_2 such that $F_2 = k \cdot F_1$.

Then,

$$\frac{\text{Perimeter of } F_2}{\text{Perimeter of } F_1} = k$$

and

$$\frac{\text{Area of } F_2}{\text{Area of } F_1} = k^2$$

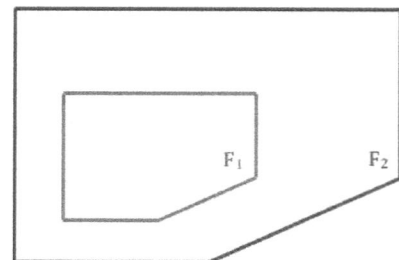

Geometry
Circle Lengths and Areas

Circumference and Area

$C = 2\pi \cdot r$ is the circumference (i.e., the perimeter) of the circle.

$A = \pi r^2$ is the area of the circle.

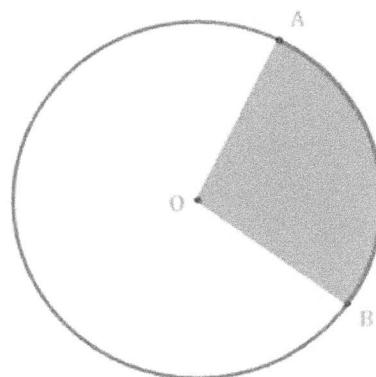

where: r is the radius of the circle.

Length of an Arc on a Circle

A common problem in the geometry of circles is to measure the length of an arc on a circle.

Definition: An arc is a segment along the circumference of a circle.

$$arc\ length = \frac{m\widehat{AB}}{360} \cdot C$$

where: $m\angle\widehat{AB}$ is the measure (in degrees) of the arc. Note that this is also the measure of the central angle $\angle AOB$.

C is the circumference of the circle.

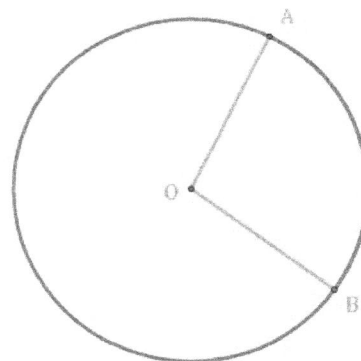

Area of a Sector of a Circle

Another common problem in the geometry of circles is to measure the area of a sector a circle.

Definition: A sector is a region in a circle that is bounded by two radii and an arc of the circle.

$$sector\ area = \frac{m\widehat{AB}}{360} \cdot A$$

where: $m\angle\widehat{AB}$ is the measure (in degrees) of the arc. Note that this is also the measure of the central angle $\angle AOB$.

A is the area of the circle.

Geometry
Area of Composite Figures

To calculate the area of a figure that is a composite of shapes, consider each shape separately.

Example 1:

Calculate the area of the blue region in the figure to the right.

To solve this:

- Recognize that the figure is the composite of a rectangle and two triangles.
- Disassemble the composite figure into its components.
- Calculate the area of the components.
- Subtract to get the area of the composite figure.

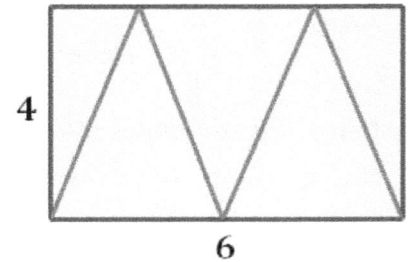

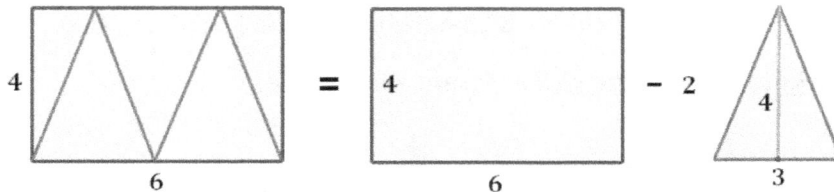

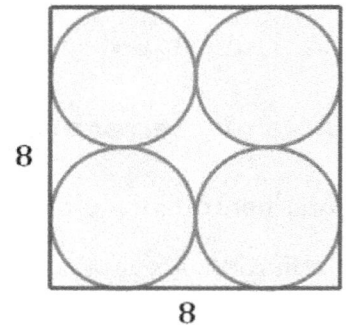

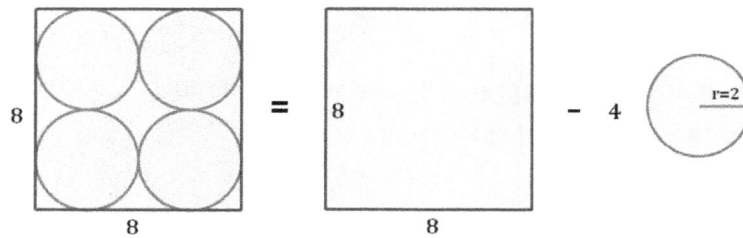

$$Area\ of\ Region\ = (4 \cdot 6) - 2\left(\frac{1}{2} \cdot 4 \cdot 3\right)\ =\ 24 - 12\ =\ 12$$

Example 2:

Calculate the area of the blue region in the figure to the right.

To solve this:

- Recognize that the figure is the composite of a square and a circle.
- Disassemble the composite figure into its components.
- Calculate the area of the components.
- Subtract to get the area of the composite figure.

$$Area\ of\ Region\ =\ 8^2 - 4(\pi \cdot 2^2)\ =\ 64 - 16\pi\ \sim\ 13.73$$

Geometry
Polyhedra

Definitions

- A Polyhedron is a 3-dimensional solid bounded by a series of polygons.

- Faces are the polygons that bound the polyhedron.

- An Edge is the line segment at the intersection of two faces.

- A Vertex is a point at the intersection of two edges.

- A Regular polyhedron is one in which all of the faces are the same regular polygon.

- A Convex Polyhedron is one in which all diagonals are contained within the interior of the polyhedron. A Concave polyhedron is one that is not convex.

- A Cross Section is the intersection of a plane with the polyhedron.

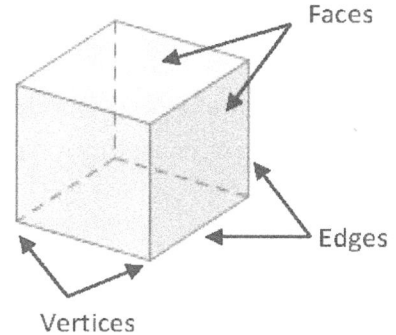

Euler's Theorem

Let: F = the number of faces of a polyhedron.

V = the number of vertices of a polyhedron.

E = the number of edges of a polyhedron.

Then, for any polyhedron that does not intersect itself,

$$F + V = E + 2$$

Euler's Theorem Example:

The cube above has ...

- 6 faces
- 8 vertices
- 12 edges

$6 + 8 = 12 + 2$ ✔

Calculating the Number of Edges

The number of edges of a polyhedron is one-half the number of sides in the polygons it comprises. Each side that is counted in this way is shared by two polygons; simply adding all the sides of the polygons, therefore, double counts the number of edges on the polyhedron.

Example: Consider a soccer ball. It is polyhedron made up of 20 hexagons and 12 pentagons. Then the number of edges is:

$$E = \frac{1}{2} \cdot [(20 \cdot 6) + (12 \cdot 5)] = 90$$

Geometry
A Hole in Euler's Theorem

Topology is a branch of mathematics that studies the properties of objects that are preserved through manipulation that does not include tearing. An object may be stretched, twisted and otherwise deformed, but not torn. In this branch of mathematics, a donut is equivalent to a coffee cup because both have one hole; you can deform either the cup or the donut and create the other, like you are playing with clay.

All of the usual polyhedra have no holes in them, so Euler's Equation holds. What happens if we allow the polyhedra to have holes in them? That is, what if we consider topological shapes different from the ones we normally consider?

Euler's Characteristic

When Euler's Equation is rewritten as $F - E + V = 2$, the left hand side of the equation is called the **Euler Characteristic**.

> *The Euler Characteristic of a shape is:* $F - E + V$

Generalized Euler's Theorem

Let: F = the number of faces of a polyhedron.

V = the number of vertices of a polyhedron.

E = the number of edges of a polyhedron.

g = the number of holes in the polyhedron. g is called the **genus** of the shape.

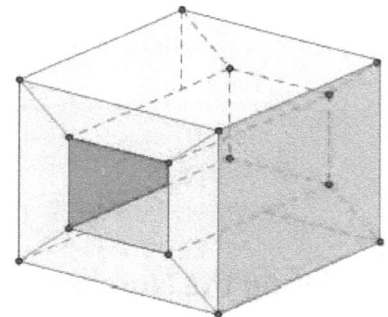

Then, for any polyhedron that does not intersect itself,

$$F - E + V = 2 - 2g$$

Note that the value of Euler's Characteristic can be negative if the shape has more than one hole in it (i.e., if $g \geq 2$)!

Example:

The cube with a tunnel in it has ...

$$V = 16$$
$$E = 32$$
$$F = 16$$
$$\text{so,} \quad V - E + F = 0$$

Geometry
Platonic Solids

A **Platonic Solid** is a convex regular polyhedron with faces composed of congruent convex regular polygons. There five of them:

Tetrahedron Cube Octohedron Dodecahedron Icosohedron

Key Properties of Platonic Solids

It is interesting to look at the key properties of these regular polyhedra.

Name	Faces	Vertices	Edges	Type of Face
Tetrahedron	4	4	6	Triangle
Cube	6	8	12	Square
Octahedron	8	6	12	Triangle
Dodecahedron	12	20	30	Pentagon
Icosahedron	20	12	30	Triangle

Notice the following patterns in the table:

- All of the numbers of faces are even. Only the cube has a number of faces that is not a multiple of 4.
- All of the numbers of vertices are even. Only the octahedron has a number of faces that is not a multiple of 4.
- The number of faces and vertices seem to alternate (e.g., cube 6-8 vs. octahedron 8-6).
- All of the numbers of edges are multiples of 6.
- There are only three possibilities for the numbers of edges – 6, 12 and 30.
- The faces are one of: regular triangles, squares or regular pentagons.

Geometry
Prisms

Definitions

- A Prism is a polyhedron with two congruent polygonal faces that lie in parallel planes.

- The Bases are the parallel polygonal faces.

- The Lateral Faces are the faces that are not bases.

- The Lateral Edges are the edges between the lateral faces.

- The Slant Height is the length of a lateral edge. Note that all lateral edges are the same length.

- The Height is the perpendicular length between the bases.

- A Right Prism is one in which the angles between the bases and the lateral edges are right angles. Note that in a right prism, the height and the slant height are the same.

- An Oblique Prism is one that is not a right prism.

- The Surface Area of a prism is the sum of the areas of all its faces.

- The Lateral Area of a prism is the sum of the areas of its lateral faces.

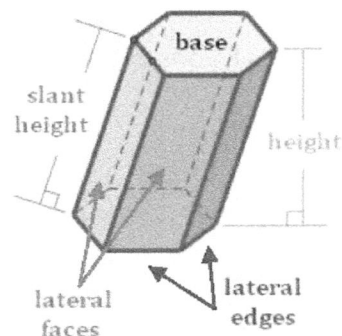

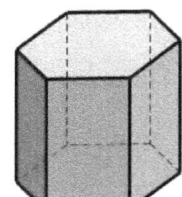

Right Hexagonal Prism

Surface Area and Volume of a Right Prism

Surface Area: $SA = Ph + 2B$ where, $P = the\ perimeter\ of\ the\ base$
Lateral SA: $SA = Ph$ $h = the\ height\ of\ the\ prism$
Volume: $V = Bh$ $B = the\ area\ of\ the\ base$

Cavalieri's Principle

If two solids have the same height and the same cross-sectional area at every level, then they have the same volume. This principle allows us to derive a formula for the volume of an oblique prism from the formula for the volume of a right prism.

Surface Area and Volume of an Oblique Prism

Surface Area: $SA = LSA + 2B$ where, $LSA = the\ lateral\ surface\ area$
Volume: $V = Bh$ $h = the\ height\ of\ the\ prism$
                                                          $B = the\ area\ of\ the\ base$

> The lateral surface area of an oblique prism is the sum of the areas of the faces, which must be calculated individually.

Geometry
Cylinders

Definitions

- A Cylinder is a figure with two congruent circular bases in parallel planes.
- The Axis of a cylinder is the line connecting the centers of the circular bases.
- A cylinder has only one Lateral Surface. When deconstructed, the lateral surface of a cylinder is a rectangle with length equal to the circumference of the base.
- There are no Lateral Edges in a cylinder.
- The Slant Height is the length of the lateral side between the bases. Note that all lateral distances are the same length. The slant height has applicability only if the cylinder is oblique.
- The Height is the perpendicular length between the bases.

- A Right Cylinder is one in which the angles between the bases and the lateral side are right angles. Note that in a right cylinder, the height and the slant height are the same.
- An Oblique Cylinder is one that is not a right cylinder.

- The Surface Area of a cylinder is the sum of the areas of its bases and its lateral surface.
- The Lateral Area of a cylinder is the areas of its lateral surface.

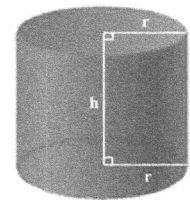

Surface Area and Volume of a Right Cylinder

Surface Area: $SA = Ch + 2B$
$\qquad\qquad = 2\pi rh + 2\pi r^2$

Lateral SA: $SA = Ch = 2\pi rh$

Volume: $V = Bh = \pi r^2 h$

where, $C = $ the circumference of the base
$h = $ the height of the cylinder
$B = $ the area of the base
$r = $ the radius of the base

Surface Area and Volume of an Oblique Cylinder

Surface Area: $SA = Pl + 2B$

Volume: $V = Bh = \pi r^2 h$

where, $P = $ the perimeter of a right section*
$\qquad\qquad$ of the cylinder
$l = $ the slant height of the cylinder
$h = $ the height of the cylinder
$B = $ the area of the base
$r = $ the radius of the base

> * A **right section** of an oblique cylinder is a cross section perpendicular to the axis of the cylinder.

Geometry
Surface Area by Decomposition

Sometimes the student is asked to calculate the surface are of a prism that does not quite fit into one of the categories for which an easy formula exists. In this case, the answer may be to decompose the prism into its component shapes, and then calculate the areas of the components. Note: this process also works with cylinders and pyramids.

Decomposition of a Prism

To calculate the surface area of a prism, decompose it and look at each of the prism's faces individually.

Example: Calculate the surface area of the triangular prism at right.

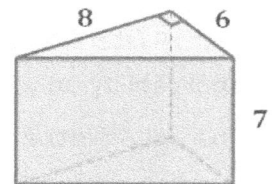

To do this, first notice that we need the value of the hypotenuse of the base. Use the Pythagorean Theorem or Pythagorean Triples to determine the missing value is **10**. Then, decompose the figure into its various faces:

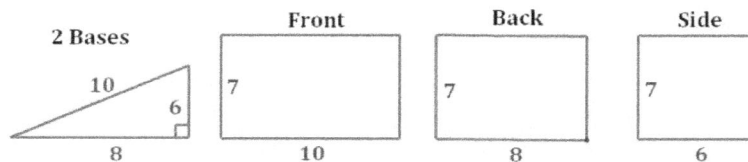

The surface area, then, is calculated as:

$$SA = (2\ Bases) + (Front) + (Back) + (Side)$$

$$SA = 2 \cdot \left(\frac{1}{2} \cdot 6 \cdot 8\right) + (10 \cdot 7) + (8 \cdot 7) + (6 \cdot 7) = 216$$

Decomposition of a Right Cylinder

The cylinder at right is decomposed into two circles (the bases) and a rectangle (the lateral face).

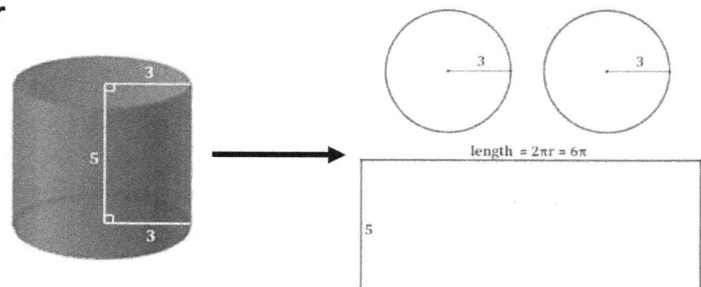

The surface area, then, is calculated as:

$$SA = (2\ tops) + (lateral\ face)$$

$$SA = 2 \cdot (\pi \cdot 3^2) + (6\pi \cdot 5) = 48\pi \sim 150.80$$

Geometry
Pyramids

Pyramids

- A Pyramid is a polyhedron in which the base is a polygon and the lateral sides are triangles with a common vertex.

- The Base is a polygon of any size or shape.

- The Lateral Faces are the faces that are not the base.

- The Lateral Edges are the edges between the lateral faces.

- The Apex of the pyramid is the intersection of the lateral edges. It is the point at the top of the pyramid.

- The Slant Height of a regular pyramid is the altitude of one of the lateral faces.

- The Height is the perpendicular length between the base and the apex.

- A Regular Pyramid is one in which the lateral faces are congruent triangles. The height of a regular pyramid intersects the base at its center.

- An Oblique Pyramid is one that is not a right pyramid. That is, the apex is not aligned directly above the center of the base.

- The Surface Area of a pyramid is the sum of the areas of all its faces.

- The Lateral Area of a pyramid is the sum of the areas of its lateral faces.

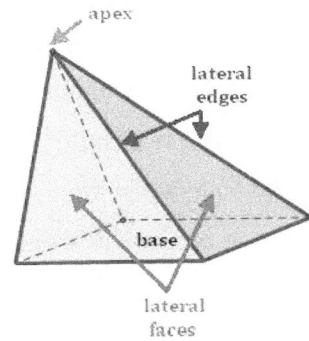

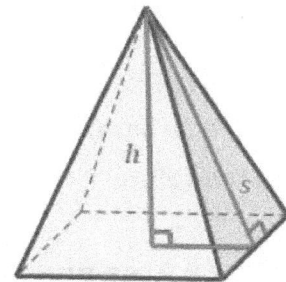

Surface Area and Volume of a Regular Pyramid

Surface Area: $SA = \frac{1}{2}Ps + B$

Lateral SA: $SA = \frac{1}{2}Ps$

Volume: $V = \frac{1}{3}Bh$

where, $P = the\ perimeter\ of\ the\ base$
$s = the\ slant\ height\ of\ the\ pyramid$
$h = the\ height\ of\ the\ pyramid$
$B = the\ area\ of\ the\ base$

Surface Area and Volume of an Oblique Pyramid

Surface Area: $SA = LSA + B$

Volume: $V = \frac{1}{3}Bh$

where, $LSA = the\ lateral\ surface\ area$
$h = the\ height\ of\ the\ pyramid$
$B = the\ area\ of\ the\ base$

> The lateral surface area of an oblique pyramid is the sum of the areas of the faces, which must be calculated individually.

Geometry
Cones

Definitions

- A Circular Cone is a 3-dimensional geometric figure with a circular base which tapers smoothly to a vertex (or apex). The apex and base are in different planes. Note: there is also an elliptical cone that has an ellipse as a base, but that will not be considered here.

- The Base is a circle.

- The Lateral Surface is area of the figure between the base and the apex.

- There are no Lateral Edges in a cone.

- The Apex of the cone is the point at the top of the cone.

- The Slant Height of a cone is the length along the lateral surface from the apex to the base.

- The Height is the perpendicular length between the base and the apex.

- A Right Cone is one in which the height of the cone intersects the base at its center.

- An Oblique Cone is one that is not a right cone. That is, the apex is not aligned directly above the center of the base.

- The Surface Area of a cone is the sum of the area of its lateral surface and its base.

- The Lateral Area of a cone is the area of its lateral surface.

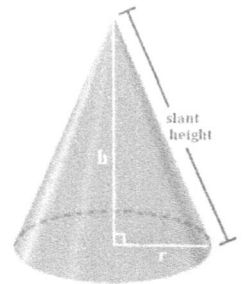

Surface Area and Volume of a Right Cone

Surface Area: $SA = \pi r s + \pi r^2$

Lateral SA: $SA = \pi r s$

Volume: $V = \frac{1}{3}Bh = \frac{1}{3}\pi r^2 h$

where, r = the radius of the base
s = the slant height of the cone
h = the height of the cone
B = the area of the base

Surface Area and Volume of an Oblique Cone

Surface Area: $SA = LSA + \pi r^2$

Volume: $V = \frac{1}{3}Bh = \frac{1}{3}\pi r^2 h$

where, LSA = the lateral surface area
r = the radius of the base
h = the height of the cone

> There is no easy formula for the lateral surface area of an oblique cone.

Geometry
Spheres

Definitions

- A Sphere is a 3-dimensional geometric figure in which all points are a fixed distance from a point. A good example of a sphere is a ball.

- Center – the middle of the sphere. All points on the sphere are the same distance from the center.

- Radius – a line segment with one endpoint at the center and the other endpoint on the sphere. The term "radius" is also used to refer to the distance from the center to the points on the sphere.

- Diameter – a line segment with endpoints on the sphere that passes through the center.

- Great Circle – the intersection of a plane and a sphere that passes through the center.

- **Hemisphere** – half of a sphere. A great circle separates a plane into two hemispheres.

- Secant Line – a line that intersects the sphere in exactly two points.

- **Tangent Line**– a line that intersects the sphere in exactly one point.

- Chord – a line segment with endpoints on the sphere that does not pass through the center.

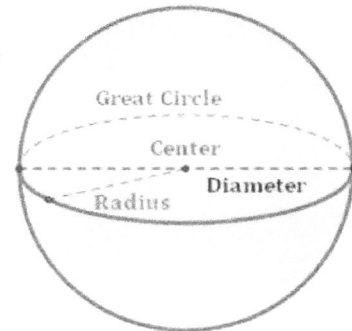

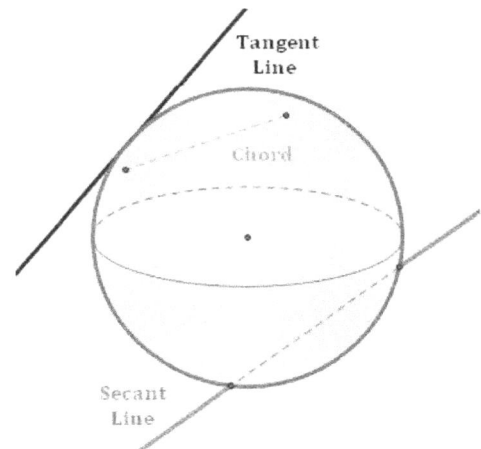

Surface Area and Volume of a Sphere

Surface Area: $SA = 4\pi r^2$

Volume: $V = \frac{4}{3}\pi r^3$

where, $r = the\ radius\ of\ the\ sphere$

Geometry
Similar Solids

Similar Solids have equal ratios of corresponding linear measurements (e.g., edges, radii). So, all of their key dimensions are proportional.

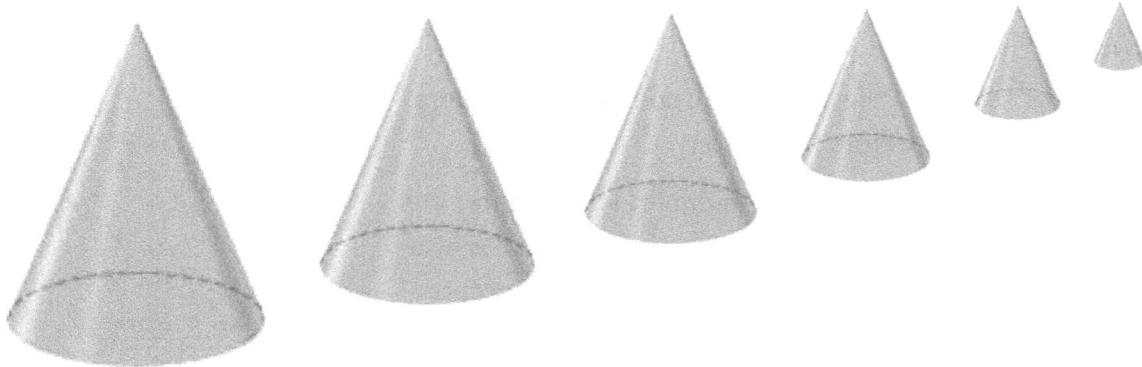

Edges, Surface Area and Volume of Similar Figures

Let k be the scale factor relating two similar geometric solids F_1 and F_2 such that $F_2 = k \cdot F_1$. Then, for corresponding parts of F_1 and F_2,

$$\frac{\text{Edge of } F_2}{\text{Edge of } F_1} = k$$

and

$$\frac{\text{Surface Area of } F_2}{\text{Surface Area of } F_1} = k^2$$

And

$$\frac{\text{Volume of } F_2}{\text{Volume of } F_1} = k^3$$

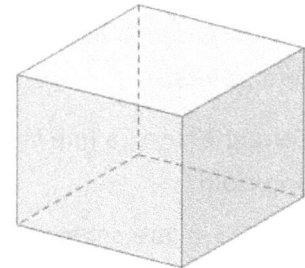

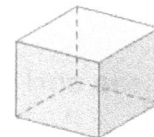

These formulas hold true for any corresponding portion of the figures. So, for example:

$$\frac{\text{Total Edge Length of } F_2}{\text{Total Edge Length of } F_1} = k \qquad \frac{\text{Area of a Face of } F_2}{\text{Area of a Face of } F_1} = k^2$$

Geometry
Summary of Perimeter and Area Formulas – 2D Shapes

Shape	Figure	Perimeter	Area
Kite		$P = 2b + 2c$ $b, c = sides$	$A = \dfrac{1}{2}(d_1 d_2)$ $d_1, d_2 = diagonals$
Trapezoid		$P = b_1 + b_2 + c + d$ $b_1, b_2 = bases$ $c, d = sides$	$A = \dfrac{1}{2}(b_1 + b_2)h$ $b_1, b_2 = bases$ $h = height$
Parallelogram		$P = 2b + 2c$ $b, c = sides$	$A = bh$ $b = base$ $h = height$
Rectangle		$P = 2b + 2c$ $b, c = sides$	$A = bh$ $b = base$ $h = height$
Rhombus		$P = 4s$ $s = side$	$A = bh = \dfrac{1}{2}(d_1 d_2)$ $d_1, d_2 = diagonals$
Square		$P = 4s$ $s = side$	$A = s^2 = \dfrac{1}{2}(d_1 d_2)$ $d_1, d_2 = diagonals$
Regular Polygon		$P = ns$ $n = number\ of\ sides$ $s = side$	$A = \dfrac{1}{2}\, a \cdot P$ $a = apothem$ $P = perimeter$
Circle		$C = 2\pi r = \pi d$ $r = radius$ $d = diameter$	$A = \pi r^2$ $r = radius$
Ellipse		$P \approx 2\pi \sqrt{\dfrac{1}{2}(r_1{}^2 + r_2{}^2)}$ $r_1 = major\ axis\ radius$ $r_2 = minor\ axis\ radius$	$A = \pi r_1 r_2$ $r_1 = major\ axis\ radius$ $r_2 = minor\ axis\ radius$

Geometry
Summary of Surface Area and Volume Formulas – 3D Shapes

Shape	Figure	Surface Area	Volume
Sphere		$SA = 4\pi r^2$ $r = radius$	$V = \dfrac{4}{3}\pi r^3$ $r = radius$
Right Cylinder		$SA = 2\pi rh + 2\pi r^2$ $h = height$ $r = radius\ of\ base$	$V = \pi r^2 h$ $h = height$ $r = radius\ of\ base$
Cone		$SA = \pi rl + \pi r^2$ $l = slant\ height$ $r = radius\ of\ base$	$V = \dfrac{1}{3}\pi r^2 h$ $h = height$ $r = radius\ of\ base$
Square Pyramid		$SA = 2sl + s^2$ $s = base\ side\ length$ $l = slant\ height$	$V = \dfrac{1}{3}s^2 h$ $s = base\ side\ length$ $h = height$
Rectangular Prism		$SA = 2 \cdot (lw + lh + wh)$ $l = length$ $w = width$ $h = height$	$V = lwh$ $l = length$ $w = width$ $h = height$
Cube		$SA = 6s^2$ $s = side\ length\ (all\ sides)$	$V = s^3$ $s = side\ length\ (all\ sides)$
General Right Prism		$SA = Ph + 2B$ $P = Perimeter\ of\ Base$ $h = height\ (or\ length)$ $B = area\ of\ Base$	$V = Bh$ $B = area\ of\ Base$ $h = height$

CHAPTER 1

Lines, Angles, and Triangles

1.1 Historical Background of Geometry

The word *geometry* is derived from the Greek words *geos* (meaning *earth*) and *metron* (meaning *measure*). The ancient Egyptians, Chinese, Babylonians, Romans, and Greeks used geometry for surveying, navigation, astronomy, and other practical occupations.

The Greeks sought to systematize the geometric facts they knew by establishing logical reasons for them and relationships among them. The work of men such as Thales (600 B.C.), Pythagoras (540 B.C.), Plato (390 B.C.), and Aristotle (350 B.C.) in systematizing geometric facts and principles culminated in the geometry text *Elements*, written in approximately 325 B.C. by Euclid. This most remarkable text has been in use for over 2000 years.

1.2 Undefined Terms of Geometry: Point, Line, and Plane

1.2A Point, Line, and Plane are Undefined Terms

These undefined terms underlie the definitions of all geometric terms. They can be given meanings by way of descriptions. However, these descriptions, which follow, are not to be thought of as definitions.

1.2B Point

A *point* has position only. It has no length, width, or thickness.

A point is represented by a dot. Keep in mind, however, that the dot *represents* a point but *is not* a point, just as a dot on a map may represent a locality but is not the locality. A dot, unlike a point, has size.

A point is designated by a capital letter next to the dot, thus point A is represented: *A*.

1.2C Line

A line has length but has no width or thickness.

A line may be represented by the path of a piece of chalk on the blackboard or by a stretched rubber band.

A line is designated by the capital letters of any two of its points or by a small letter, thus:

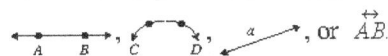

$$\underset{A \quad\quad B}{\longleftrightarrow}, \quad \underset{C \quad\quad D}{\frown}, \quad \underset{a}{\nearrow}, \quad \text{or} \quad \overleftrightarrow{AB}.$$

A *line* may be straight, curved, or a combination of these. To understand how lines differ, think of a line as being generated by a moving point. A *straight line*, such as $\longleftrightarrow$, is generated by a point moving always in the same direction. A *curved line*, such as $\curvearrowright$, is generated by a point moving in a continuously changing direction.

Two lines intersect in a point.

A straight line is unlimited in extent. It may be extended in either direction indefinitely.

A *ray* is the part of a straight line beginning at a given point and extending limitlessly in one direction:

$\overrightarrow{AB}$ and $\nearrow$ designate rays.

In this book, the word *line* will mean "straight line" unless otherwise stated.

1.2D Surface

A *surface* has length and width but no thickness. It may be represented by a blackboard, a side of a box, or the outside of a sphere; remember, however, that these are representations of a surface but are not surfaces.

A plane surface (or *plane*) is a surface such that a straight line connecting any two of its points lies entirely in it. A plane is a flat surface.

Plane geometry is the geometry of plane figures—those that may be drawn on a plane. Unless otherwise stated, the word *figure* will mean "plane figure" in this book.

SOLVED PROBLEMS

1.1 Illustrating undefined terms

Point, line, and plane are undefined terms. State which of these terms is illustrated by (a) the top of a desk; (b) a projection screen; (c) a ruler's edge; (d) a stretched thread; (e) the tip of a pin.

Solutions

(a) surface; (b) surface; (c) line; (d) line; (e) point.

1.3 Line Segments

A straight line segment is the part of a straight line between two of its points, including the two points, called *endpoints*. It is designated by the capital letters of these points with a bar over them or by a small letter. Thus, $\overline{AB}$ or r represents the straight line segment $A \overset{r}{\longrightarrow} B$ between A and B.

The expression *straight line segment* may be shortened to *line segment* or to *segment*, if the meaning is clear. Thus, $\overline{AB}$ and *segment AB* both mean "the straight line segment AB."

1.3A Dividing a Line Segment into Parts

If a line segment is divided into parts:

1. The length of the whole line segment equals the sum of the lengths of its parts. Note that the length of $\overline{AB}$ is designated AB. A number written beside a line segment designates its length.

2. The length of the whole line segment is greater than the length of any part.
   Suppose $\overline{AB}$ is divided into three parts of lengths a, b, and c; thus $A \overset{a}{\bullet} \overset{b}{\bullet} \overset{c}{\longrightarrow} B$. Then $AB = a + b + c$. Also, AB is greater than a; this may be written as $AB > a$.

If a line segment is divided into two equal parts:

1. The point of division is the *midpoint* of the line segment.

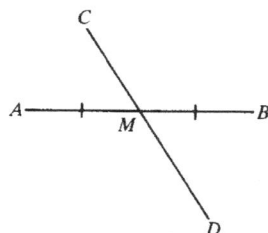

Fig. 1-1

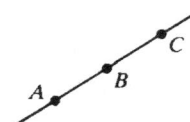

Fig. 1-2

2. A line that crosses at the midpoint is said to *bisect* the segment.
   Because $AM = MB$ in Fig. 1-1, M is the midpoint of $\overline{AB}$, and $\overline{CD}$ bisects $\overline{AB}$. Equal line segments may be shown by crossing them with the same number of strokes. Note that $\overline{AM}$ and $\overline{MB}$ are crossed with a single stroke.

3. If three points A, B, and C lie on a line, then we say they are *collinear*. If A, B, and C are collinear and $AB + BC = AC$, then B is between A and C (see Fig. 1-2).

1.3B Congruent Segments

Two line segments having the same length are said to be *congruent*. Thus, if $AB = CD$, then $\overline{AB}$ is congruent to $\overline{CD}$, written $\overline{AB} \cong \overline{CD}$.

SOLVED PROBLEMS

1.2 Naming line segments and points

See Fig. 1-3.

(a) Name each line segment shown.

(b) Name the line segments that intersect at A.

(c) What other line segment can be drawn using points A, B, C, and D?

(d) Name the point of intersection of $\overline{CD}$ and $\overline{AD}$.

(e) Name the point of intersection of $\overline{BC}, \overline{AC}$, and $\overline{CD}$.

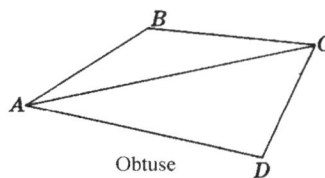

Obtuse

Fig. 1-3

Solutions

(a) $\overline{AB}, \overline{BC}, \overline{CD}, \overline{AC}$, and $\overline{AD}$. These segments may also be named by interchanging the letters; thus, $\overline{BA}, \overline{CB}, \overline{DC}, \overline{CA}$, and $\overline{DA}$ are also correct.

(b) $\overline{AB}, \overline{AC}$, and $\overline{AD}$

(c) $\overline{BD}$

(d) D

(e) C

1.3 Finding lengths and points of line segments

See Fig. 1-4.

(a) State the lengths of $\overline{AB}, \overline{AC}$, and $\overline{AF}$.

(b) Name two midpoints.

(c) Name two bisectors.

(d) Name all congruent segments.

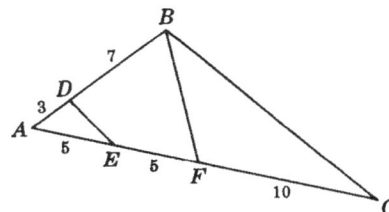

Fig. 1-4

Solutions

(a) $AB = 3 + 7 = 10; AC = 5 + 5 + 10 = 20; AF = 5 + 5 = 10.$

(b) E is midpoint of $\overline{AF}$; F is midpoint of $\overline{AC}$.

(c) $\overline{DE}$ is bisector of $\overline{AF}$; $\overline{BF}$ is bisector of $\overline{AC}$.

(d) $\overline{AB}, \overline{AF}$, and $\overline{FC}$ (all have length 10); $\overline{AE}$ and $\overline{EF}$ (both have length 5).

1.4 Circles

A *circle* is the set of all points in a plane that are the same distance from the *center*. The symbol for circle is $\odot$; for circles, $\circledS$. Thus, $\odot O$ stands for the circle whose center is O.

The *circumference* of a circle is the distance around the circle. It contains 360 *degrees* (360°).

A *radius* is a segment joining the center of a circle to a point on the circle (see Fig. 1-5). From the definition of a circle, it follows that the radii of a circle are congruent. Thus, $\overline{OA}, \overline{OB}$, and OC of Fig. 1-5 are radii of $\odot O$ and $\overline{OA} \cong \overline{OB} \cong \overline{OC}$.

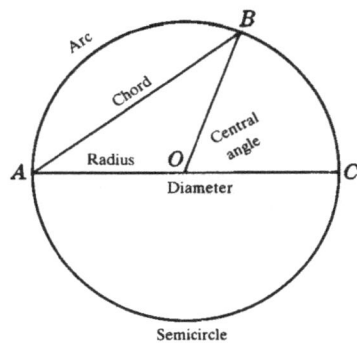
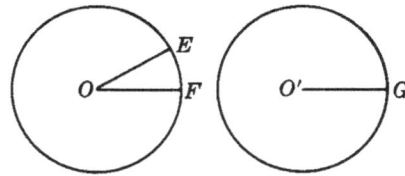

Fig. 1-5   Fig. 1-6

A *chord* is a segment joining any two points on a circle. Thus, $\overline{AB}$ and $\overline{AC}$ are chords of $\odot O$.

A *diameter* is a chord through the center of the circle; it is the longest chord and is twice the length of a radius. $\overline{AC}$ is a diameter of $\odot O$.

An *arc* is a continuous part of a circle. The symbol for arc is $\frown$, so that $\overset{\frown}{AB}$ stands for arc *AB*. An arc of measure 1° is 1/360th of a circumference.

A *semicircle* is an arc measuring one-half of the circumference of a circle and thus contains 180°. A diameter divides a circle into two semicircles. For example, diameter $\overline{AC}$ cuts $\odot O$ of Fig. 1-5 into two semicircles.

A *central angle* is an angle formed by two radii. Thus, the angle between radii $\overline{OB}$ and $\overline{OC}$ is a central angle. A central angle measuring 1° cuts off an arc of 1°; thus, if the central angle between $\overline{OE}$ and $\overline{OF}$ in Fig. 1-6 is 1°, then $\overset{\frown}{EF}$ measures 1°.

Congruent circles are circles having congruent radii. Thus, if $\overline{OE} \cong \overline{O'G}$, then $\odot O \cong \odot O'$.

SOLVED PROBLEMS

1.4   Finding lines and arcs in a circle

In Fig. 1-7 find (a) *OC* and *AB*; (b) the number of degrees in $\overset{\frown}{AD}$; (c) the number of degrees in $\overset{\frown}{BC}$.

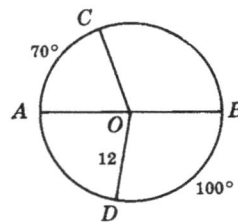

Fig. 1-7

Solutions

(a) Radius *OC* = radius *OD* = 12. Diameter *AB* = 24.

(b) Since semicircle *ADB* contains 180°, $\overset{\frown}{AD}$ contains $180° - 100° = 80°$.

(c) Since semicircle *ACB* contains 180°, $\overset{\frown}{BC}$ contains $180° - 70° = 110°$.

1.5   Angles

An *angle* is the figure formed by two rays with a common end point. The rays are the *sides* of the angle, while the end point is its *vertex*. The symbol for angle is $\angle$ or $\measuredangle$; the plural is $\measuredangle$.

Thus, $\overrightarrow{AB}$ and $\overrightarrow{AC}$ are the sides of the angle shown in Fig. 1-8(a), and *A* is its vertex.

1.5A Naming an Angle

An angle may be named in any of the following ways:

1. With the vertex letter, if there is only one angle having this vertex, as ∠B in Fig. 1-8(b).
2. With a small letter or a number placed between the sides of the angle and near the vertex, as ∠a or ∠1 in Fig. 1-8(c).
3. With three capital letters, such that the vertex letter is between two others, one from each side of the angle. In Fig. 1-8(d), ∠E may be named ∠DEG or ∠GED.

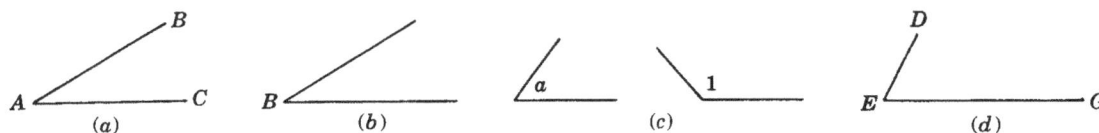

Fig. 1-8

1.5B Measuring the Size of an Angle

The size of an angle depends on the extent to which one side of the angle must be rotated, or turned about the vertex, until it meets the other side. We choose degrees to be the unit of measure for angles. The measure of an angle is the number of degrees it contains. We will write $m\angle A = 60°$ to denote that "angle A measures 60°."

The protractor in Fig. 1-9 shows that ∠A measures of 60°. If $\vec{AC}$ were rotated about the vertex A until it met $\vec{AB}$, the amount of turn would be 60°.

In using a protractor, be sure that the vertex of the angle is at the center and that one side is along the $0°-180°$ diameter.

The size of an angle *does not* depend on the lengths of the sides of the angle.

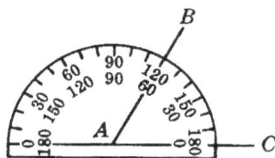

Fig. 1-9

Fig. 1-10

The size of ∠B in Fig. 1-10 would not be changed if its sides $\vec{AB}$ and $\vec{BC}$ were made larger or smaller.

No matter how large or small a clock is, the angle formed by its hands at 3 o'clock measures 90°, as shown in Figs. 1-11 and 1-12.

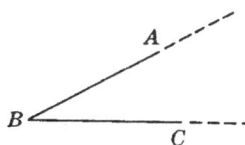

Fig. 1-11

Fig. 1-12

Angles that measure less than 1° are usually represented as fractions or decimals. For example, one-thousandth of the way around a circle is either $\frac{360°}{1000}$ or 0.36°.

In some fields, such as navigation and astronomy, small angles are measured in *minutes* and *seconds*. One degree is comprised of 60 minutes, written $1° = 60'$. A minute is 60 seconds, written $1' = 60''$. In this notation, one-thousandth of a circle is $21'36''$ because $\frac{21}{60} + \frac{36}{3600} = \frac{1296}{3600} = \frac{360}{1000}$.

1.5C Kinds of Angles

1. *Acute angle*: An acute angle is an angle whose measure is less than 90°.

   Thus, in Fig. 1-13 $a°$ is less than 90°; this is symbolized as $a° < 90°$.

2. *Right angle*: A right angle is an angle that measures 90°.

   Thus, in Fig. 1-14, m(rt. $\angle A$) = 90°. The square corner denotes a right angle.

3. *Obtuse angle*: An obtuse angle is an angle whose measure is more than 90° and less than 180°.

   Thus, in Fig. 1-15, 90° is less than $b°$ and $b°$ is less than 180°; this is denoted by 90° < $b°$ < 180°.

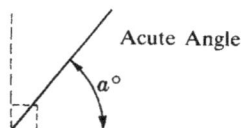
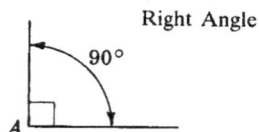
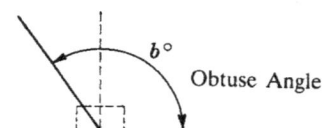

| Fig. 1-13 | Fig. 1-14 | Fig. 1-15 |

4. *Straight angle*: A straight angle is an angle that measures 180°.

   Thus, in Fig. 1-16, m(st. $\angle B$) = 180°. Note that the sides of a straight angle lie in the same straight line. But do not confuse a straight angle with a straight line!

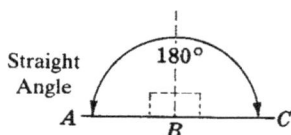
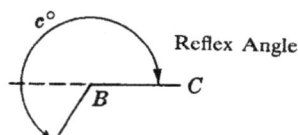

| Fig. 1-16 | Fig. 1-17 |

5. *Reflex angle*: A reflex angle is an angle whose measure is more than 180° and less than 360°.

   Thus, in Fig. 1-17, 180° is less than $c°$ and $c°$ is less than 360°; this is symbolized as 180° < $c°$ < 360°.

1.5D Additional Angle Facts

1. *Congruent angles* are angles that have the same number of degrees. In other words, if $m\angle A = m\angle B$, then $\angle A \cong \angle B$.

   Thus, in Fig. 1-18, rt. $\angle A \cong$ rt. $\angle B$ since each measures 90°.

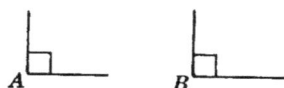
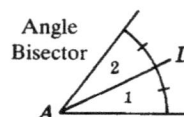

| Fig. 1-18 | Fig. 1-19 |

2. A line that *bisects* an angle divides it into two congruent parts.

   Thus, in Fig. 1-19, if $\overline{AD}$ bisects $\angle A$, then $\angle 1 \cong \angle 2$. (Congruent angles may be shown by crossing their arcs with the same number of strokes. Here the arcs of ∡ 1 and 2 are crossed by a single stroke.)

3. *Perpendiculars* are lines or rays or segments that meet at right angles.

   The symbol for perpendicular is $\perp$; for perpendiculars, ⊥s. In Fig. 1-20, $\overline{CD} \perp \overline{AB}$, so right angles 1 and 2 are formed.

4. A *perpendicular bisector* of a given segment is perpendicular to the segment and bisects it.

   In Fig. 1-21, $\overleftrightarrow{GH}$ is the bisector of $\overline{EF}$; thus, $\angle 1$ and $\angle 2$ are right angles and M is the midpoint of $\overline{EF}$.

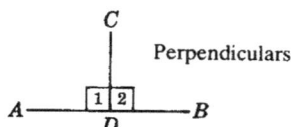
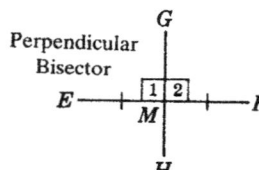

| Fig. 1-20 | Fig. 1-21 |

SOLVED PROBLEMS

1.5 Naming an angle

Name the following angles in Fig. 1-22: (a) two obtuse angles; (b) a right angle; (c) a straight angle; (d) an acute angle at D; (e) an acute angle at B.

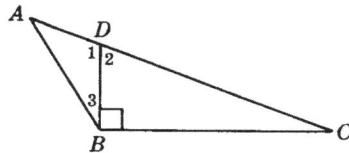

Fig. 1-22

Solutions

(a) $\angle ABC$ and $\angle ADB$ (or $\angle 1$). The angles may also be named by reversing the order of the letters: $\angle CBA$ and $\angle BDA$.

(b) $\angle DBC$

(c) $\angle ADC$

(d) $\angle 2$ or $\angle BDC$

(e) $\angle 3$ or $\angle ABD$

1.6 Adding and subtracting angles

In Fig. 1-23, find (a) $m\angle AOC$; (b) $m\angle BOE$; (c) the measure of obtuse $\angle AOE$.

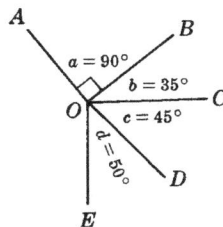

Fig. 1-23

Solutions

(a) $m\angle AOC = m\angle a + m\angle b = 90° + 35° = 125°$

(b) $m\angle BOE = m\angle b + m\angle c + m\angle d = 35° + 45° + 50° = 130°$

(c) $m\angle AOE = 360° - (m\angle a + m\angle b + m\angle c + m\angle d) = 360° - 220° = 140°$

1.7 Finding parts of angles

Find (a) $\frac{2}{5}$ of the measure of a rt. $\angle$; (b) $\frac{2}{3}$ of the measure of a st. $\angle$; (c) $\frac{1}{2}$ of $31°$; (d) $\frac{1}{10}$ of $70°20'$.

Solutions

(a) $\frac{2}{5}(90°) = 36°$

(b) $\frac{2}{3}(180°) = 120°$

(c) $\frac{1}{2}(31°) = 15\frac{1}{2}° = 15°30'$

(d) $\frac{1}{10}(70°20') = \frac{1}{10}(70°) + \frac{1}{10}(20') = 7°2'$

1.8 Finding rotations

In a half hour, what turn or rotation is made (a) by the minute hand, and (b) by the hour hand of a clock? What rotation is needed to turn (c) from north to southeast in a clockwise direction, and (d) from northwest to southwest in a counterclockwise direction (see Fig. 1-24)?

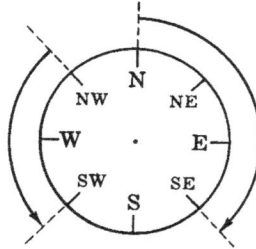

Fig. 1-24

Solutions

(a) In 1 hour, a minute hand completes a full circle of $360°$. Hence, in a half hour it turns $180°$.

(b) In 1 hour, an hour hand turns $\frac{1}{12}$ of $360°$ or $30°$. Hence, in a half hour it turns $15°$.

(c) Add a turn of $90°$ from north to east and a turn of $45°$ from east to southeast to get $90° + 45° = 135°$.

(d) The turn from northwest to southwest is $\frac{1}{4}(360°) = 90°$.

1.9 Finding angles

Find the measure of the angle formed by the hands of the clock in Fig. 1-25, (a) at 8 o'clock; (b) at 4:30.

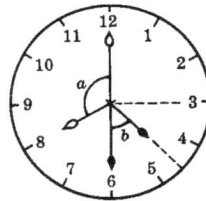

Fig. 1-25

Solutions

(a) At 8 o'clock, $m\angle a = \frac{1}{3}(360°) = 120°$.

(b) At 4:30, $m\angle b = \frac{1}{2}(90°) = 45°$.

1.10 Applying angle facts

In Fig. 1-26, (a) name two pairs of perpendicular segments; (b) find $m\angle a$ if $m\angle b = 42°$; (c) find $m\angle AEB$ and $m\angle CED$.

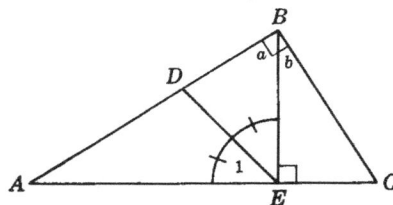

Fig. 1-26

Solutions

(a) Since $\angle ABC$ is a right angle, $\overline{AB} \perp \overline{BC}$. Since $\angle BEC$ is a right angle, $\overline{BE} \perp \overline{AC}$.

(b) $m\angle a = 90° - m\angle b = 90° - 42° = 48°$.

(c) $m\angle AEB = 180° - m\angle BEC = 180° - 90° = 90°$. $m\angle CED = 180° - m\angle 1 = 180° - 45° = 135°$.

1.6   Triangles

A *polygon* is a closed plane figure bounded by straight line segments as sides. Thus, Fig. 1-27 is a polygon of five sides, called a *pentagon*; it is named pentagon *ABCDE*, using its letters in order.

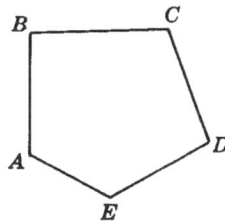

Fig. 1-27

A *quadrilateral* is a polygon having four sides.

A *triangle* is a polygon having three sides. A *vertex* of a triangle is a point at which two of the sides meet. (*Vertices* is the plural of vertex.) The symbol for triangle is $\triangle$; for triangles, $\triangle\!\!\!\triangle$.

A triangle may be named with its three letters in any order or with a Roman numeral placed inside of it. Thus, the triangle shown in Fig. 1-28 is $\triangle ABC$ *or* $\triangle$I; its sides are $\overline{AB}, \overline{AC}$, and $\overline{BC}$; its vertices are *A*, *B*, and *C*; its angles are $\angle A$, $\angle B$, and $\angle C$.

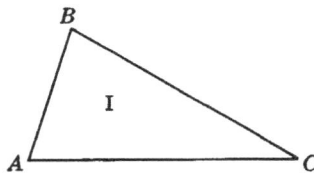

Fig. 1-28

1.6A   Classifying Triangles

Triangles are classified according to the equality of the lengths of their sides or according to the kind of angles they have.

Triangles According to the Equality of the Lengths of their Sides (Fig. 1-29)

1. *Scalene triangle*: A scalene triangle is a triangle having no congruent sides.

   Thus in scalene triangle *ABC*, $a \neq b \neq c$. The small letter used for the length of each side agrees with the capital letter of the angle opposite it. Also, $\neq$ means "is not equal to."

2. *Isosceles triangle*: An isosceles triangle is a triangle having at least two congruent sides.

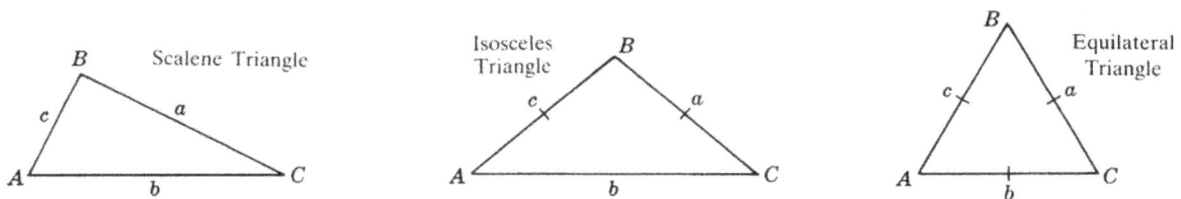

Fig. 1-29

Thus in isosceles triangle ABC, $a = c$. These equal sides are called the *legs* of the isosceles triangle; the remaining side is the *base*, b. The angles on either side of the base are the *base angles*; the angle opposite the base is the *vertex angle*.

3. *Equilateral triangle*: An equilateral triangle is a triangle having three congruent sides.

    Thus in equilateral triangle ABC, $a = b = c$. Note that an equilateral triangle is also an isosceles triangle.

Triangles According to the Kind of Angles (Fig. 1-30)

1. *Right triangle*: A right triangle is a triangle having a right angle.

    Thus in right triangle ABC, $\angle C$ is the right angle. Side c opposite the right angle is the *hypotenuse*. The perpendicular sides, a and b, are the *legs* or *arms* of the right triangle.

2. *Obtuse triangle*: An obtuse triangle is a triangle having an obtuse angle.

    Thus in obtuse triangle DEF, $\angle D$ is the obtuse angle.

3. *Acute triangle*: An acute triangle is a triangle having three acute angles.

    Thus in acute triangle HJK, $\angle H$, $\angle J$, and $\angle K$ are acute angles.

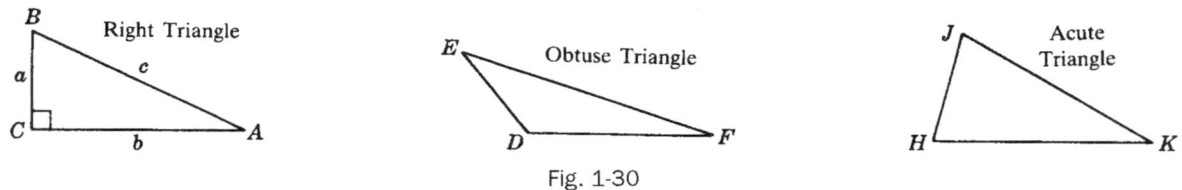

Fig. 1-30

1.6B   Special Lines in a Triangle

1. *Angle bisector of a triangle*: An angle bisector of a triangle is a segment or ray that bisects an angle and extends to the opposite side.

    Thus $\overrightarrow{BD}$, the angle bisector of $\angle B$ in Fig. 1-31, bisects $\angle B$, making $\angle 1 \cong \angle 2$.

2. *Median of a triangle*: A median of a triangle is a segment from a vertex to the midpoint of the opposite side.

    Thus $\overline{BM}$, the median to $\overline{AC}$, in Fig. 1-32, bisects $\overline{AC}$, making $AM = MC$.

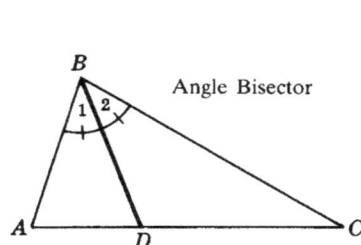 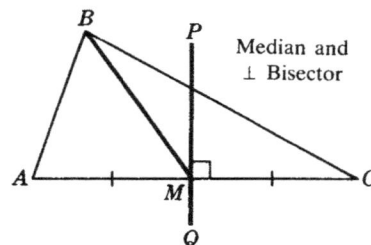

Fig. 1-31   Fig. 1-32

3. *Perpendicular bisector of a side*: A perpendicular bisector of a side of a triangle is a line that bisects and is perpendicular to a side.

    Thus $\overleftrightarrow{PQ}$, the perpendicular bisector of $\overline{AC}$ in Fig. 1-32, bisects $\overline{AC}$ and is perpendicular to it.

4. *Altitude to a side of a triangle*: An altitude of a triangle is a segment from a vertex perpendicular to the opposite side.

    Thus $\overline{BD}$, the altitude to $\overline{AC}$ in Fig. 1-33, is perpendicular to $\overline{AC}$ and forms right angles 1 and 2. Each angle bisector, median, and altitude of a triangle extends from a vertex to the opposite side.

5. *Altitudes of obtuse triangle*: In an obtuse triangle, the altitude drawn to either side of the obtuse angle falls outside the triangle.

Thus in obtuse triangle *ABC* (shaded) in Fig. 1-34, altitudes $\overline{BD}$ and $\overline{CE}$ fall outside the triangle. In each case, a side of the obtuse angle must be extended.

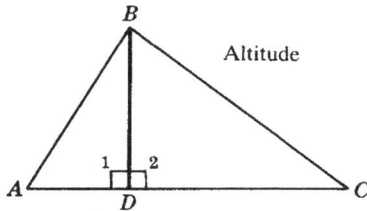

Fig. 1-33

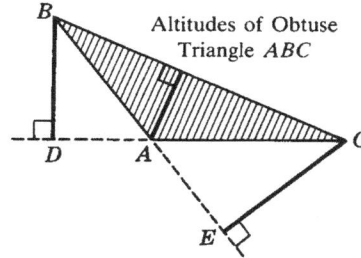

Altitudes of Obtuse Triangle *ABC*

Fig. 1-34

SOLVED PROBLEMS

1.11 Naming a triangle and its parts

In Fig. 1-35, name (a) an obtuse triangle, and (b) two right triangles and the hypotenuse and legs of each. (c) In Fig. 1-36, name two isosceles triangles; also name the legs, base, and vertex angle of each.

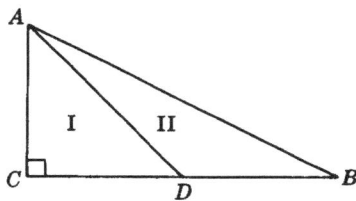

Fig. 1-35

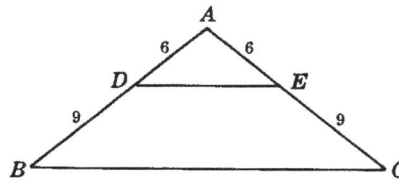

Fig. 1-36

Solutions

(a) Since ∠*ADB* is an obtuse angle, ∠*ADB or* △II is obtuse.

(b) Since ∠*C* is a right angle, △I and △*ABC* are right triangles. In △I, $\overline{AD}$ is the hypotenuse and $\overline{AC}$ and $\overline{CD}$ are the legs. In △*ABC*, *AB* is the hypotenuse and $\overline{AC}$ and $\overline{BC}$ are the legs.

(c) Since *AD* = *AE*, △*ADE* is an isosceles triangle. In △*ADE*, $\overline{AD}$ and $\overline{AE}$ are the legs, $\overline{DE}$ is the base, and ∠*A* is the vertex angle.

Since *AB* = *AC*, △*ABC* is an isosceles triangle. In △*ABC*, $\overline{AB}$ and $\overline{AC}$ are the legs, $\overline{BC}$ is the base, and ∠*A* is the vertex angle.

1.12 Special lines in a triangle

Name the equal segments and congruent angles in Fig. 1-37, (a) if $\overline{AE}$ is the altitude to $\overline{BC}$; (b) if $\overline{CG}$ bisects ∠*ACB*; (c) if $\overline{KL}$ is the perpendicular bisector of $\overline{AD}$; (d) if $\overline{DF}$ is the median to $\overline{AC}$.

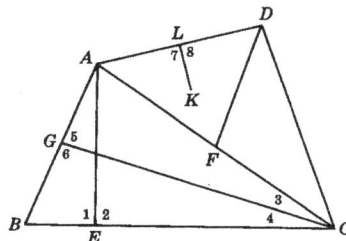

Fig. 1-37

Solutions

(a) Since $\overline{AE} \perp \overline{BC}, \angle 1 \cong \angle 2$.

(b) Since $\overline{CG}$ bisects $\angle ACB, \angle 3 \cong \angle 4$.

(c) Since $\overline{LK}$ is the $\perp$ bisector of $\overline{AD}, AL = LD$ and $\angle 7 \cong \angle 8$.

(d) Since $\overline{DF}$ is median to $\overline{AC}, AF = FC$.

1.7 Pairs of Angles

1.7A Kinds of Pairs of Angles

1. *Adjacent angles*: Adjacent angles are two angles that have the same vertex and a common side between them.

   Thus, the entire angle of $c°$ in Fig. 1-38 has been cut into two adjacent angles of $a°$ and $b°$. These adjacent angles have the same vertex A, and a common side $\overrightarrow{AD}$ between them. Here, $a° + b° = c°$.

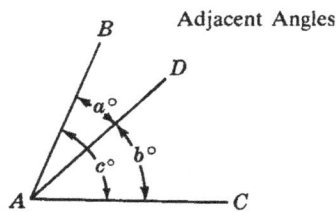
Fig. 1-38

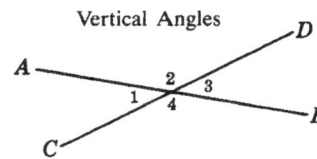
Fig. 1-39

2. *Vertical angles*: Vertical angles are two nonadjacent angles formed by two intersecting lines.

   Thus, $\angle 1$ and $\angle 3$ in Fig. 1-39 are vertical angles formed by intersecting lines $\overleftrightarrow{AB}$ and $\overleftrightarrow{CD}$. Also, $\angle 2$ and $\angle 4$ are another pair of vertical angles formed by the same lines.

3. *Complementary angles*: Complementary angles are two angles whose measures total $90°$.

   Thus, in Fig. 1-40(a) the angles of $a°$ and $b°$ are adjacent complementary angles. However, in (b) the complementary angles are nonadjacent. In each case, $a° + b° = 90°$. Either of two complementary angles is said to be the *complement* of the other.

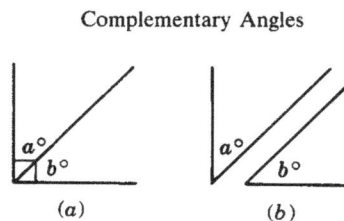
Fig. 1-40

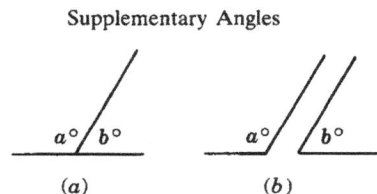
Fig. 1-41

4. *Supplementary angles*: Supplementary angles are two angles whose measures total $180°$.

   Thus, in Fig. 1-41(a) the angles of $a°$ and $b°$ are adjacent supplementary angles. However, in Fig. 1-41(b) the supplementary angles are nonadjacent. In each case, $a° + b° = 180°$. Either of two supplementary angles is said to be the *supplement* of the other.

1.7B Principles of Pairs of Angles

PRINCIPLE 1: *If an angle of $c°$ is cut into two adjacent angles of $a°$ and $b°$, then $a° + b° = c°$.*

Thus if $a° = 25°$ and $b° = 35°$ in Fig. 1-42, then $c° + 25° + 35° = 60°$.

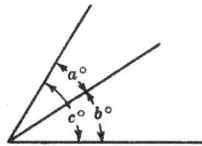

Fig. 1-42

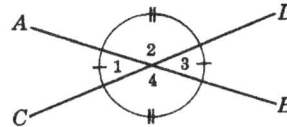

Fig. 1-43

PRINCIPLE 2: *Vertical angles are congruent.*

Thus if $\overleftrightarrow{AB}$ and $\overleftrightarrow{CD}$ are straight lines in Fig. 1-43, then $\angle 1 \cong \angle 3$ and $\angle 2 \cong \angle 4$. Hence, if $m\angle 1 = 40°$, then $m\angle 3 = 40°$; in such a case, $m\angle 2 = m\angle 4 = 140°$.

PRINCIPLE 3: *If two complementary angles contain $a°$ and $b°$, then $a° + b° = 90°$.*

Thus if angles of $a°$ and $b°$ are complementary and $a° = 40°$, then $b° = 50°$ [Fig. 1-44(a) or (b)].

PRINCIPLE 4: *Adjacent angles are complementary if their exterior sides are perpendicular to each other.*

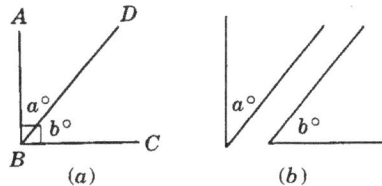

Fig. 1-44

Thus in Fig. 1-44(a), $a°$ and $b°$ are complementary since their exterior sides $\overline{AB}$ and $\overline{BC}$ are perpendicular to each other.

PRINCIPLE 5: *If two supplementary angles contain $a°$ and $b°$, then $a° + b° = 180°$.*

Thus if angles of $a°$ and $b°$ are supplementary and $a° = 140°$, then $b° = 40°$ [Fig. 1-45(a) or (b)].

PRINCIPLE 6: *Adjacent angles are supplementary if their exterior sides lie in the same straight line.*

Thus in Fig. 1-45(a) $a°$ and $b°$ are supplementary angles since their exterior sides $\overrightarrow{AB}$ and $\overrightarrow{BC}$ lie in the same straight line $\overrightarrow{AC}$.

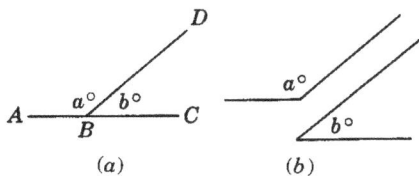

Fig. 1-45

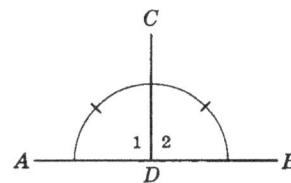

Fig. 1-46

PRINCIPLE 7: *If supplementary angles are congruent, each of them is a right angle. (Equal supplementary angles are right angles.)*

Thus if $\angle 1$ and $\angle 2$ in Fig. 1-46 are both congruent and supplementary, then each of them is a right angle.

SOLVED PROBLEMS

1.13 Naming pairs of angles

(a) In Fig. 1-47(a), name two pairs of supplementary angles.

(b) In Fig. 1-47(b), name two pairs of complementary angles.

(c) In Fig. 1-47(c), name two pairs of vertical angles.

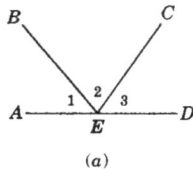 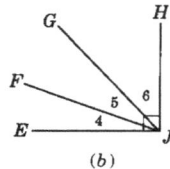 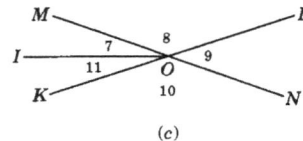

(a) (b) (c)

Fig. 1-47

Solutions

(a) Since their sum is 180°, the supplementary angles are (1) $\angle 1$ and $\angle BED$; (2) $\angle 3$ and $\angle AEC$.

(b) Since their sum is 90°, the complementary angles are (1) $\angle 4$ and $\angle FJH$; (2) $\angle 6$ and $\angle EJG$.

(c) Since $\overleftrightarrow{KL}$ and $\overleftrightarrow{MN}$ are intersecting lines, the vertical angles are (1) $\angle 8$ and $\angle 10$; (2) $\angle 9$ and $\angle MOK$.

1.14 Finding pairs of angles

Find two angles such that:

(a) The angles are supplementary and the larger is twice the smaller.

(b) The angles are complementary and the larger is 20° more than the smaller.

(c) The angles are adjacent and form an angle of 120°. The larger is 20° less than three times the smaller.

(d) The angles are vertical and complementary.

Solutions

In each solution, x is a number only. This number indicates the number of degrees contained in the angle. Hence, if $x = 60$, the angle measures 60°.

(a) Let $x = m$ (smaller angle) and $2x = m$ (larger angle), as in Fig. 1-48(a).
   Principle 5: $x + 2x = 180$, so $3x = 180$; $x = 60$.
   $$2x = 120. \qquad Ans. \quad 60° \text{ and } 120°$$

(b) Let $x = m$ (smaller angle) and $x + 20 = m$ (larger angle), as in Fig. 1-48(b).
   Principle 3: $x + (x + 20) = 90$, or $2x + 20 = 90$; $x = 35$.
   $$x + 20 = 55. \qquad Ans. \quad 35° \text{ and } 55°$$

(c) Let $x = m$ (smaller angle) and $3x - 20 = m$ (larger angle) as in Fig. 1-48(c).
   Principle 1: $x + (3x - 20) = 120$, or $4x - 20 = 120$; $x = 35$.
   $$3x - 20 = 85. \qquad Ans. \quad 35° \text{ and } 85°$$

(d) Let $x = m$ (each vertical angle), as in Fig. 1-48(d). They are congruent by Principle 2.
   Principle 3: $x + x = 90°$, or $2x = 90$; $x = 45$. $\qquad Ans. \quad 45°$ each.

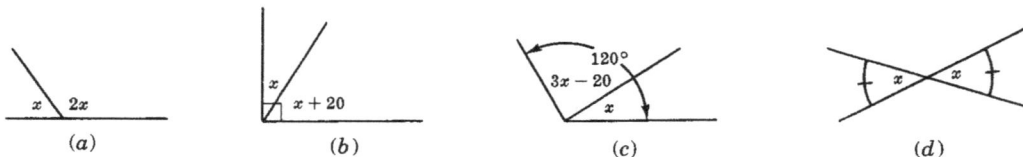

(a) (b) (c) (d)

Fig. 1-48

1.15 Finding a pair of angles using two unknowns

For each of the following, be represented by a and b. Obtain two equations for each case, and then find the angles.

(a) The angles are adjacent, forming an angle of 88°. One is 36° more than the other.

(b) The angles are complementary. One is twice as large as the other.

(c) The angles are supplementary. One is 60° less than twice the other.

(d) The angles are supplementary. The difference of the angles is 24°.

Solutions

(a) $a + b = 88$
$a = b + 36$ *Ans.*   62° and 26°

(b) $a + b = 90$
$a = 2b$ *Ans.*   60° and 30°

(c) $a + b = 180$
$a = 2b - 60$ *Ans.*   100° and 80°

(d) $a + b = 180°$
$a - b = 24°$ *Ans.*   78° and 102°

SUPPLEMENTARY PROBLEMS

1.1. Point, line, and plane are undefined terms. Which of these is illustrated by (a) the tip of a sharpened pencil; (b) the shaving edge of a blade; (c) a sheet of paper; (d) a side of a box; (e) the crease of a folded paper; (f) the junction of two roads on a map? (1.1)

1.2. (a) Name the line segments that intersect at E in Fig. 1-49. (1.2)

(b) Name the line segments that intersect at D.

(c) What other line segments can be drawn using points A, B, C, D, E, and F?

(d) Name the point of intersection of $\overline{AC}$ and $\overline{BD}$.

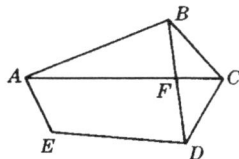
Fig. 1-49

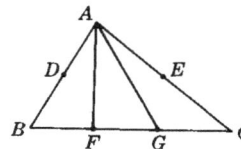
Fig. 1-50

1.3. (a) Find the length of $\overline{AB}$ in Fig. 1-50 if AD is 8 and D is the midpoint of $\overline{AB}$ (1.3)

(b) Find the length of $\overline{AE}$ if AC is 21 and E is the midpoint of $\overline{AC}$.

1.4. (a) Find OB in Fig. 1-51 if diameter $AD = 36$. (1.4)

(b) Find the number of degrees in $\overarc{AE}$ if E is the midpoint of semicircle $\overarc{AED}$. Find the number of degrees in (c) $\overarc{CD}$; (d) $\overarc{AC}$; (e) $\overarc{AEC}$.

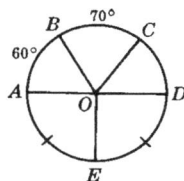
Fig. 1-51

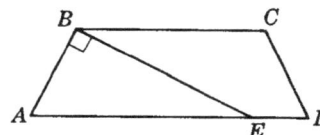
Fig. 1-52

1.5. Name the following angles in Fig. 1-52 (a) an acute angle at *B*; (b) an acute angle at *E*; (c) a right angle; (d) three obtuse angles; (e) a straight angle. (1.5)

1.6. (a) Find *m∠ADC* if *m∠c* = 45° and *m∠d* = 85° in Fig. 1-53. (1.6)

   (b) Find *m∠AEB* if *m∠e* = 60°.

   (c) Find *m∠EBD* if *m∠a* = 15°.

   (d) Find *m∠ABC* if *m∠b* = 42°.

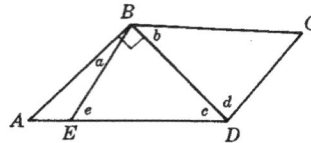

Fig. 1-53

1.7. Find (a) $\frac{5}{6}$ of a rt. ∠; (b) $\frac{2}{9}$ of a st. ∠; (c) $\frac{1}{3}$ of 31°; (d) $\frac{1}{5}$ of 45°55′. (1.7)

1.8. What turn or rotation is made (a) by an hour hand in 3 hours; (b) by the minute hand in $\frac{1}{3}$ of an hour? What rotation is needed to turn from (c) west to northeast in a clockwise direction; (d) east to south in a counterclockwise direction; (e) southwest to northeast in either direction? (1.8)

1.9. Find the angle formed by the hand of a clock (a) at 3 o'clock; (b) at 10 o'clock; (c) at 5:30 AM; (d) at 11:30 PM. (1.9)

1.10. In Fig. 1-54: (1.10)

   (a) Name two pairs of perpendicular lines.

   (b) Find *m∠BCD* if *m∠4* is 39°.

   If *m∠1* = 78°, find (c) *m∠BAD*; (d) *m∠2*; (e) *m∠CAE*.

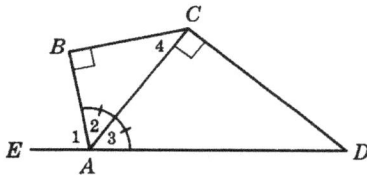   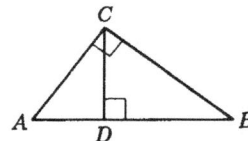   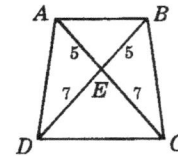

Fig. 1-54   Fig. 1-55

1.11. (a) In Fig. 1-55(a), name three right triangles and the hypotenuse and legs of each. (1.11)

   In Fig. 1-55(b), (b) name two obtuse triangles and (c) name two isosceles triangles, also naming the legs, base, and vertex angle of each.

1.12. In Fig. 1-56, name the congruent lines and angles (a) if $\overline{PR}$ is a ⊥ bisector of $\overline{AB}$; (b) if $\overline{BF}$ bisects ∠*ABC*; (c) if $\overline{CG}$ is an altitude to $\overline{AD}$; (d) if $\overline{EM}$ is a median to $\overline{AD}$. (1.12)

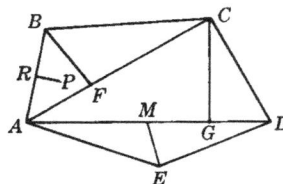

Fig. 1-56

1.13. In Fig. 1-57, state the relationship between: (1.13)

    (a) $\angle 1$ and $\angle 4$ (d) $\angle 4$ and $\angle 5$

    (b) $\angle 3$ and $\angle 4$ (e) $\angle 1$ and $\angle 3$

    (c) $\angle 1$ and $\angle 2$ (f) $\angle AOD$ and $\angle 5$

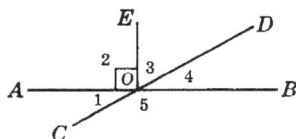

Fig. 1-57

1.14. Find two angles such that: (1.14)

    (a) The angles are complementary and the measure of the smaller is $40°$ less than the measure of the larger.

    (b) The angles are complementary and the measure of the larger is four times the measure of the smaller.

    (c) The angles are supplementary and the measure of the smaller is one-half the measure of the larger.

    (d) The angles are supplementary and the measure of the larger is $58°$ more than the measure of the smaller.

    (e) The angles are supplementary and the measure of the larger is $20°$ less than three times the measure of the smaller.

    (f) The angles are adjacent and form an angle measuring $140°$. The measure of the smaller is $28°$ less than the measure of the larger.

    (g) The angles are vertical and supplementary.

1.15. For each of the following, let the two angles be represented by a and b. Obtain two equations for each case, and then find the angles. (1.15)

    (a) The angles are adjacent and form an angle measuring $75°$. Their difference is $21°$.

    (b) The angles are complementary. One measures $10°$ less than three times the other.

    (c) The angles are supplementary. One measures $20°$ more than four times the other.

CHAPTER 2

Methods of Proof

2.1 Proof By Deductive Reasoning

2.1A Deductive Reasoning is Proof

Deductive reasoning enables us to derive true or acceptably true conclusions from statements which are true or accepted as true. It consists of three steps as follows:

1. Making a *general statement* referring to a whole set or class of things, such as the class of dogs: *All dogs are quadrupeds (have four feet)*.
2. Making a *particular statement* about one or some of the members of the set or class referred to in the general statement: *All greyhounds are dogs*.
3. Making a *deduction* that follows logically when the general statement is applied to the particular statement: *All greyhounds are quadrupeds*.

Deductive reasoning is called *syllogistic reasoning* because the three statements together constitute a syllogism. In a syllogism the general statement is called the major premise, the particular statement is the minor premise, and the deduction is the conclusion. Thus, in the above syllogism:

1. The major premise is: *All dogs are quadrupeds*.
2. The minor premise is: *All greyhounds are dogs*.
3. The conclusion is: *All greyhounds are quadrupeds*.

Using a circle, as in Fig. 2-1, to represent each set or class will help you understand the relationships involved in deductive reasoning.

1. Since the major premise or general statement states that all dogs are quadrupeds, the circle representing dogs must be inside that for quadrupeds.
2. Since the minor premise or particular statement states that all greyhounds are dogs, the circle representing greyhounds must be inside that for dogs.

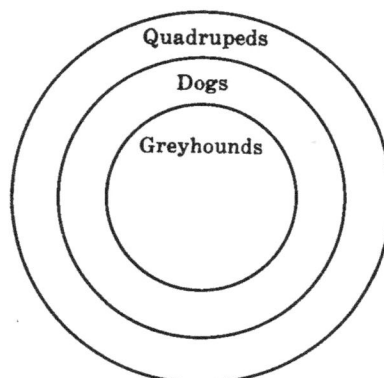

Fig. 2-1

3. The conclusion is obvious. Since the circle of greyhounds must be inside the circle of quadrupeds, the only possible conclusion is that greyhounds are quadrupeds.

2.1B Observation, Measurement, and Experimentation are not Proof

Observation cannot serve as proof. Eyesight, as in the case of a color-blind person, may be defective. Appearances may be misleading. Thus, in each part of Fig. 2-2, AB does not seem to equal CD although it actually does.

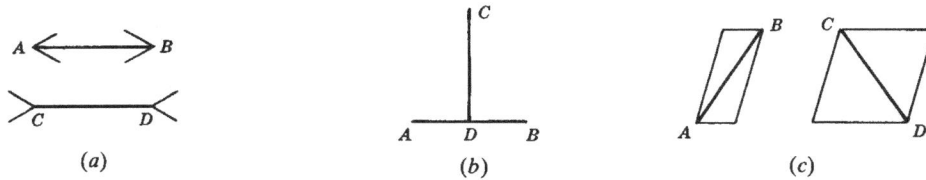

Fig. 2-2

Measurement cannot serve as proof. Measurement applies only to the limited number of cases involved. The conclusion it provides is not exact but approximate, depending on the precision of the measuring instrument and the care of the observer. In measurement, allowance should be made for possible error equal to half the smallest unit of measurement used. Thus if an angle is measured to the nearest degree, an allowance of half a degree of error should be made.

Experiment cannot serve as proof. Its conclusions are only probable ones. The degree of probability depends on the particular situations or instances examined in the process of experimentation. Thus, it is probable that a pair of dice are loaded if ten successive 7s are rolled with the pair, and the probability is much greater if twenty successive 7s are rolled; however, neither probability is a certainty.

SOLVED PROBLEMS

2.1 Using circles to determine group relationships

In (a) to (e) each letter, such as A, B, and R, represents a set or group. Complete each statement. Show how circles may be used to represent the sets or groups.

(a) If A is B and B is C, then __?__ .

(b) If A is B and B is E and E is R, then __?__ .

(c) If X is Y and __?__ , then X is M.

(d) If C is D and E is C, then __?__ .

(e) If squares (S) are rectangles (R) and rectangles are parallelograms (P), then __?__ .

Solutions

(a) A is C (b) A is R (c) Y is M (d) E is D (e) Squares are parallelograms

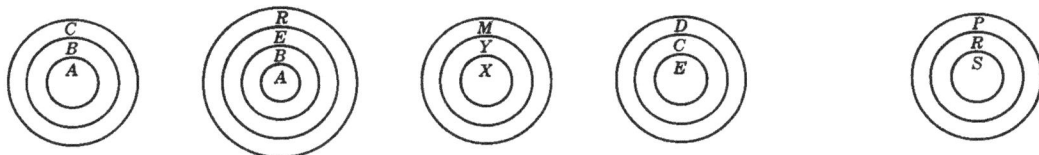

2.2 Completing a syllogism

Write the statement needed to complete each syllogism:

Major Premise (General Statement)	Minor Premise (Particular Statement)	Conclusion (Deducted Statement)
(a) A cat is a domestic animal.	Fluffy is a cat.	?
(b) All people must die.	?	Jan must die.
(c) Vertical angles are congruent.	$\angle c$ and $\angle d$ are vertical angles.	?
(d) ?	A square is a rectangle.	A square has congruent diagonals.
(e) An obtuse triangle has only one obtuse angle.	?	$\triangle ABC$ has only one obtuse angle.

Solutions

(a) Fluffy is a domestic animal.   (d) A rectangle has congruent diagonals.

(b) Jan is a person.   (e) $\triangle ABC$ is an obtuse triangle.

(c) $\angle c \cong \angle d$.

2.2 Postulates (Assumptions)

The entire structure of proof in geometry rests upon, or begins with, some unproved general statements called *postulates*. These are statements which we must willingly assume or accept as true so as to be able to deduce other statements.

2.2A Algebraic Postulates

POSTULATE 1: *Things equal to the same or equal things are equal to each other; if $a = b$ and $c = b$, then $a = c$.* (Transitive Postulate)

Thus the total value of a dime is equal to the value of two nickels because each is equal to the value of ten pennies.

POSTULATE 2: *A quantity may be substituted for its equal in any expression or equation.* (Substitution Postulate)

Thus if $x = 5$ and $y = x + 3$, we may substitute 5 for x and find $y = 5 + 3 = 8$.

POSTULATE 3: *The whole equals the sum of its parts.* (Partition Postulate)

Thus the total value of a dime, a nickel, and a penny is 16 cents.

POSTULATE 4: *Any quantity equals itself.* (Reflexive Postulate or Identity Postulate)

Thus $x = x$, $m\angle A = m\angle A$, and $AB = AB$.

POSTULATE 5: *If equals are added to equals, the sums are equal; if $a = b$ and $c = d$, then $a + c = b + d$.* (Addition Postulate)

If	7 dimes = 70 cents	If	$x + y = 12$	
and	2 dimes = 20 cents	and	$x - y = 8$	
then	9 dimes = 90 cents	then	$2x = 20$	

POSTULATE 6: *If equals are subtracted from equals, the differences are equal; if a = b and c = d, then a − c = b − d.* (Subtraction Postulate)

If	7 dimes = 70 cents	If	$x + y = 12$
and	2 dimes = 20 cents	and	$x - y = 8$
then	5 dimes = 50 cents	then	$2y = 4$

POSTULATE 7: *If equals are multiplied by equals, the products are equal; if a = b and c = d, then ac = bd.* (Multiplication Postulate)

Thus if the price of one book is $2, the price of three books is $6.

Special multiplication axiom: Doubles of equals are equal.

POSTULATE 8: *If equals are divided by equals, the quotients are equal; if a = b and c = d, then a/c = b/d, where c, d ≠ 0.* (Division Postulate)

Thus if the price of 1 lb of butter is 80 cents then, at the same rate, the price of $\frac{1}{4}$ lb is 20 cents.

POSTULATE 9: *Like powers of equals are equal; if a = b, then $a^n = b^n$.* (Powers Postulate)

Thus if $x = 5$, then $x^2 = 5^2$ or $x^2 = 25$.

POSTULATE 10: *Like roots of equals are equal; if a = b then $\sqrt[n]{a} = \sqrt[n]{b}$.*

Thus if $y^3 = 27$, then $y = \sqrt[3]{27} = 3$.

2.2B Geometric Postulates

POSTULATE 11: *One and only one straight line can be drawn through any two points.*

Thus, $\overset{\leftrightarrow}{AB}$ is the only line that can be drawn between A and B in Fig. 2-3.

Fig. 2-3 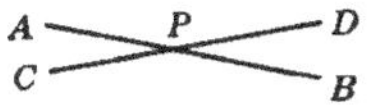   Fig. 2-4

POSTULATE 12: *Two lines can intersect in one and only one point.*

Thus, only P is the point of intersection of $\overset{\leftrightarrow}{AB}$ and $\overset{\leftrightarrow}{CD}$ in Fig. 2-4.

POSTULATE 13: *The length of a segment is the shortest distance between two points.*

Thus, $\overline{AB}$ is shorter than the curved or broken line segment between A and B in Fig. 2-5.

Fig. 2-5 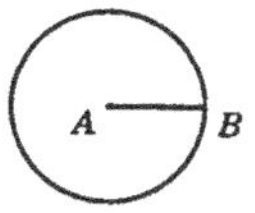   Fig. 2-6

POSTULATE 14: *One and only one circle can be drawn with any given point as center and a given line segment as a radius.*

Thus, only circle A in Fig. 2-6 can be drawn with A as center and $\overline{AB}$ as a radius.

POSTULATE 15: *Any geometric figure can be moved without change in size or shape.*

Thus, △I in Fig. 2-7 can be moved to a new position without changing its size or shape.

Fig. 2-7

Fig. 2-8

POSTULATE 16: *A segment has one and only one midpoint.*

Thus, only M is the midpoint of $\overline{AB}$ in Fig. 2-8.

POSTULATE 17: *An angle has one and only one bisector.*

Thus, only $\overrightarrow{AD}$ is the bisector of $\angle A$ in Fig. 2-9.

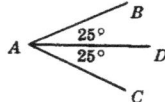

Fig. 2-9

POSTULATE 18: *Through any point on a line, one and only one perpendicular can be drawn to the line.*

Thus, only $\overrightarrow{PC} \perp \overleftrightarrow{AB}$ at point P on $\overleftrightarrow{AB}$ in Fig. 2-10.

Fig. 2-10

POSTULATE 19: *Through any point outside a line, one and only one perpendicular can be drawn to the given line.*

Thus, only $\overline{PC}$ can be drawn $\perp \overleftrightarrow{AB}$ from point P outside $\overleftrightarrow{AB}$ in Fig. 2-11.

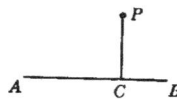

Fig. 2-11

SOLVED PROBLEMS

2.3 Applying postulate 1

In each part, what conclusion follows when Postulate 1 is applied to the given data from Figs. 2-12 and 2-13?

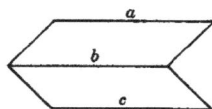

Fig. 2-12

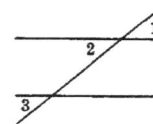

Fig. 2-13

(a) Given: $a = 10, b = 10, c = 10$

(b) Given: $a = 25, a = c$

(c) Given: $a = b, c = b$

(d) Given: $m\angle 1 = 40°, m\angle 2 = 40°, m\angle 3 = 40°$

(e) Given: $m\angle 1 = m\angle 2, m\angle 3 = m\angle 1$

(f) Given: $m\angle 3 = m\angle 1, m\angle 2 = m\angle 3$

Solutions

(a) Since a, b, and c each equal 10, $a = b = c$.

(b) Since c and 25 each equal $a, c = 25$.

(c) Since a and c each equal b, $a = c$.

(d) Since $\angle 1$, $\angle 2$, and $\angle 3$ each measures $40°$, $\angle 1 \cong \angle 2 \cong \angle 3$.

(e) Since $\angle 2$ and $\angle 3$ each $\cong \angle 1$, $\angle 2 \cong \angle 3$.

(f) Since $\angle 1$ and $\angle 2$ each $\cong \angle 3$, $\angle 1 \cong \angle 2$.

2.4 Applying postulate 2

In each part, what conclusion follows when Postulate 2 is applied to the given data?

(a) Evaluate $2a + 2b$ when $a = 4$ and $b = 8$.

(b) Find x if $3x + 4y = 35$ and $y = 5$.

(c) Given: $m\angle 1 + m\angle B + m\angle 2 = 180°$, $\angle 1 \cong \angle A$, and $\angle 2 \cong \angle C$ in Fig. 2-14.

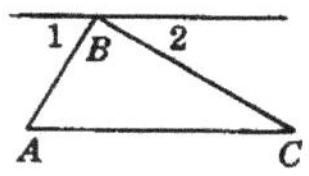

Fig. 2-14

Solutions

(a) Substitute 4 for a and 8 for b:
$2a + 2b$
$2(4) + 2(8)$
$8 + 16 = 24$ *Ans.*

(b) Substitute 5 for y:
$3x + 4y = 35$
$3x + 4(5) = 35$
$3x + 20 = 35$
$3x = 15, \quad x = 5$ *Ans.*

(c) Substitute $\angle A$ for $\angle 1$ and $\angle C$ for $\angle 2$:
$m\angle 1 + m\angle B + m\angle 2 = 180°$
$m\angle A + m\angle B + m\angle C = 180°$ *Ans.*

2.5 Applying postulate 3

State the conclusions that follow when Postulate 3 is applied to the data in (a) Fig. 2.15(a) and (b) Fig. 2-15(b).

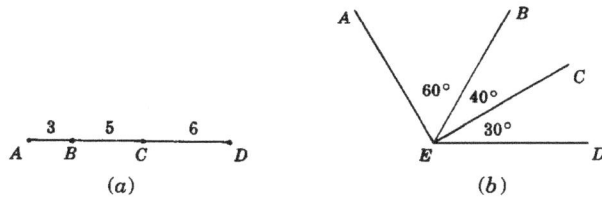

Fig. 2-15

Solutions

(a) $AC = 3 + 5 = 8$
$BD = 5 + 6 = 11$
$AD = 3 + 5 + 6 = 14$

(b) $m\angle AEC = 60° + 40° = 100°$
$m\angle BED = 40° + 30° = 70°$
$m\angle AED = 60° + 40° + 30° = 130°$

2.6 Applying postulates 4, 5, and 6

In each part, state a conclusion that follows when Postulates 4, 5, and 6 are applied to the given data.

(a) Given: $a = e$ (Fig. 2-16)

(b) Given: $a = c, b = d$ (Fig. 2-16)

(c) Given: $m\angle BAC = m\angle DAE$ (Fig. 2-17)

(d) Given: $m\angle BAC = m\angle BCA, m\angle 1 = m\angle 3$ (Fig. 2-17)

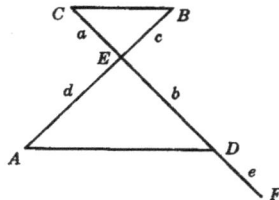   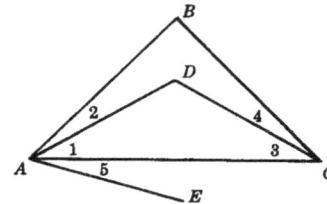

Fig. 2-16   Fig. 2-17

Solutions

(a)
$$a = e \qquad \text{Given}$$
$$\underline{b = b} \qquad \text{Identity}$$
$$a + b = b + e \qquad \text{Add. Post}$$
$$CD = EF \qquad \text{Subst.}$$

(b)
$$a = c \qquad \text{Given}$$
$$\underline{b = d} \qquad \text{Given}$$
$$a + b = c + d \qquad \text{Add. Post.}$$
$$CD = AB \qquad \text{Subst.}$$

(c)
$$m\angle BAC = m\angle DAE \qquad \text{Given}$$
$$\underline{m\angle 1 = m\angle 1} \qquad \text{Given}$$
$$m\angle BAC - m\angle 1 = m\angle DAE - m\angle 1 \qquad \text{Subt. Post.}$$
$$m\angle 2 = m\angle 5 \qquad \text{Subst.}$$

(d)
$$m\angle BAC = m\angle BCA \qquad \text{Given}$$
$$\underline{m\angle 1 = m\angle 3} \qquad \text{Given}$$
$$m\angle BAC - m\angle 1 = m\angle BCA - m\angle 3 \qquad \text{Subt. Post.}$$
$$m\angle 2 = m\angle 4 \qquad \text{Subst.}$$

2.7 Applying postulates 7 and 8

State the conclusions that follow when the multiplication and division axioms are applied to the data in (a) Fig. 2-18 and (b) Fig. 2-19.

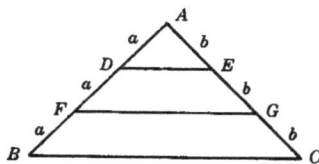   **Given:** $a = b$
                          $\overline{AB}$ and $\overline{AC}$ are trisected.

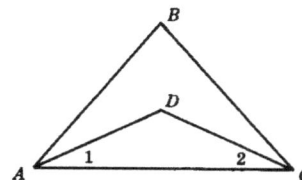   **Given:** $m\angle A = m\angle C$
                          $m\angle 1 = \frac{1}{2}m\angle A$
                          $m\angle 2 = \frac{1}{2}m\angle C$

Fig. 2-18   Fig. 2-19

Solutions

(a) If $a = b$, then $2a = 2b$ since doubles of equals are equal. Hence, $AF = DB = AG = EC$. Also, $3a = 3b$, using the Multiplication Postulate. Hence, $AB = AC$.

(b) If $m\angle A = m\angle C$, then $\frac{1}{2}m\angle A = \frac{1}{2}m\angle C$ since halves of equals are equal. Hence, $m\angle 1 = m\angle 2$.

2.8   Applying postulates to statements

Complete each sentence and state the postulate that applies.

(a) If Harry and Alice are the same age today, then in 10 years __?__ .

(b) Since 32°F and 0°C both name the temperature at which water freezes, we know that __?__ .

(c) If Henry and John are the same weight now and each loses 20 lb, then __?__ .

(d) If two stocks of equal value both triple in value, then __?__ .

(e) If two ribbons of equal size are cut into five equal parts, then __?__ .

(f) If Joan and Agnes are the same height as Anne, then __?__ .

(g) If two air conditioners of the same price are each discounted 10 percent, then __?__ .

Solutions

(a) They will be the same age. (Add. Post.)

(b) 32°F = 0°C. (Trans. Post.)

(c) They will be the same weight. (Subt. Post.)

(d) They will have the same value. (Mult. Post.)

(e) Their parts will be of the same size. (Div. Post.)

(f) Joan and Agnes are of the same height. (Trans. Post.)

(g) They will have the same price. (Subt. Post.)

2.9   Applying geometric postulates

State the postulate needed to correct each diagram and accompanying statement in Fig. 2-20.

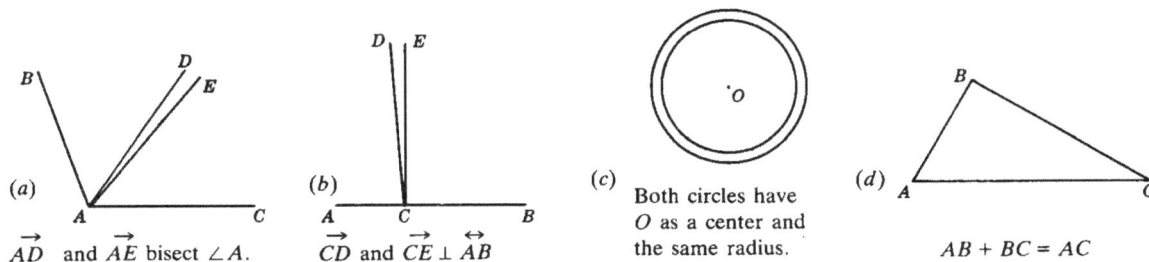

(a) $\overrightarrow{AD}$ and $\overrightarrow{AE}$ bisect $\angle A$.

(b) $\overrightarrow{CD}$ and $\overrightarrow{CE} \perp \overleftrightarrow{AB}$.

(c) Both circles have O as a center and the same radius.

(d) $AB + BC = AC$

Fig. 2-20

Solutions

(a) Postulate 17. (b) Postulate 18. (c) Postulate 14. (d) Postulate 13. (AC is less than the sum of AB and BC.)

2.3   Basic Angle Theorems

A *theorem* is a statement, which, when proved, can be used to prove other statements or derive other results. Each of the following basic theorems requires the use of definitions and postulates for its proof.

   Note: We shall use the term *principle* to include important geometric statements such as theorems, postulates, and definitions.

PRINCIPLE 1: *All right angles are congruent.*

Thus, $\angle A \cong \angle B$ in Fig. 2-21.

Fig. 2-21

PRINCIPLE 2: *All straight angles are congruent.*

Thus, $\angle C \cong \angle D$ in Fig. 2-22.

Fig. 2-22

PRINCIPLE 3: *Complements of the same or of congruent angles are congruent.*

This is a combination of the following two principles:

1. *Complements of the same angle are congruent.* Thus, $\angle a \cong \angle b$ in Fig. 2.23 and each is the complement of $\angle x$.

2. *Complements of congruent angles are congruent.* Thus, $\angle c \cong \angle d$ in Fig. 2-24 and their complements are the congruent ∕ₛ x and y.

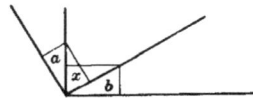

Fig. 2-23

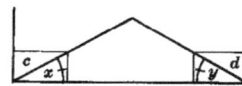

Fig. 2-24

PRINCIPLE 4: *Supplements of the same or of congruent angles are congruent.*

This is a combination of the following two principles:

1. *Supplements of the same angle are congruent.* Thus, $\angle a \cong \angle b$ in Fig. 2-25 and each is the supplement of $\angle x$.

2. *Supplements of congruent angles are congruent.* Thus, $\angle c \cong \angle d$ in Fig. 2-26 and their supplements are the congruent angles x and y.

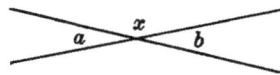

Fig. 2-25

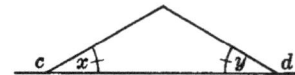

Fig. 2-26

PRINCIPLE 5: *Vertical angles are congruent.*

Thus, in Fig. 2-27, $\angle a \cong \angle b$; this follows from Principle 4, since $\angle a$ and $\angle b$ are supplements of the same angle, $\angle c$.

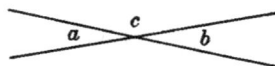

Fig. 2-27

2.10 Applying basic theorems: principles 1 to 5

State the basic angle theorem needed to prove $\angle a \cong \angle b$ in each part of Fig. 2-28.

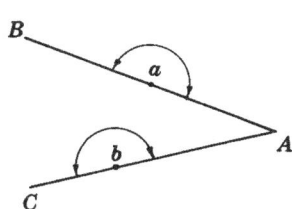

(a) **Given:**
$\overrightarrow{AB}$ and $\overrightarrow{AC}$ are straight lines.

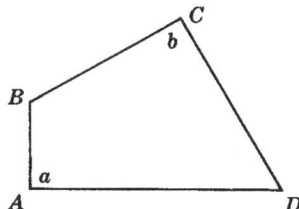

(b) **Given:** $\overline{BA} \perp \overline{AD}$,
$\overline{BC} \perp \overline{CD}$

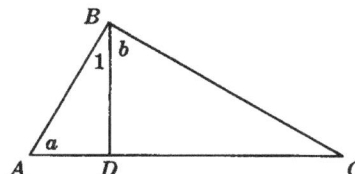

(c) **Given:** $\overline{AB} \perp \overline{BC}$
$\angle a$ comp. $\angle 1$

Fig. 2-28

Solutions

(a) Since $\overleftrightarrow{AB}$ and $\overleftrightarrow{AC}$ are straight lines, $\angle a$ and $\angle b$ are straight $\angle$s. Hence, $\angle a \cong \angle b$. *Ans.* All straight angles are congruent.

(b) Since $\overline{BA} \perp \overline{AD}$ and $\overline{BC} \perp \overline{CD}$, $\angle a$ and $\angle b$ are rt. $\angle$s. Hence, $\angle a \cong \angle b$. *Ans.* All right angles are congruent.

(c) Since $\overline{AB} \perp \overline{BC}$, $\angle B$ is a rt. $\angle$, making $\angle b$ the complement of $\angle 1$. Since $\angle a$ is the complement of $\angle 1$, $\angle a \cong \angle b$. *Ans.* Complements of the same angle are congruent.

2.4 Determining the Hypothesis and Conclusion

2.4A Statement Forms: Subject-Predicate Form and If-Then Form

The statements "A heated metal expands" and "If a metal is heated, then it expands" are two forms of the same idea. The following table shows how each form may be divided into its two important parts, the *hypothesis*, which tells *what is given*, and the *conclusion*, which tells *what is to be proved*. Note that in the if-then form, the word *then* may be omitted.

Form	Hypothesis (What is given)	Conclusion (What is to be proved)
Subject-predicate form: *A heated metal expands.*	**Hypothesis is subject:** *A heated metal*	**Conclusion is predicate:** *expands*
If-then form: *If a metal is heated, then it expands.*	**Hypothesis is if clause:** *If a metal is heated*	**Conclusion is then clause:** *then is expands*

2.4B Converse of a Statement

The converse of a statement is formed by interchanging the hypothesis and conclusion. Hence to form the converse of an if-then statement, interchange the if and then clauses. In the case of the subject-predicate form, interchange the subject and the predicate.

Thus, the converse of "triangles are polygons" is "polygons are triangles." Also, the converse of "if a metal is heated, then it expands" is "if a metal expands, then it is heated." Note in each of these cases that the statement is true but its converse need not necessarily be true.

PRINCIPLE 1: *The converse of a true statement is not necessarily true.*

Thus, the statement "triangles are polygons" is true. Its converse need not be true.

PRINCIPLE 2: *The converse of a definition is always true.*

Thus, the converse of the definition "a triangle is a polygon of three sides" is "a polygon of three sides is a triangle." Both the definition and its converse are true.

SOLVED PROBLEMS

2.11 Determining the hypothesis and conclusion in subject-predicate form

Determine the hypothesis and conclusion of each statement.

Statements	Solutions	
	Hypothesis (subject)	Conclusion (predicate)
(a) Perpendiculars form right angles.	Perpendiculars	form right angles
(b) Complements of the same angle are congruent.	Complements of the same angle	are congruent
(c) An equilateral triangle is equiangular.	An equilateral triangle	is equiangular
(d) A right triangle has only one right angle.	A right triangle	has only one right angle
(e) A triangle is not a quadrilateral.	A triangle	is not a quadrilateral

2.12 Determining the hypothesis and conclusion in if-then form

Determine the hypothesis and conclusion of each statement.

Statements	Solutions	
	Hypothesis (if clause)	Conclusion (then clause)
(a) If a line bisects an angle, then it divides the angle into two congruent parts.	If a line bisects an angle	then it divides the angle into two congruent parts
(b) A triangle has an obtuse angle if it is an obtuse triangle.	If it is an obtuse triangle	(then) a triangle has an obtuse angle
(c) If a student is sick, she should not go to school.	If a student is sick	(then) she should not go to school
(d) A student, if he wishes to pass, must study regularly.	If he wishes to pass	(then) a student must study regularly

2.13 Forming converses and determining their truth

State whether the given statement is true. Then form its converse and state whether this is necessarily true.

(a) A quadrilateral is a polygon.

(b) An obtuse angle has greater measure than a right angle.

(c) Florida is a state of the United States.

(d) If you are my pupil, then I am your teacher.

(e) An equilateral triangle is a triangle that has all congruent sides.

Solutions

(a) Statement is true. Its converse, "a polygon is a quadrilateral," is not necessarily true; it might be a triangle.

(b) Statement is true. Its converse, "an angle with greater measure than a right angle is an obtuse angle," is not necessarily true; it might be a straight angle.

(c) Statement is true. Its converse, "a state of the United States is Florida," is not necessarily true; it might be any one of the other 49 states.

(d) Statement is true. Its converse, "if I am your teacher, then you are my pupil," is also true.

(e) The statement, a definition, is true. Its converse, "a triangle that has all congruent sides is an equilateral triangle," is also true.

2.5   Proving a Theorem

Theorems should be proved using the following step-by-step procedure. The form of the proof is shown in the example that follows the procedure. Note that accepted symbols and abbreviations may be used.

1. Divide the theorem into its hypothesis (what is given) and its conclusion (what is to be proved). Underline the hypothesis with a single line, and the conclusion with a double line.

2. On one side, make a marked diagram. Markings on the diagram should include such helpful symbols as square corners for right angles, cross marks for equal parts, and question marks for parts to be proved equal.

3. On the other side, next to the diagram, state what is given and what is to be proved. The "Given" and "To Prove" must refer to the parts of the diagram.

4. Present a plan. Although not essential, a plan is very advisable. It should state the major methods of proof to be used.

5. On the left, present statements in successively numbered steps. The last statement must be the one to be proved. All the statements must refer to parts of the diagram.

6. On the right, next to the statements, provide a reason for each statement. Acceptable reasons in the proof of a theorem are given facts, definitions, postulates, assumed theorems, and previously proven theorems.

Step 1: **Prove:**   All right angles are equal
                              in measure.
Steps 2 **Given:** $\angle A$ and $\angle B$ are rt. $\angle$s
and 3:   **To Prove:** $m\angle A = m\angle B$
Step 4: **Plan:**   Since each angle equals 90°,
                the angles are equal in measure,
                using Post. 1: Things equal to the
                same thing are equal to each other.

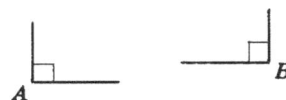

Steps 5
and 6:

Statements	Reasons
1. $\angle A$ and $\angle B$ are rt. $\angle$s.	1. Given
2. $m\angle A$ and $m\angle B$ each $= 90°$.	2. m(rt. $\angle$) $= 90°$
3. $m\angle A = m\angle B$	3. Things $=$ to same thing $=$ each other.

SOLVED PROBLEM

2.14   Proving a theorem

Use the proof procedure to prove that supplements of angles of equal measure have equal measure.

Step 1:   **Prove:**   Supplements of angles of
                        equal measure have equal measure.
Steps 2 **Given:** $\angle a$ sup. $\angle 1$, $\angle b$ sup. $\angle 2$
and 3: $m\angle 1 = m\angle 2$
              To Prove: $m\angle a = m\angle b$
Step 4: **Plan:**   Using the subtraction postulate,
                the equal angle measures may be
                subtracted from the equal sums of
                measures of pairs of supplementary
                angles. The equal remainders are the
                measures of the supplements.

*Steps 5
and 6:*

Statements	Reasons
1. $\angle a$ sup. $\angle 1, \angle b$ sup. $\angle 2$	1. Given
2. $m\angle a + m\angle 1 = 180°$ $m\angle b + m\angle 2 = 180°$	2. Sup. ∡s are ∡ the sum of whose measures $= 180°$.
3. $m\angle a + m\angle 1 = m\angle b + m\angle 2$	3. Things $=$ to the same thing $=$ each other.
4. $m\angle 1 = m\angle 2$	4. Given
5. $m\angle a = m\angle b$	5. If $=$s are subtracted from $=$s, the differences are $=$.

SUPPLEMENTARY PROBLEMS

2.1. Complete each statement. In (a) to (e), each letter, such as C, D, or R, represents a set or group. (2.2)

(a) If A is B and B is H, then ___?___ .

(b) If C is D and P is C, then ___?___ .

(c) If ___?___ and B is R, then B is S.

(d) If E is F, F is G, and G is K, then ___?___ .

(e) If G is H, H is R, and ___?___ , then A is R.

(f) If triangles are polygons and polygons are geometric figures, then ___?___ .

(g) If a rectangle is a parallelogram and a parallelogram is a quadrilateral, then ___?___ .

2.2. State the conclusions which follow when Postulate 1 is applied to the given data, which refer to Fig. 2-29. (2.3)

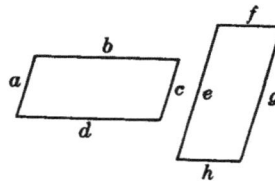
Fig. 2-29

(a) $a = 7, c = 7, f = 7$

(b) $b = 15, b = g$

(c) $f = h, h = a$

(d) $a = c, c = f, f = h$

(e) $b = d, d = g, g = e$

2.3. State the conclusions which follow when Postulate 2 is applied in each case. (2.4)

(a) Evaluate $a^2 + 3a$ when $a = 10$.

(b) Evaluate $x^2 - 4y$ when $x = 4$ and $y = 3$.

(c) Does $b^2 - 8 = 17$ when $b = 5$?

(d) Find x if $x + y = 20$ and $y = x + 3$.

(e) Find y if $x + y = 20$ and $y = 3x$.

(f) Find x if $5x - 2y = 24$ and $y = 3$.

(g) Find x if $x^2 + 3y = 45$ and $y = 3$.

2.4. State the conclusions that follow when Postulate 3 is applied to the data in Fig. 2-30(a) and (b).   (2.5)

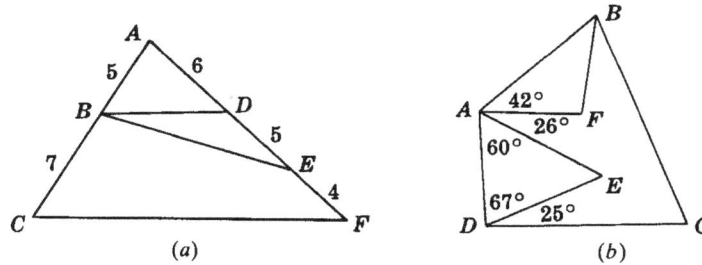

Fig. 2-30

2.5. State a conclusion involving two new equals that follows when Postulate 4, 5, or 6 is applied to the given data. (2.6)

   (a) Given: $b = e$ (Fig. 2-31).

   (b) Given: $b = c, a = d$ (Fig. 2-31).

   (c) Given: $\angle 4 \cong \angle 5$ (Fig. 2-32).

   (d) Given: $\angle 1 \cong \angle 3, \angle 2 \cong \angle 4$ (Fig. 2-32).

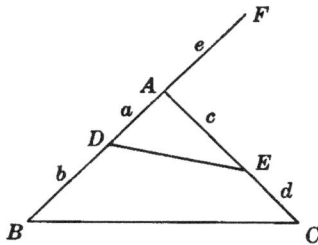

Fig. 2-31

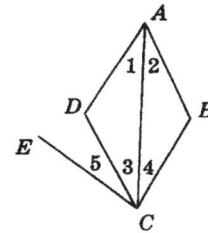

Fig. 2-32

2.6. In Fig. 2-33 $\overline{AD}$ and $\overline{BC}$ are trisected (divided into 3 equal parts).   (2.7)

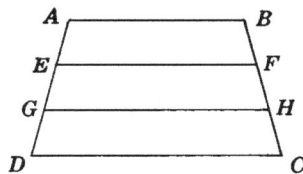

Fig. 2-33

   (a) If $\overline{AD} \cong \overline{BC}$, why is $\overline{AE} \cong \overline{BF}$?

   (b) If $\overline{EG} \cong \overline{FH}$, why is $\overline{AG} \cong \overline{BH}$?

   (c) If $\overline{GD} \cong \overline{HC}$, why is $\overline{AD} \cong \overline{BC}$?

   (d) If $\overline{ED} \cong \overline{FC}$, why is $\overline{EG} \cong \overline{FH}$?

2.7. In Fig. 2-34 $\angle BCD$ and $\angle ADC$ are trisected.

   (a) If $m\angle BCD = m\angle ADC$, why does $m\angle FCD = m\angle FDC$?

   (b) If $m\angle 1 = m\angle 2$, why does $m\angle BCD = m\angle ADC$?

(c) If $m\angle 1 = m\angle 2$, why does $m\angle ADF = m\angle BCF$?

(d) If $m\angle EDC = m\angle ECD$, why does $m\angle 1 = m\angle 2$?

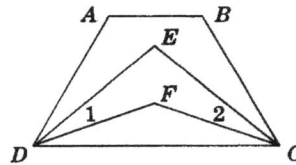

Fig. 2-34

2.8. Complete each statement, and name the postulate that applies. (2.8)

(a) If Bill and Helen earn the same amount of money each hour and their rate of pay is increased by the same amount, then __?__ .

(b) In the past year, those stocks have tripled in value. If they had the same value last year, then __?__ .

(c) A week ago, there were two classes that had the same register. If the same number of pupils were dropped in each, then __?__ .

(d) Since 100°C and 212°F are the boiling temperatures of water, then __?__ .

(e) If two boards have the same length and each is cut into four equal parts, then __?__ .

(f) Since he has $2000 in Bank A, $3000 in Bank B and $5000 in Bank C, then __?__ .

(g) If three quarters and four nickels are compared with three quarters and two dimes, __?__ .

2.9. Answer each of the following by stating the basic angle theorem needed. The questions refer to Fig. 2-35. (2.10)

(a) Why does $m\angle 1 = m\angle 2$?

(b) Why does $m\angle DBC = m\angle ECB$?

(c) If $m\angle 3 = m\angle 4$, why does $m\angle 5 = m\angle 6$?

(d) If $\overrightarrow{AF} \perp \overline{DE}$ and $\overrightarrow{GC} \perp \overline{DE}$, why does $m\angle 7 = m\angle 8$?

(e) If $\overrightarrow{AF} \perp \overline{DE}$, $\overrightarrow{GC} \perp \overline{DE}$, and $m\angle 11 = m\angle 12$, why does $m\angle 9 = m\angle 10$?

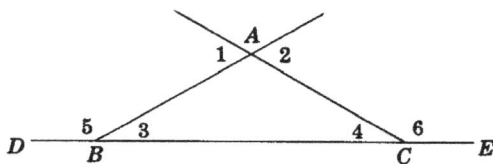   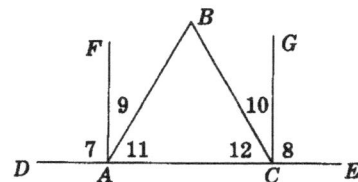

Fig. 2-35

2.10. Determine the hypothesis and conclusion of each statement. (2.11 and 2.12)

(a) Stars twinkle.

(b) Jet planes are the speediest.

(c) Water boils at 212°F.

(d) If it is the American flag, its colors are red, white, and blue.

(e) You cannot learn geometry if you fail to do homework in the subject.

(f) A batter goes to first base if the umpire calls a fourth ball.

(g) If *A* is *B*'s brother and *C* is *B*'s son, then *A* is *C*'s uncle.

(h) An angle bisector divides the angle into two equal parts.

(i) A segment is trisected if it is divided into three congruent parts.

(j) A pentagon has five sides and five angles.

(k) Some rectangles are squares.

(l) Angles do not become larger if their sides are made longer.

(m) Angles, if they are congruent and supplementary, are right angles.

(n) The figure cannot be a polygon if one of its sides is not a straight line segment.

2.11. State the converse of each of the following true statements. State whether the converse is necessarily true. (2.13)

(a) Half a right angle is an acute angle.

(b) An obtuse triangle is a triangle having one obtuse angle.

(c) If the umpire called a third strike, then the batter is out.

(d) If I am taller than you, then you are shorter than I.

(e) If I am heavier than you, then our weights are unequal.

2.12. Prove each of the following. (2.14)

(a) Straight angles are congruent.

(b) Complements of congruent angles are congruent.

(c) Vertical angles are congruent.

Congruent Triangles

CHAPTER 3

3.1 Congruent Triangles

Congruent figures are figures that have the same size and the same shape; they are the exact duplicates of each other. Such figures can be moved on top of one another so that their corresponding parts line up exactly. For example, two circles having the same radius are congruent circles.

Congruent triangles are triangles that have the same size and the same shape.

If two triangles are congruent, their corresponding sides and angles must be congruent. Thus, congruent triangles ABC and $A'B'C'$ in Fig. 3-1 have congruent corresponding sides ($\overline{AB} \cong \overline{A'C'}$, $\overline{BC} \cong \overline{B'C'}$, and $\overline{AC} \cong \overline{A'C'}$) and congruent corresponding angles ($\angle A \cong \angle A'$, $\angle B \cong \angle B'$, and $\angle C \cong \angle C'$).

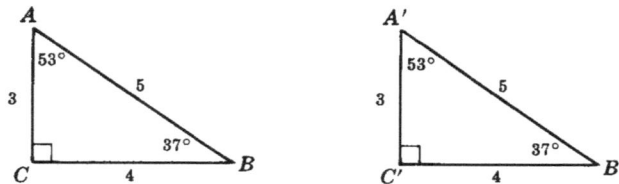

Fig. 3-1

(Read $\triangle ABC \cong \triangle A'B'C'$ as "Triangle ABC is congruent to triangle A-prime, B-prime, C-prime.")

Note in the congruent triangles how corresponding equal parts may be located. Corresponding sides lie opposite congruent angles, and corresponding angles lie opposite congruent sides.

3.1A Basic Principles of Congruent Triangles

PRINCIPLE 1:   *If two triangles are congruent, then their corresponding parts are congruent.* (Corresponding parts of congruent triangles are congruent.)

Thus if $\triangle ABC \cong \triangle A'B'C'$ in Fig. 3-2, then $\angle A \cong \angle A'$, $\angle B \cong \angle B'$, $\angle C \cong \angle C'$, $a = a'$, $b = b'$, and $c = c'$.

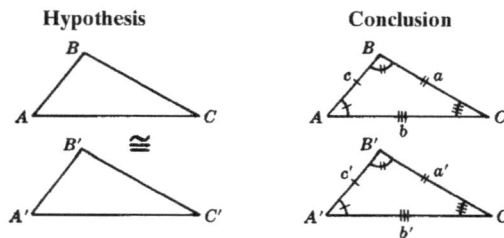

Fig. 3-2

Methods of Proving that Triangles are Congruent

PRINCIPLE 2:   (Side-Angle-Side, SAS) *If two sides and the included angle of one triangle are congruent to the corresponding parts of another, then the triangles are congruent.*

Thus if $b = b'$, $c = c'$, and $\angle A \cong \angle A'$ in Fig. 3-3, then $\triangle ABC \cong \triangle A'B'C'$.

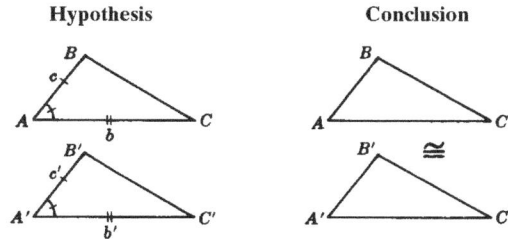

Fig. 3-3

PRINCIPLE 3:   (Angle-Side-Angle, ASA) *If two angles and the included side of one triangle are congruent to the corresponding parts of another, then the triangles are congruent.*

Thus if $\angle A \cong \angle A'$, $\angle C \cong \angle C'$, and $b = b'$ in Fig. 3-4, then $\triangle ABC \cong \triangle A'B'C'$.

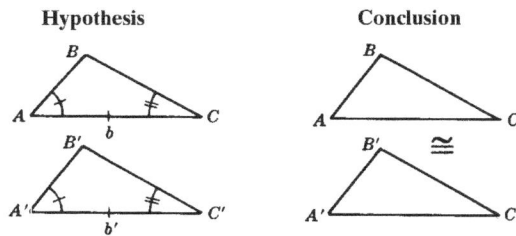

Fig. 3-4

PRINCIPLE 4:   (Side-Side-Side, SSS) *If three sides of one triangle are congruent to three sides of another, then the triangles are congruent.*

Thus if $a = a'$, $b = b'$, and $c = c'$ in Fig. 3-5, then $\triangle ABC \cong \triangle A'B'C'$.

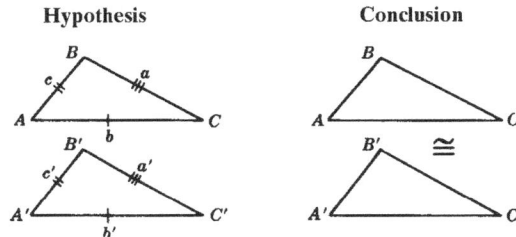

Fig. 3-5

SOLVED PROBLEMS

3.1   Selecting congruent triangles

From each set of three triangles in Fig. 3-6 select the congruent triangles and state the congruency principle that is involved.

Solutions

(a) $\triangle I \cong \triangle II$, by SAS. In $\triangle III$, the right angle is not between 3 and 4.

(b) $\triangle II \cong \triangle III$, by ASA. In $\triangle I$, side 10 is not between 70° and 30°.

(c) $\triangle I \cong \triangle II \cong \triangle III$ by SSS.

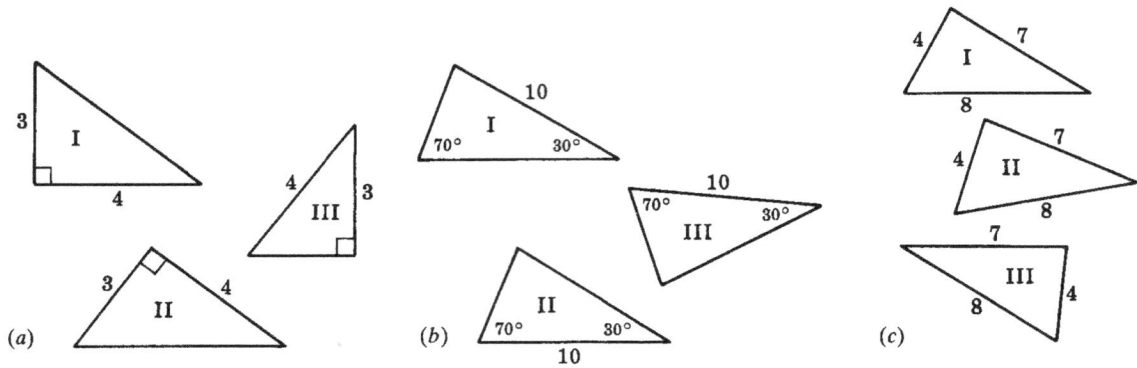

Fig. 3-6

3.2 Determining the reason for congruency of triangles

In each part of Fig. 3-7, $\triangle I$ can be proved congruent of $\triangle II$. Make a diagram showing the equal parts of both triangles and state the congruency principle that is involved.

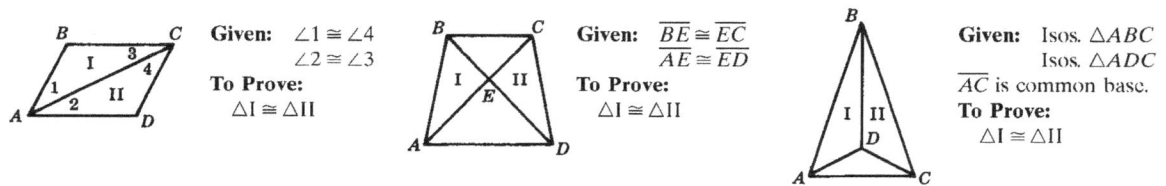

Given: $\angle 1 \cong \angle 4$
            $\angle 2 \cong \angle 3$
To Prove:
            $\triangle I \cong \triangle II$

Given: $\overline{BE} \cong \overline{EC}$
            $\overline{AE} \cong \overline{ED}$
To Prove:
            $\triangle I \cong \triangle II$

Given: Isos. $\triangle ABC$
            Isos. $\triangle ADC$
$\overline{AC}$ is common base.
To Prove:
            $\triangle I \cong \triangle II$

Fig. 3-7

Solutions

(a) AC is a common side of both $\triangle$ [Fig. 3-8(a)]. $\triangle I \cong \triangle II$ by ASA.

(b) $\angle 1$ and $\angle 2$ are vertical angles [Fig. 3-8(b)]. $\triangle I \cong \triangle II$ by SAS.

(c) BD is a common side of both $\triangle$ [Fig. 3-8(c)]. $\triangle I \cong \triangle II$ by SSS.

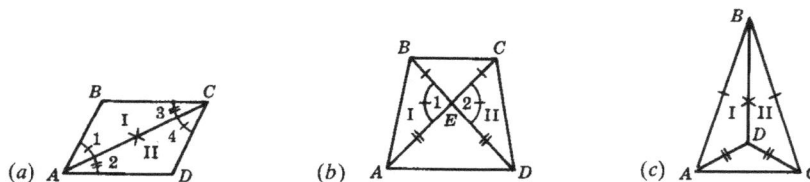

Fig. 3-8

3.3 Finding parts needed to prove triangles congruent

State the additional parts needed to prove $\triangle I \cong \triangle II$ in the given figure by the given congruency principle.

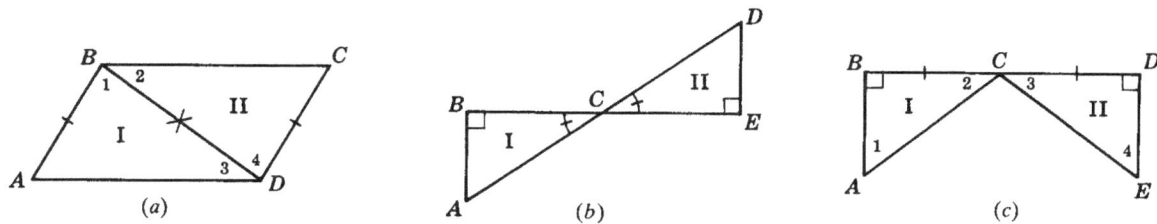

Fig. 3-9

(a) In Fig. 3-9(a) by SSS.

(b) In Fig. 3-9(a) by SAS.

(c) In Fig. 3-9(b) by ASA.

(d) In Fig. 3-9(c) by ASA.

(e) In Fig. 3-9(c) by SAS.

Solutions

(a) If $\overline{AD} \cong \overline{BC},$ then $\triangle I \cong \triangle II$ by SSS.

(b) If $\angle 1 \cong \angle 4,$ then $\triangle I \cong \triangle II$ by SAS.

(c) If $\overline{BC} \cong \overline{CE},$ then $\triangle I \cong \triangle II$ by ASA.

(d) If $\angle 2 \cong \angle 3,$ then $\triangle I \cong \triangle II$ by ASA.

(e) If $\overline{AB} \cong \overline{DE},$ then $\triangle I \cong \triangle II$ by SAS.

3.4 Selecting corresponding parts of congruent triangles

In each part of Fig. 3-10, the equal parts needed to prove $\triangle I \cong \triangle II$ are marked. List the remaining parts that are congruent.

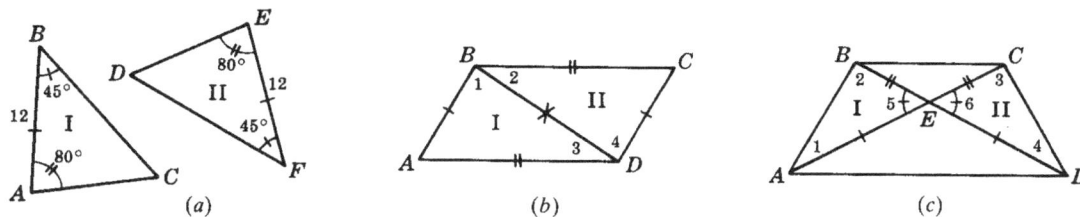

Fig. 3-10

Solutions

Congruent corresponding sides lie opposite congruent angles. Congruent corresponding angles lie opposite congruent sides.

(a) Opposite 45°, $\overline{AC} \cong \overline{DE}$, Opposite 80°, $\overline{BC} \cong \overline{DF}$. Opposite the side of length 12; $\angle C \cong \angle D$.

(b) Opposite $\overline{AB}$ and $\overline{CD}$, $\angle 3 \cong \angle 2$. Opposite $\overline{BC}$ and $\overline{AD}$, $\angle 1 \cong \angle 4$. Opposite common side $\overline{BD}$, $\angle A \cong \angle C$.

(c) Opposite $\overline{AE}$ and $\overline{ED}$, $\angle 2 \cong \angle 3$. Opposite $\overline{BE}$ and $\overline{EC}$, $\angle 1 \cong \angle 4$. Opposite $\angle 5$ and $\angle 6$, $\overline{AB} \cong \overline{CD}$.

3.5   Applying algebra to congruent triangles

In each part of Fig. 3-11, find x and y.

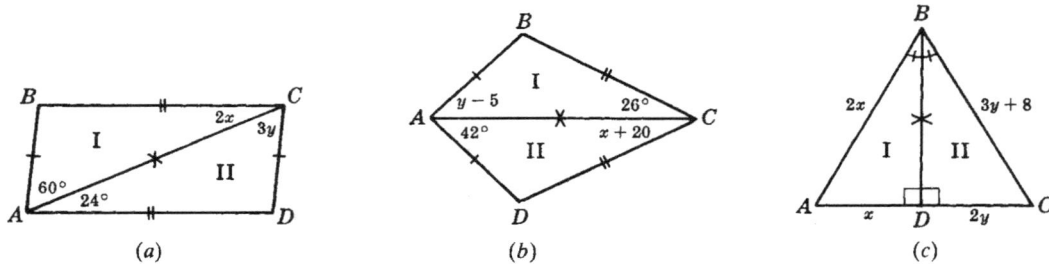

Fig. 3-11

Solutions

(a) Since $\triangle I \cong \triangle II$, by SSS, corresponding angles are congruent. Hence, $2x = 24$ or $x = 12$, and $3y = 60$ or $y = 20$.

(b) Since $\triangle I \cong \triangle II$, by SSS, corresponding angles are congruent. Hence, $x + 20 = 26$ or $x = 6$, and $y - 5 = 42$ or $y = 47$.

(c) Since $\triangle I \cong \triangle II$, by ASA, corresponding sides are congruent. Then $2x = 3y + 8$ and $x = 2y$. Substituting $2y$ for x in the first of these equations, we obtain $2(2y) = 3y + 8$ or $y = 8$. Then $x = 2y = 16$.

3.6   Proving a congruency problem

Given: $\overline{BF} \perp \overline{DE}$
$\overline{BF} \perp \overline{AC}$
$\angle 3 \cong \angle 4$
To Prove: $\overline{AF} \cong \overline{FC}$
Prove: Prove $\triangle I \cong \triangle II$

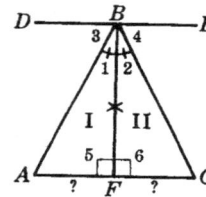

PROOF:

Statements	Reasons
1. $\overline{BF} \perp \overline{AC}$	1. Given
2. $\angle 5 \cong \angle 6$	2. ⊥s form rt. ∡s; rt. ∡s are ≅
3. $\overline{BF} \cong \overline{BF}$	3. Reflexive property
4. $\overline{BF} \perp \overline{DE}$	4. Given
5. $\angle 1$ is the complement of $\angle 3$. $\angle 2$ is the complement of $\angle 4$.	5. Adjacent angles are complementary if exterior sides are ⊥ to each other.
6. $\angle 3 \cong \angle 4$	6. Given
7. $\angle 1 \cong \angle 2$	7. Complements of ≅ ∡s are =
8. $\triangle I \cong \triangle II$	8. ASA
9. $\overline{AF} \cong \overline{FC}$	9. Corresponding parts of congruent ▲ are ≅

3.7   Proving a congruency problem stated in words

Prove that if the opposite sides of a quadrilateral are equal and a diagonal is drawn, equal angles are formed between the diagonal and the sides.

Solution

If the opposite sides of a quadrilateral are congruent and a diagonal is drawn, congruent angles are formed between the diagonal and the sides.

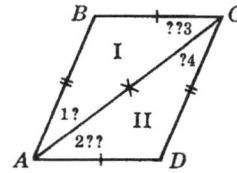

Given: Quadrilateral $ABCD$
$\overline{AB} \cong \overline{CD}, \overline{BC} \cong \overline{AD}$
$\overline{AC}$ is a diagonal.
To Prove: $\angle 1 \cong \angle 4, \angle 2 \cong \angle 3$
Plan: Prove $\triangle I \cong \triangle II$

PROOF:

Statements	Reasons
1. $\overline{AB} \cong \overline{CD}, \overline{BC} \cong \overline{AD}$	1. Given
2. $\overline{AC} \cong \overline{AC}$	2. Reflexive property
3. $\triangle I \cong \triangle II$	3. SSS
4. $\angle 1 \cong \angle 4, \angle 2 \cong \angle 3$	4. Corresponding parts of $\cong \triangle$ are $\cong$.

3.2 Isosceles and Equilateral Triangles

3.2A Principles of Isosceles and Equilateral Triangles

PRINCIPLE 1:   *If two sides of a triangle are congruent, the angles opposite these sides are congruent.* (Base angles of an isosceles triangle are congruent.)

Thus in $\triangle ABC$ in Fig. 3-12, if $\overline{AB} \cong \overline{BC}$, then $\angle A \cong \angle C$.

A proof of Principle 1 is given in Chapter 16.

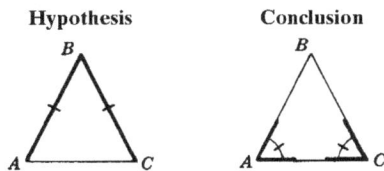

Fig. 3-12

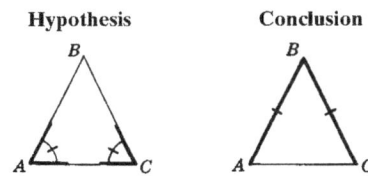

Fig. 3-13

PRINCIPLE 2:   *If two angles of a triangle are congruent, the sides opposite these angles are congruent.*

Thus in $\triangle ABC$ in Fig. 3-13, if $\angle A \cong \angle C$, then $\overline{AB} \cong \overline{BC}$.

Principle 2 is the converse of Principle 1. A proof of Principle 2 is given in Chapter 16.

PRINCIPLE 3:   *An equilateral triangle is equiangular.*

Thus in $\triangle ABC$ in Fig. 3-14, if $\overline{AB} \cong \overline{BC} \cong \overline{CA}$, then $\angle A \cong \angle B \cong \angle C$.

Principle 3 is a corollary of Principle 1. A *corollary* of a theorem is another theorem whose statement and proof follow readily from the theorem.

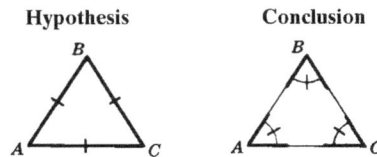

Fig. 3-14

PRINCIPLE 4:   *An equiangular triangle is equilateral.*

Thus in $\triangle ABC$ in Fig. 3-15, if $\angle A \cong \angle B \cong \angle C$, then $\overline{AB} \cong \overline{BC} \cong \overline{CA}$.

Principle 4 is the converse of Principle 3 and a corollary of Principle 2.

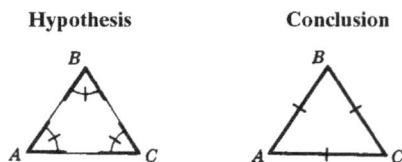

Fig. 3-15

SOLVED PROBLEMS

3.8 Applying principles 1 and 3

In each part of Fig. 3-16, name the congruent angles that are opposite congruent sides of a triangle.

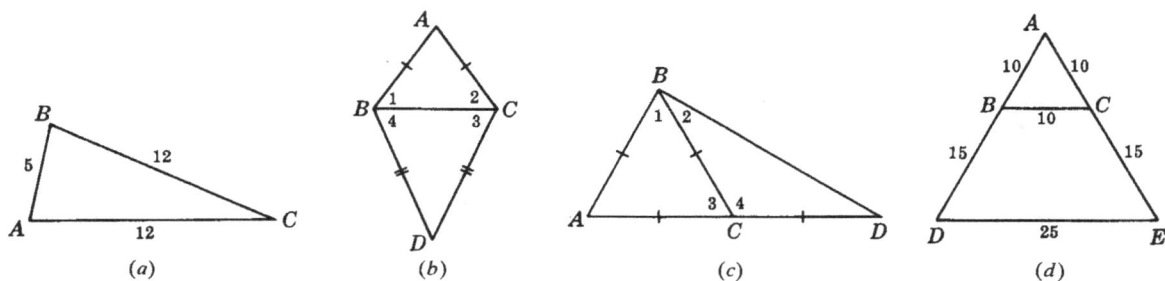

Fig. 3-16

Solutions

(a) Since $\overline{AC} \cong \overline{BC}$, $\angle A \cong \angle B$.

(b) Since $\overline{AB} \cong \overline{AC}$, $\angle 1 \cong \angle 2$. Since $\overline{BD} \cong \overline{CD}$, $\angle 3 \cong \angle 4$.

(c) Since $\overline{AB} \cong \overline{AC} \cong \overline{BC}$, $\angle A \cong \angle 1 \cong \angle 3$. Since $\overline{BC} \cong \overline{CD}$, $\angle 2 \cong \angle D$.

(d) Since $\overline{AB} \cong \overline{BC} \cong \overline{AC}$, $\angle A \cong \angle ACB \cong \angle ABC$. Since $\overline{AE} \cong \overline{AD} \cong \overline{DE}$, $\angle A \cong \angle D \cong \angle E$.

3.9 Applying principles 2 and 4

In each part of Fig. 3-17, name the congruent sides that are opposite congruent angles of a triangle.

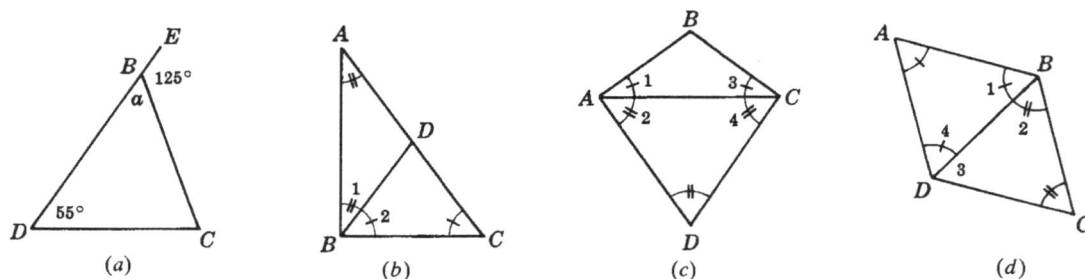

Fig. 3-17

Solutions

(a) Since $m\angle a = 55°$, $\angle a \cong \angle D$. Hence, $\overline{BC} \cong \overline{CD}$.

(b) Since $\angle A \cong \angle 1$, $\overline{AD} \cong \overline{BD}$. Since $\angle 2 \cong \angle C$, $\overline{BD} \cong \overline{CD}$.

(c) Since $\angle 1 \cong \angle 3$, $\overline{AB} \cong \overline{BC}$. Since $\angle 2 \cong \angle 4 \cong \angle D$, $\overline{CD} \cong \overline{AD} \cong \overline{AC}$.

(d) Since $\angle A \cong \angle 1 \cong \angle 4$, $\overline{AB} \cong \overline{BD} \cong \overline{AD}$. Since $\angle 2 \cong \angle C$, $\overline{BD} \cong \overline{CD}$.

3.10 Applying isosceles triangle principles

In each of Fig. 3-18(a) and (b), $\triangle$I can be proved congruent to $\triangle$II. Make a diagram showing the congruent parts of both triangles and state the congruency principle involved.

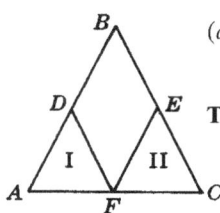

(a) **Given:** $\overline{AB} \cong \overline{BC}$
$\overline{AD} \cong \overline{EC}$
F is midpoint of $\overline{AC}$.
To Prove: $\triangle$I $\cong \triangle$II

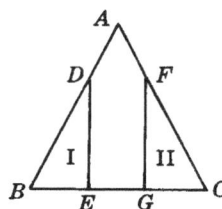

(b) **Given:** $\overline{AB} \cong \overline{AC}$
$\overline{BC}$ is trisected at E and G.
$\overline{DE} \perp \overline{BC}$
$\overline{FG} \perp \overline{BC}$
To Prove: $\triangle$I $\cong \triangle$II

Fig. 3-18

Solutions

(a) Since $\overline{AB} \cong \overline{BC}$, $\angle A \cong \angle C$. $\triangle$I $\cong \triangle$II by SAS [see Fig. 3-19(a)].

(b) Since $\overline{AB} \cong \overline{AC}$, $\angle B \cong \angle C$. $\triangle$I $\cong \triangle$II by ASA [see Fig. 3-19(b)].

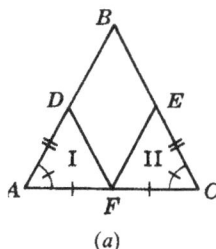

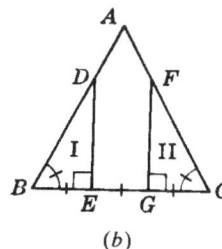

Fig. 3-19

3.11 Proving an isosceles triangle problem

Given: $\overline{AB} \cong \overline{BC}$
$\overline{AC}$ is trisected at D and E.
To Prove: $\angle 1 \cong \angle 2$
Plan: Prove $\triangle$I $\cong \triangle$II to obtain $\overline{BD} \cong \overline{BE}$.

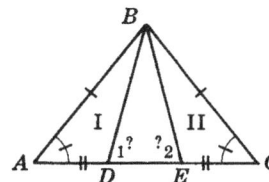

PROOF:

Statements	Reasons
1. AC is trisected at D and E	1. Given
2. $\overline{AD} \cong \overline{EC}$	2. To trisect is to divide into three congruent parts
3. $\overline{AB} \cong \overline{BC}$	3. Given
4. $\angle A \cong \angle C$	4. In a $\triangle$, $\angle$ opposite $\cong$ sides are $\cong$
5. $\triangle$I $\cong \triangle$II	5. SAS
6. $\overline{BD} \cong \overline{BE}$	6. Corresponding parts of $\triangle$ are $\cong$
7. $\angle 1 \cong \angle 2$	7. Same as 4

3.12 Proving an isosceles triangle problem stated in words

Prove that the bisector of the vertex angle of an isosceles triangle is a median to the base.

Solution

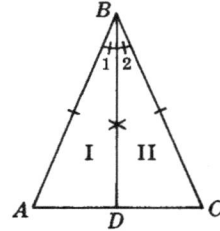

<u>The bisector of the vertex angle of an isosceles triangle</u> <u>is a median to the base.</u>

Given: Isosceles $\triangle ABC$ ($\overline{AB} \cong \overline{BC}$)
$\overline{BD}$ bisects $\angle B$

To Prove: $\overline{BD}$ is a median to $\overline{AC}$

Plan: Prove $\triangle I \cong \triangle II$ to obtain $\overline{AD} \cong \overline{DC}$.

PROOF:

Statements	Reasons
1. $\overline{AB} \cong \overline{BC}$	1. Given
2. $\overline{BD}$ bisects $\angle B$.	2. Given
3. $\angle 1 \cong \angle 2$	3. To bisect is to divide into two congruent parts.
4. $\overline{BD} \cong \overline{BD}$	4. Reflexive property.
5. $\triangle I \cong \triangle II$	5. SAS
6. $\overline{AD} \cong \overline{DC}$	6. Corresponding parts of $\cong \triangle$ are $\cong$.
7. $\overline{BD}$ is a median to $\overline{AC}$.	7. A line from a vertex of $\triangle$ bisecting opposite side is a median.

SUPPLEMENTARY PROBLEMS

3.1. Select the congruent triangles in (a) Fig. 3-20, (b) Fig. 3-21, and (c) Fig. 3-22, and state the congruency principle in each case. (3.1)

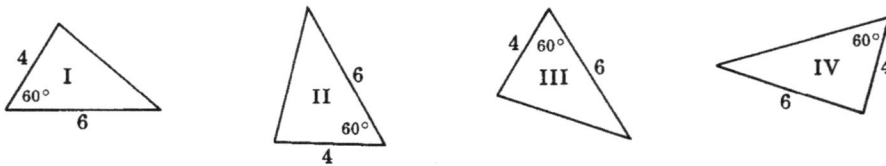

Fig. 3-20

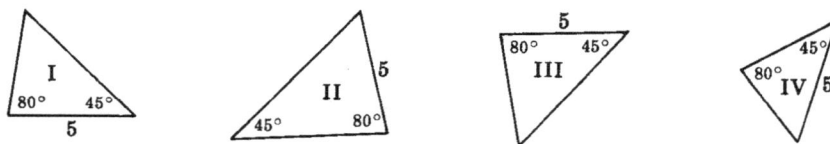

Fig. 3-21

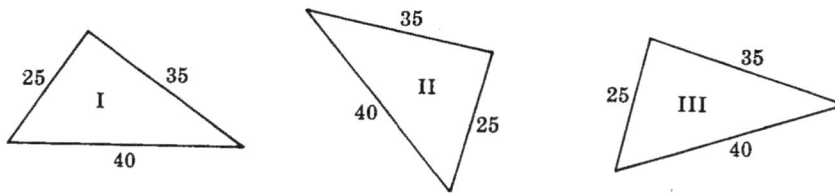

Fig. 3-22

3.2. In each figure below, △I can be proved congruent to △II. State the congruency principle involved. (3.2)

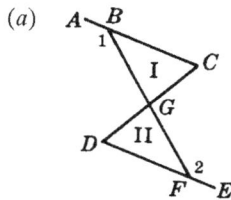

(a)
Given: ∠1 ≅ ∠2
G is midpoint of $\overline{BF}$.
To Prove: △I ≅ △II

(e)
Given: ∠EAB ≅ ∠EAC
$\overline{AD} \perp \overline{BC}$
To Prove: △I ≅ △II

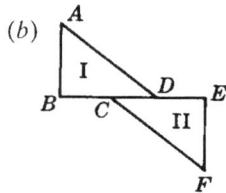

(b)
Given: $\overline{AB} \perp \overline{BE}$
$\overline{EF} \perp \overline{BE}$
$\overline{BC} \cong \overline{DE}$
$\overline{AB} \cong \overline{EF}$
To Prove: △I ≅ △II

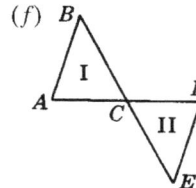

(f)
Given: $\overline{AD}$ and $\overline{BE}$ bisect each other.
To Prove: △I ≅ △II

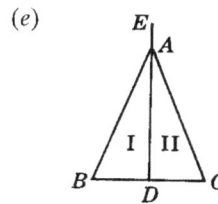

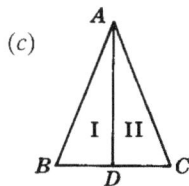

(c)
Given: $\overline{AB} \cong \overline{AC}$
$\overline{AD}$ is median to $\overline{BC}$.
To Prove: △I ≅ △II

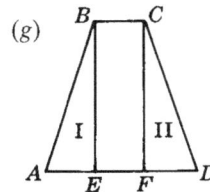

(g)
Given: $\overline{BE} \perp \overline{AD}$
$\overline{CF} \perp \overline{AD}$
$\overline{BE} \cong \overline{CF}$
$\overline{AD}$ is trisected.
To Prove: △I ≅ △II

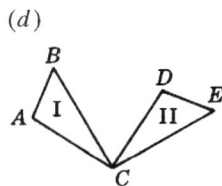

(d)
Given: $\overline{BC} \perp \overline{CE}$
$\overline{AC} \perp \overline{CD}$
$\overline{AC} \cong \overline{CD}$
$\overline{BC} \cong \overline{CE}$
To Prove: △I ≅ △II

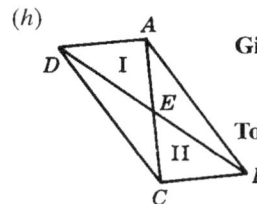

(h)
Given: $\overline{AD} \perp \overline{AC}$
$\overline{BC} \perp \overline{AC}$
$\overline{BD}$ bisects $\overline{AC}$.
To Prove: △I ≅ △II

3.3. State the additional parts needed to prove △I ≅ △II in the given figure by the given congruency principle. (3.3)

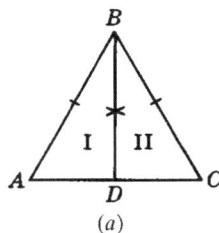

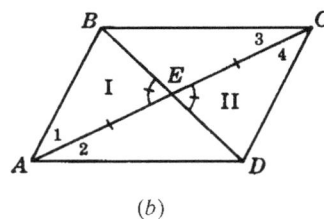

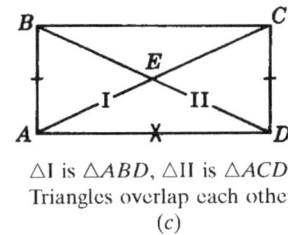

(a)

(b)

(c)
△I is △ABD, △II is △ACD.
Triangles overlap each other.

Fig. 3-23

(a) In Fig. 3-23(a) by SSS.

(b) In Fig. 3-23(a) by SAS.

(c) In Fig. 3-23(b) by ASA.

(d) In Fig. 3-23(b) by SAS.

(e) In Fig. 3-23(c) by SSS.

(f) In Fig. 3-23(c) by SAS.

3.4. In each part of Fig. 3-24, the congruent parts needed to prove $\triangle$I $\cong$ $\triangle$II are marked. Name the remaining parts that are congruent.   (3.4)

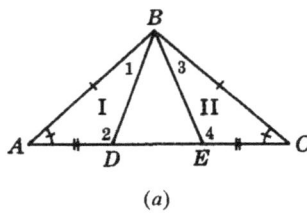   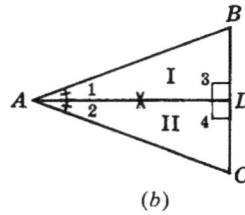   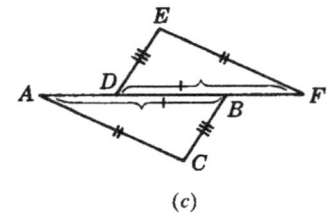

(a)   (b)   (c)

Fig. 3-24

3.5. In each part of Fig. 3-25, find x and y.   (3.5)

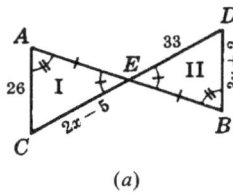   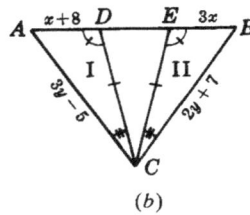   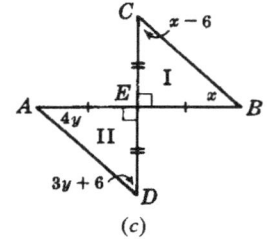

(a)   (b)   (c)

Fig. 3-25

3.6. Prove each of the following.   (3.6)

(a) In Fig. 3-26: **Given:** $\overline{BD} \perp \overline{AC}$
                              D is midpoint of $\overline{AC}$.
              To Prove: $\overline{AB} \cong \overline{BC}$

(b) In Fig. 3-26: **Given:** $\overline{BD}$ is altitude to $\overline{AC}$.
                              $\overline{BD}$ bisects $\angle B$.
              To Prove: $\angle A \cong \angle C$

(c) In Fig. 3-27: **Given:** $\angle 1 \cong \angle 2, \overline{BF} \cong \overline{DE}$
                              $\overline{BF}$ bisects $\angle B$.
                              $\overline{DE}$ bisects $\angle D$.
                              $\angle B$ and $\angle D$ are rt. $\angle$s.
              To Prove: $\overline{AB} \cong \overline{CD}$

(d) In Fig. 3-27: **Given:** $\overline{BC} \cong \overline{AD}$
                              E is midpoint of $\overline{BC}$.
                              F is midpoint of $\overline{AD}$.
                              $\overline{AB} \cong \overline{CD}, \overline{BF} \cong \overline{DE}$
              To Prove: $\angle A \cong \angle C$

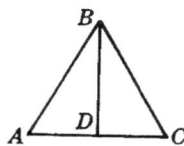   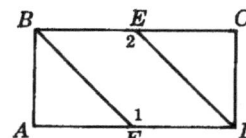

Fig. 3-26   Fig. 3-27

(e) In Fig. 3-28: **Given:** $\angle 1 \cong \angle 2$
$\overline{CE}$ bisects $\overline{BF}$.
To Prove: $\angle C \cong \angle E$

(f) In Fig. 3-28: **Given:** $\overline{BF}$ and $\overline{CE}$ bisect each other.
To Prove: $\overline{BC} \cong \overline{EF}$

(g) In Fig. 3-29: **Given:** $\overline{CD} \cong \overline{C'D'}, \overline{AD} \cong \overline{A'D'}$
$\overline{CD}$ is altitude to $\overline{AB}$.
$\overline{C'D'}$ is altitude to $\overline{A'B'}$.
To Prove: $\angle A \cong \angle A'$

(h) In Fig. 3-29: **Given:** $\overline{CD}$ bisects $\angle C$.
$\overline{C'D'}$ bisects $\angle C'$
$\angle C \cong \angle C'$,
$\angle B \cong \angle B'$,
$BC \cong \overline{B'C'}$.
To Prove: $\overline{CD} \cong \overline{C'D'}$

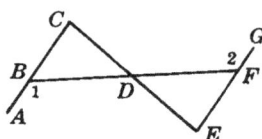

Fig. 3-28

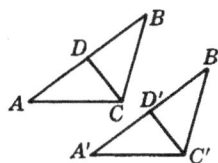

Fig. 3-29

3.7. Prove each of the following: (3.7)

(a) If a line bisects an angle of a triangle and is perpendicular to the opposite side, then it bisects that side.

(b) If the diagonals of a quadrilateral bisect each other, then its opposite sides are congruent.

(c) If the base and a leg of one isosceles triangle are congruent to the base and a leg of another isosceles triangle, then their vertex angles are congruent.

(d) Lines drawn from a point on the perpendicular bisector of a given line to the ends of the given line are congruent.

(e) If the legs of one right triangle are congruent respectively to the legs of another, their hypotenuses are congruent.

3.8. In each part of Fig. 3-30, name the congruent angles that are opposite sides of a triangle. (3.8)

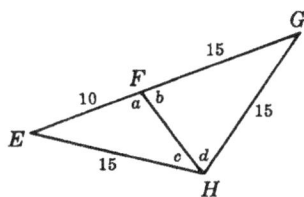

(a)

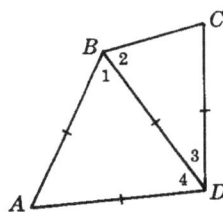

(b)

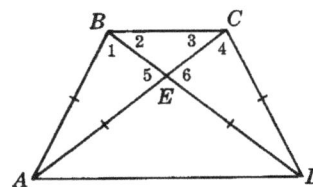

(c)

Fig. 3-30

3.9. In each part of Fig. 3-31, name the congruent sides that are opposite congruent angles of a triangle.   (3.9)

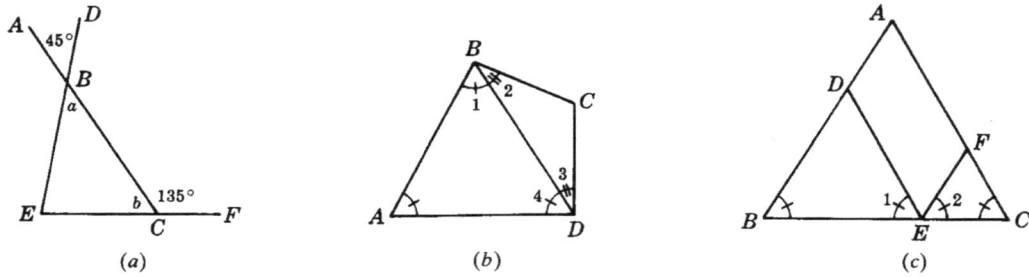

Fig. 3-31

3.10. In each part of Fig. 3-32, two triangles are to be proved congruent. Make a diagram showing the congruent parts of both triangles and state the reason for congruency.   (3.10)

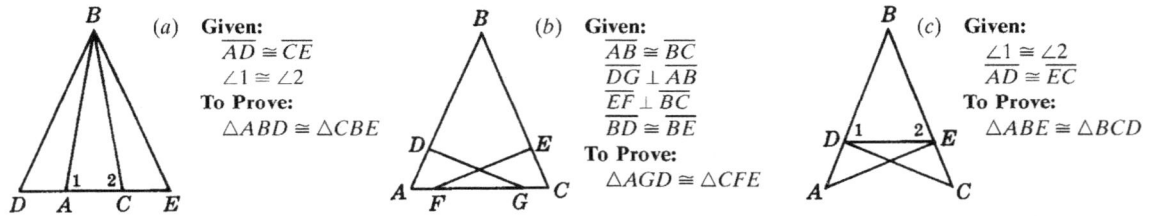

(a) **Given:**
$\overline{AD} \cong \overline{CE}$
$\angle 1 \cong \angle 2$
To Prove:
$\triangle ABD \cong \triangle CBE$

(b) **Given:**
$\overline{AB} \cong \overline{BC}$
$\overline{DG} \perp \overline{AB}$
$\overline{EF} \perp \overline{BC}$
$\overline{BD} \cong \overline{BE}$
To Prove:
$\triangle AGD \cong \triangle CFE$

(c) **Given:**
$\angle 1 \cong \angle 2$
$\overline{AD} \cong \overline{EC}$
To Prove:
$\triangle ABE \cong \triangle BCD$

Fig. 3-32

3.11. In each part of Fig. 3-33, $\triangle$I, $\triangle$II, and $\triangle$III can be proved congruent. Make a diagram showing the congruent parts and state the reason for congruency.   (3.10)

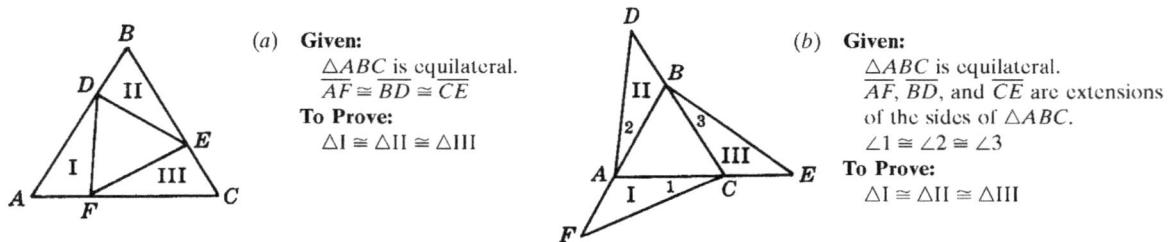

(a) **Given:**
$\triangle ABC$ is equilateral.
$\overline{AF} \cong \overline{BD} \cong \overline{CE}$
To Prove:
$\triangle I \cong \triangle II \cong \triangle III$

(b) **Given:**
$\triangle ABC$ is equilateral.
$\overline{AF}, \overline{BD},$ and $\overline{CE}$ are extensions of the sides of $\triangle ABC$.
$\angle 1 \cong \angle 2 \cong \angle 3$
To Prove:
$\triangle I \cong \triangle II \cong \triangle III$

Fig. 3-33

3.12. Prove each of the following: (3.11)

(a) In Fig. 3-34:   **Given:** $\overline{AB} \cong \overline{AC}$
                     F is midpoint of $\overline{BC}$.
                     $\angle 1 \cong \angle 2$
         To Prove: $\overline{FD} \cong \overline{FE}$

(b) In Fig. 3-34:   **Given:** $\overline{AB} \cong \overline{AC}$
                     $\overline{AD} \cong \overline{AE}$
                     $\overline{FD} \perp \overline{AB}, \overline{FE} \perp \overline{AC}$
         To Prove: $\overline{BF} \cong \overline{FC}$

(c) In Fig. 3-35:   **Given:** $\overline{AB} \cong \overline{AC}$
                     $\angle A$ is trisected.
         To Prove: $\overline{AD} \cong \overline{AE}$

(d) In Fig. 3-35:   **Given:** $\overline{AB} \cong \overline{AC}$
                     $\overline{DB} \cong \overline{BC}$
                     $\overline{CE} \cong \overline{BC}$
         To Prove: $\overline{AD} \cong \overline{AE}$

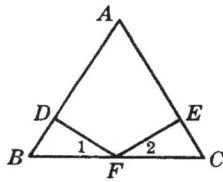

Fig. 3-34

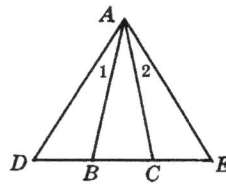

Fig. 3-35

3.13.   Prove each of the following: (3.12)

(a) The median to the base of an isosceles triangle bisects the vertex angle.

(b) If the bisector of an angle of a triangle is also an altitude to the opposite side, then the other two sides of the triangle are congruent.

(c) If a median to a side of a triangle is also an altitude to that side, then the triangle is isosceles.

(d) In an isosceles triangle, the medians to the legs are congruent.

(e) In an isosceles triangle, the bisectors of the base angles are congruent.

CHAPTER 4

Parallel Lines, Distances, and Angle Sums

4.1 Parallel Lines

Parallel lines are straight lines which lie in the same plane and do not intersect however far they are extended. The symbol for parallel is ∥; thus, $\overset{\leftrightarrow}{AB} \parallel \overset{\leftrightarrow}{CD}$ is read "line $\overset{\leftrightarrow}{AB}$ is parallel to line $\overset{\leftrightarrow}{CD}$." In diagrams, arrows are used to indicate that lines are parallel (see Fig. 4-1).

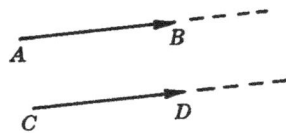

Fig. 4-1

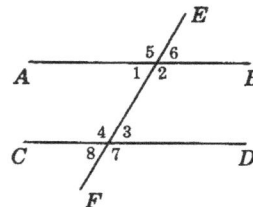

Fig. 4-2

A *transversal* of two or more lines is a line that cuts across these lines. Thus, $\overset{\leftrightarrow}{EF}$ is a transversal of $\overset{\leftrightarrow}{AB}$ and $\overset{\leftrightarrow}{CD}$, in Fig. 4-2.

The *interior angles* formed by two lines cut by a transversal are the angles between the two lines, while the *exterior angles* are those outside the lines. Thus, of the eight angles formed by $\overset{\leftrightarrow}{AB}$ and $\overset{\leftrightarrow}{CD}$ cut by $\overset{\leftrightarrow}{EF}$ in Fig. 4-2, the interior angles are $\angle 1, \angle 2, \angle 3$, and $\angle 4$; the exterior angles are $\angle 5, \angle 6, \angle 7$, and $\angle 8$.

4.1A Pairs of Angles Formed by Two Lines Cut by a Transversal

Corresponding angles of two lines cut by a transversal are angles on the same side of the transversal and on the same side of the lines. Thus, $\angle 1$ and $\angle 2$ in Fig. 4-3 are corresponding angles of $\overset{\leftrightarrow}{AB}$ and $\overset{\leftrightarrow}{CD}$ cut by transversal $\overset{\leftrightarrow}{EF}$. Note that in this case the two angles are both to the right of the transversal and both below the lines.

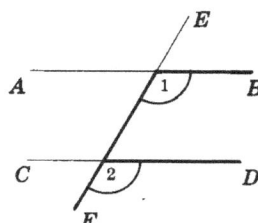

Fig. 4-3

When two parallel lines are cut by a transversal, the sides of two corresponding angles form a capital F in varying positions, as shown in Fig. 4-4.

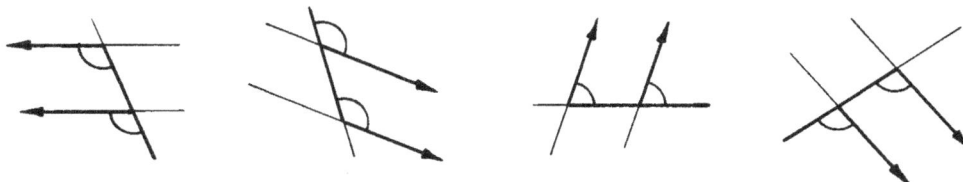

Fig. 4-4

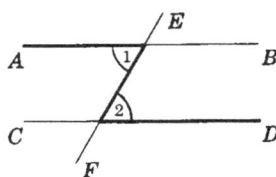

Fig. 4-5

Alternate interior angles of two lines cut by a transversal are nonadjacent angles between the two lines and on opposite sides of the transversal. Thus, $\angle 1$ and $\angle 2$ in Fig. 4-5 are alternate interior angles of $\overleftrightarrow{AB}$ and $\overleftrightarrow{CD}$ cut by $\overleftrightarrow{EF}$. When parallel lines are cut by a transversal, the sides of two alternate interior angles form a capital Z or N in varying positions, as shown in Fig. 4-6.

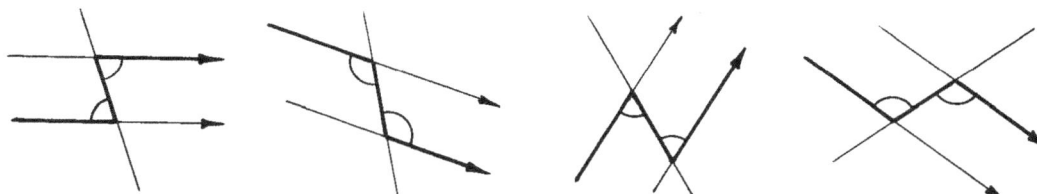

Fig. 4-6

When parallel lines are cut by a transversal, *interior angles on the same side of the transversal* can be readily located by noting the capital U formed by their sides (Fig. 4-7).

Fig. 4-7

4.1B Principles of Parallel Lines

PRINCIPLE 1: *Through a given point not on a given line, one and only one line can be drawn parallel to a given line.* (Parallel-Line Postulate)

Thus, either l_1 or l_2 but not both may be parallel to l_3 in Fig. 4-8.

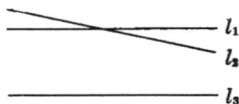

Fig. 4-8

Proving that Lines are Parallel

PRINCIPLE 2:   *Two lines are parallel if a pair of corresponding angles are congruent.*

Thus, $l_1 \parallel l_2$ if $\angle a \cong \angle b$ in Fig. 4-9.

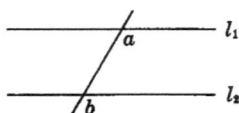

Fig. 4-9

PRINCIPLE 3:   *Two lines are parallel if a pair of alternate interior angles are congruent.*

Thus, $l_1 \parallel l_2$ if $\angle c \cong \angle d$ in Fig. 4-10.

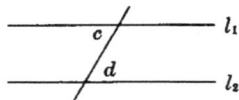

Fig. 4-10

PRINCIPLE 4:   *Two lines are parallel if a pair of interior angles on the same side of a transversal are supplementary.*

Thus, $l_1 \parallel l_2$ if $\angle e$ and $\angle f$ are supplementary in Fig. 4-11.

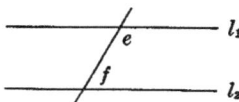

Fig. 4-11

PRINCIPLE 5:   *Lines are parallel if they are perpendicular to the same line. (Perpendiculars to the same line are parallel.)*

Thus, $l_1 \parallel l_2$ if l_1 and l_2 are each perpendicular to l_3 in Fig. 4-12.

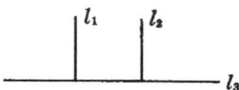

Fig. 4-12

PRINCIPLE 6:   *Lines are parallel if they are parallel to the same line. (Parallels to the same line are parallel.)*

Thus, $l_1 \parallel l_2$ if l_1 and l_2 are each parallel to l_3 in Fig. 4-13.

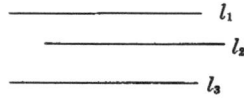

Fig. 4-13

Properties of Parallel Lines

PRINCIPLE 7: *If two lines are parallel, each pair of corresponding angles are congruent. (Corresponding angles of parallel lines are congruent.)*

Thus, if $l_1 \parallel l_2$, then $\angle a \cong \angle b$ in Fig. 4-14.

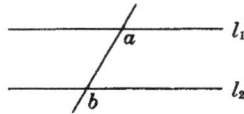

Fig. 4-14

PRINCIPLE 8: *If two lines are parallel, each pair of alternate interior angles are congruent. (Alternate interior angles of parallel lines are congruent.)*

Thus, if $l_1 \parallel l_2$, then $\angle c \cong \angle d$ in Fig. 4-15.

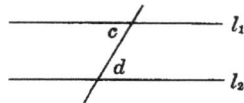

Fig. 4-15

PRINCIPLE 9: *If two lines are parallel, each pair of interior angles on the same side of the transversal are supplementary.*

Thus, if $l_1 \parallel l_2$, $\angle e$ and $\angle f$ are supplementary in Fig. 4-16.

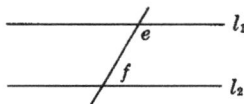

Fig. 4-16

PRINCIPLE 10: *If lines are parallel, a line perpendicular to one of them is perpendicular to the others also.*

Thus, if $l_1 \parallel l_2$ and $l_3 \perp l_1$, then $l_3 \perp l_2$ in Fig. 4-17.

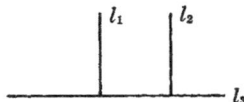

Fig. 4-17

PRINCIPLE 11: *If lines are parallel, a line parallel to one of them is parallel to the others also.*

Thus, if $l_1 \parallel l_2$ and $l_3 \parallel l_1$, then $l_3 \parallel l_2$ in Fig. 4-18.

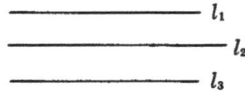

Fig. 4-18

PRINCIPLE 12: *If the sides of two angles are respectively parallel to each other, the angles are either congruent or supplementary.*

Thus, if $l_1 \parallel l_3$ and $l_2 \parallel l_4$ in Fig. 4-19, then $\angle a \cong \angle b$ and $\angle a$ and $\angle c$ are supplementary.

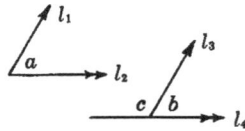

Fig. 4-19

SOLVED PROBLEMS

4.1 Numerical applications of parallel lines

In each part of Fig. 4-20, find the measure x and the measure y of the indicated angles.

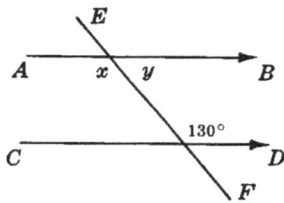

(a)

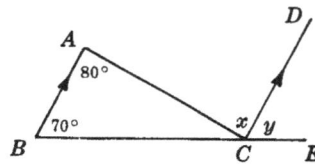

(b)

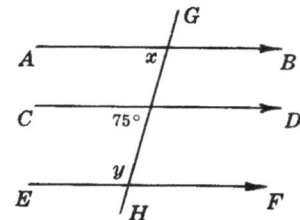

(c)

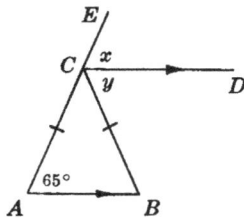

(d)

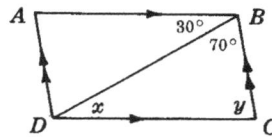

(e)

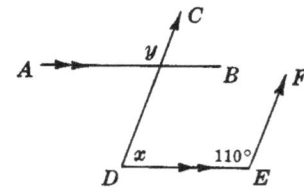

(f)

Fig. 4-20

Solutions

(a) $x = 130°$ (Principle 8). $y = 180° - 130° = 50°$ (Principle 9).

(b) $x = 80°$ (Principle 8). $y = 70°$ (Principle 7).

(c) $x = 75°$ (Principle 7). $y = 180° - 75° = 105°$ (Principle 9).

(d) $x = 65°$ (Principle 7). Since $m\angle B = m\angle A$, $m\angle B = 65°$. Hence, $y = 65°$ (Principle 8).

(e) $x = 30°$ (Principle 8). $y = 180° - (30° + 70°) = 80°$ (Principle 9).

(f) $x = 180° - 110° = 70°$ (Principle 9). $y = 110°$ (Principle 12).

4.2 Applying parallel line principles and their converses

The following short proofs refer to Fig. 4-21. In each, the first statement is given. State the parallel-line principle needed as the reason for each of the remaining statements.

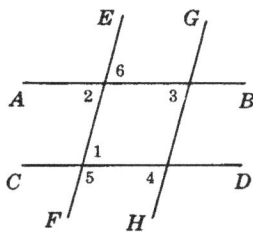

Fig. 4-21

(a) 1. $\angle 1 \cong \angle 2$ 1. Given
     2. $\overline{AB} \parallel \overline{CD}$ 2. ___?___
     3. $\angle 3 \cong \angle 4$ 3. ___?___

(b) 1. $\angle 2 \cong \angle 3$ 1. Given
     2. $\overline{EF} \parallel \overline{GH}$ 2. ___?___
     3. $\angle 4$ sup. $\angle 5$ 3. ___?___

(c) 1. $\angle 5$ sup. $\angle 4$ 1. Given
     2. $\overline{EF} \parallel \overline{GH}$ 2. ___?___
     3. $\angle 3 \cong \angle 6$ 3. ___?___

(d) 1. $\overline{EF} \perp \overline{AB}, \overline{GH} \perp \overline{AB},$ 1. Given
         $\overline{EF} \perp \overline{CD}$
     2. $\overline{EF} \parallel \overline{GH}$ 2. ___?___
     3. $\overline{CD} \perp \overline{GH}$ 3. ___?___

Solutions

(a) 2: Principle 3; 3: Principle 7.   (c) 2: Principle 4; 3: Principle 8.

(b) 2: Principle 2; 3: Principle 9.   (d) 2: Principle 5; 3: Principle 10.

4.3 Algebraic applications of parallel lines

In each part of Fig. 4-22, find x and y. Provide the reason for each equation obtained from the diagram.

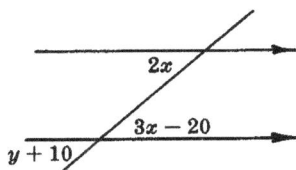

(*a*)

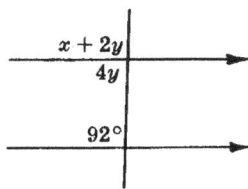

(*b*)

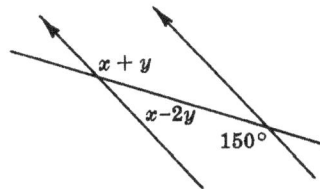

(*c*)

Fig. 4-22

Solutions

(a) $3x - 20 = 2x$ (Principle 8)
          $x = 20°$
     $y + 10 = 2x$ (Principle 7)
     $y + 10 = 40$
          $y = 30°$

(b) $4y = 180 - 92 = 88$ (Principle 9)
                  $y = 22°$
     $x + 2y = 92$ (Principle 7)
     $x + 44 = 92$
          $x = 48°$

(c) (1) $x + y = 150$ (Principle 8)

(2) $x + 2y = 30$ (Principle 9)

$3y = 120$ (Subt. Postulate)

$y = 40°$

$x + 40 = 150$

$x = 110°$

4.4 Proving a parallel-line problem

Given: $\overline{AB} \cong \overline{AC}$

$\overrightarrow{AE} \parallel \overline{BC}$

To Prove: $\overrightarrow{AE}$ bisects $\angle DAC$

Plan: Show that $\angle 1$ and $\angle 2$ are congruent to the congruent angles B and C.

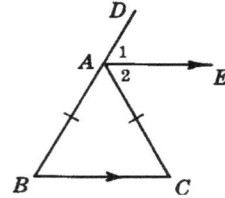

PROOF:

Statements	Reasons
1. $\overrightarrow{AE} \parallel \overline{BC}$	1. Given
2. $\angle 1 \cong \angle B$	2. Corresponding ∡ of ∥ lines are ≅.
3. $\angle 2 \cong \angle C$	3. Alternate interior ∡ of ∥ lines are ≅.
4. $\overline{AB} \cong \overline{AC}$	4. Given
5. $\angle B \cong \angle C$	5. In a △, ∡ opposite ≅ sides are ≅.
6. $\angle 1 \cong \angle 2$	6. Things ≅ to ≅ things are ≅ to each other.
7. $\overrightarrow{AE}$ bisects $\angle DAC$.	7. To divide into two congruent parts is to bisect.

4.5 Proving a parallel-line problem stated in words

Prove that if the diagonals of a quadrilateral bisect each other, the opposite sides are parallel.

Given: Quadrilateral *ABCD*

$\overline{AC}$ and $\overline{BD}$ bisect each other.

To Prove: $\overline{AB} \parallel \overline{CD}$

$\overline{AD} \parallel \overline{BC}$

Plan: Prove $\angle 1 \cong \angle 4$ by showing $\triangle I \cong \triangle II$.

Prove $\angle 2 \cong \angle 3$ by showing $\triangle III \cong \triangle IV$.

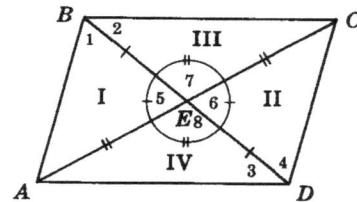

PROOF:

Statements	Reasons
1. $\overline{AC}$ and $\overline{BD}$ bisect each other.	1. Given
2. $\overline{BE} \cong \overline{ED} \; \overline{AE} \cong \overline{EC}$	2. To bisect is to divide into two congruent parts.
3. $\angle 5 \cong \angle 6, \angle 7 \cong \angle 8$	3. Vertical ∡ are ≅.
4. $\triangle I \cong \triangle II, \triangle III \cong \triangle IV$	4. SAS
5. $\angle 1 \cong \angle 4, \angle 2 \cong \angle 3$	5. Corresponding parts of congruent △ are ≅.
6. $\overline{AB} \parallel \overline{CD}, \overline{BC} \parallel \overline{AD}$	6. Lines cut by a transversal are ∥ if alternate interior ∡ are ≅.

4.2 Distances

4.2A Distances Between Two Geometric Figures

The distance between two geometric figures is the straight line segment which is the *shortest segment between the figures.*

1. The distance *between two points*, such as P and Q in Fig. 4-23(a), is the line segment $\overline{PQ}$ between them.

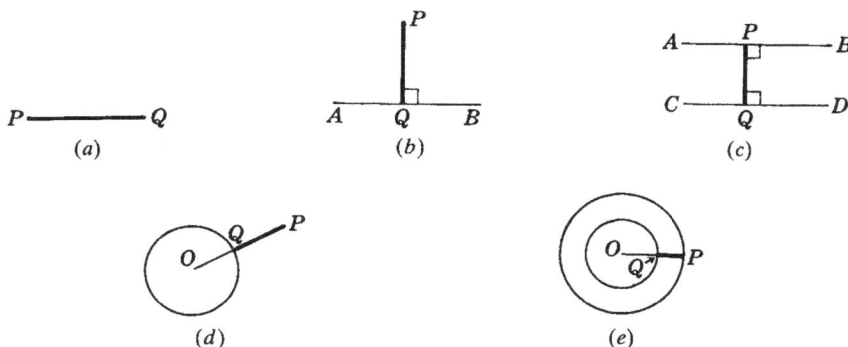

(a)   (b)   (c)

(d)   (e)

Fig. 4-23

2. The distance *between a point and a line*, such as P and $\overleftrightarrow{AB}$ in (b), is the line segment PQ, the perpendicular from the point to the line.

3. The distance *between two parallels*, such as $\overleftrightarrow{AB}$ and $\overleftrightarrow{CD}$ in (c), is the segment $\overline{PQ}$, a perpendicular between the two parallels.

4. The distance *between a point and a circle*, such as P and circle O in (d), is $\overline{PQ}$, the segment of $\overline{OP}$ between the point and the center of the circle.

5. The distance *between two concentric circles*, such as two circles whose center is O, is $\overline{PQ}$, the segment of the larger radius that lies between the two circles, as shown in (e).

4.2B Distance Principles

PRINCIPLE 1: *If a point is on the perpendicular bisector of a line segment, then it is equidistant from the ends of the line segment.*

Thus if P is on $\overleftrightarrow{CD}$, the $\perp$ bisector of $\overline{AB}$ in Fig. 4-24, then $\overline{PA} \cong \overline{PB}$.

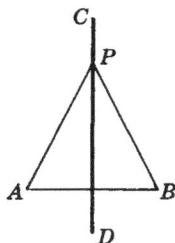

Fig. 4-24

PRINCIPLE 2:   *If a point is equidistant from the ends of a line segment, then it is on the perpendicular bi-sector of the line segment.* (Principle 2 is the converse of Principle 1.)

Thus if $\overline{PA} \cong \overline{PB}$ in Fig. 4-24, then P is on $\overset{\leftrightarrow}{CD}$, the $\perp$ bisector of $\overline{AB}$.

PRINCIPLE 3:   *If a point is on the bisector of an angle, then it is equidistant from the sides of the angle.*

Thus if P is on $\overset{\leftrightarrow}{AB}$, the bisector of $\angle A$ in Fig. 4-25, then $\overline{PQ} \cong \overline{PR}$, where PQ and PR are the distances of P from the sides of the angle.

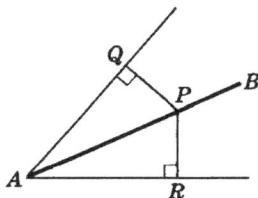

Fig. 4-25

PRINCIPLE 4:   *If a point is equidistant from the sides of an angle, then it is on the bisector of the angle.* (Principle 4 is the converse of Principle 3.)

Thus if $PQ = PR$, where PQ and PR are the distances of P from the sides of $\angle A$ in Fig. 4-25, then P is on $\overset{\leftrightarrow}{AB}$, the bi-sector of $\angle A$.

PRINCIPLE 5:   *Two points each equidistant from the ends of a line segment determine the perpendicular bisector of the line segment.* (The line joining the vertices of two isosceles triangles having a common base is the perpendicular bisector of the base.)

Thus if $\overline{PA} \cong \overline{PB}$ and $\overline{QA} \cong \overline{QB}$ in Fig. 4-26, then P and Q determine $\overset{\leftrightarrow}{CD}$, the $\perp$ bisector of $\overline{AB}$.

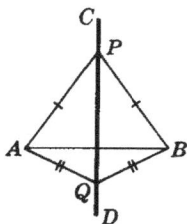

Fig. 4-26

PRINCIPLE 6:   *The perpendicular bisectors of the sides of a triangle meet in a point which is equidistant from the vertices of the triangle.*

Thus if P is the intersection of the $\perp$ bisectors of the sides of $\triangle ABC$ in Fig. 4-27, then $\overline{PA} \cong \overline{PB} \cong \overline{PC}$. P is the center of the circumscribed circle and is called the *circumcenter of* $\triangle ABC$.

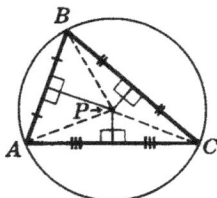

Fig. 4-27

PRINCIPLE 7:   *The bisectors of the angles of a triangle meet in a point which is equidistant from the sides of the triangle.*

Thus if Q is the intersection of the bisectors of the angles of $\triangle ABC$ in Fig. 4-28, then $\overline{QR} \cong \overline{QS} \cong \overline{QT}$, where these are the distances from Q to the sides of $\triangle ABC$. Q is the center of the inscribed circle and is called the *incenter* of $\triangle ABC$.

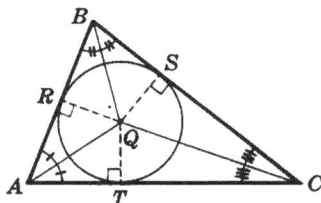

Fig. 4-28

SOLVED PROBLEMS

4.6 Finding distances

In each of the following, find the distance and indicate the kind of distance involved. In Fig. 4-29(a), find the distance (a) from P to A; (b) from P to $\overleftrightarrow{CD}$; (c) from A to $\overleftrightarrow{BC}$; (d) from $\overleftrightarrow{AB}$ to $\overleftrightarrow{CD}$. In Fig. 4-29(b), find the distance (e) from P to inner circle O; (f) from P to outer circle O; (g) between the concentric circles.

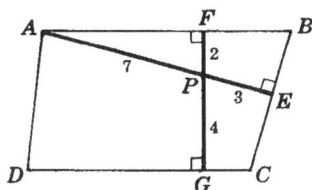

(a)

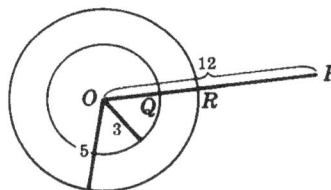

(b)

Fig. 4-29

Solutions

(a) $PA = 7$, distance between two points

(b) $PG = 4$, distance from a point to a line

(c) $AE = 10$, distance from a point to a line

(d) $FG = 6$, distance between two parallel lines

(e) $PQ = 12 - 3 = 9$, distance from a point to a circle

(f) $PR = 12 - 5 = 7$, distance from a point to a circle

(g) $QR = 5 - 3 = 2$, distance between two concentric circles

4.7 Locating a point satisfying given conditions

In Fig. 4-30.

(a) Locate P, a point on $\overleftrightarrow{BC}$ and equidistant from A and C.

(b) Locate Q, a point on $\overleftrightarrow{AB}$ and equidistant from $\overline{BC}$ and $\overline{AC}$.

(c) Locate R, the center of the circle circumscribed about $\triangle ABC$.

(d) Locate S, the center of the circle inscribed in $\triangle ABC$.

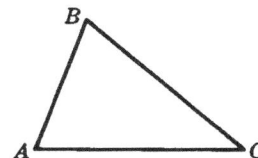

Fig. 4-30

Solutions

See Fig. 4-31.

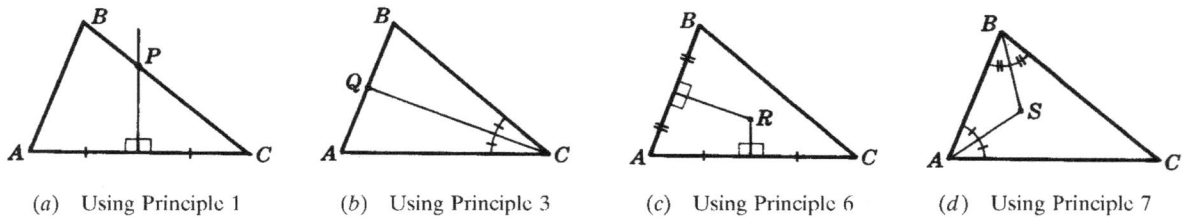

(a) Using Principle 1   (b) Using Principle 3   (c) Using Principle 6   (d) Using Principle 7

Fig. 4-31

4.8 Applying principles 2 and 4

For each $\triangle ABC$ in Fig. 4-32, describe P, Q, and R as equidistant points, and locate each on a bisector.

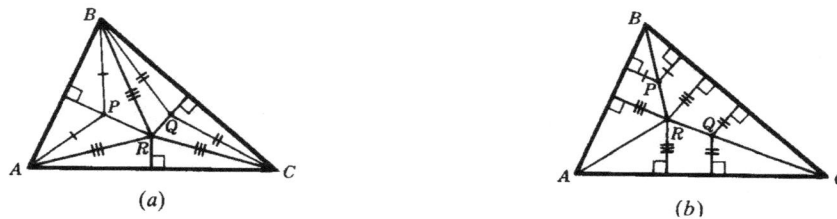

(a)   (b)

Fig. 4-32

Solutions

(a) Since P is equidistant from A and B, it is on the $\perp$ bisector of $\overline{AB}$. Since Q is equidistant from B and C, it is on the $\perp$ bisector of BC. Since R is equidistant from A, B, and C, it is on the $\perp$ bisectors of $\overline{AB}$, $\overline{BC}$, and $\overline{AC}$.

(b) Since P is equidistant from $\overleftrightarrow{AB}$ and $\overleftrightarrow{BC}$, it is on the bisector of $\angle B$. Since Q is equidistant from $\overleftrightarrow{AC}$ and $\overleftrightarrow{BC}$, it is on the bisector of $\angle C$. Since R is equidistant from $\overleftrightarrow{AB}$, $\overleftrightarrow{BC}$, and $\overleftrightarrow{AC}$, it is on the bisectors of $\angle A$, $\angle B$, and $\angle C$.

4.9 Applying principles 1, 3, 6, and 7

For each $\triangle ABC$ in Fig. 4-33, describe P, Q, and R as equidistant points. Also, describe R as the center of a circle.

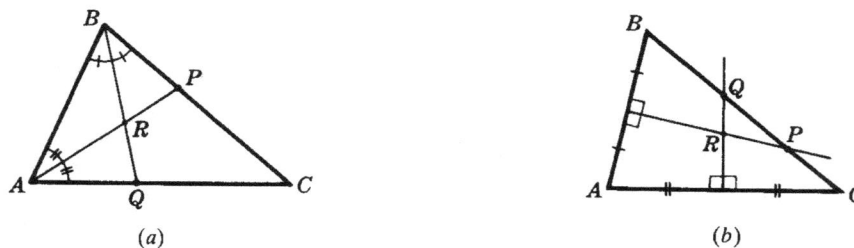

(a)   (b)

Fig. 4-33

Solutions

(a) Since P is on the bisector of $\angle A$, it is equidistant from $\overleftrightarrow{AB}$ and $\overleftrightarrow{AC}$. Since Q is on the bisectors of $\angle B$, it is equidistant from $\overleftrightarrow{AB}$ and $\overleftrightarrow{BC}$. Since R is on the bisectors of $\angle A$ and $\angle B$, it is equidistant from $\overleftrightarrow{AB}$, $\overleftrightarrow{BC}$, and $\overleftrightarrow{AC}$. R is the incenter of $\triangle ABC$, that is, the center of its inscribed circle.

(b) Since P is on the $\perp$ bisector of $\overline{AB}$, it is equidistant from A and B. Since Q is on the $\perp$ bisector of $\overline{AC}$, it is equidistant from A and C. Since R is on the $\perp$ bisectors of $\overline{AB}$ and $\overline{AC}$, it is equidistant from A, B, and C. R is the circumcenter of $\triangle ABC$, that is, the center of its circumscribed circle.

4.10 Applying principles 1, 3, 6, and 7

In each part of Fig. 4-34, find two points equidistant from the ends of a line segment, and find the perpendicular bisector determined by the two points.

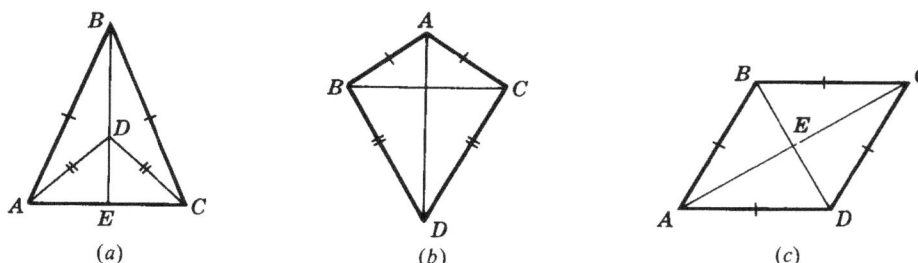

Fig. 4-34

Solutions

(a) B and D are equidistant from A and C; hence, $\overline{BE}$ is the $\perp$ bisector of $\overline{AC}$.

(b) A and D are equidistant from B and C; hence, $\overline{AD}$ is the $\perp$ bisector of $\overline{BC}$.

(c) B and D are equidistant from A and C; hence, $\overline{BD}$ is the $\perp$ bisector of $\overline{AC}$. A and C are equidistant from B and D; hence, $\overline{AC}$ is the $\perp$ bisector of $\overline{BD}$.

4.3 Sum of the Measures of the Angles of a Triangle

The angles of any triangle may be torn off, as in Fig. 4-35(a), and then fitted together as shown in (b). The three angles will form a straight angle.

We can prove that the sum of the measures of the angles of a triangle equals 180° by drawing a line through one vertex of the triangle parallel to the side opposite the vertex. In Fig. 4-35(c), $\overleftrightarrow{MN}$ is drawn through B parallel to AC. Note that the measure of the straight angle at B equals the sum of the measures of the angles of $\triangle ABC$; that is, $a° + b° + c° = 180°$. Each pair of congruent angles is a pair of alternate interior angles of parallel lines.

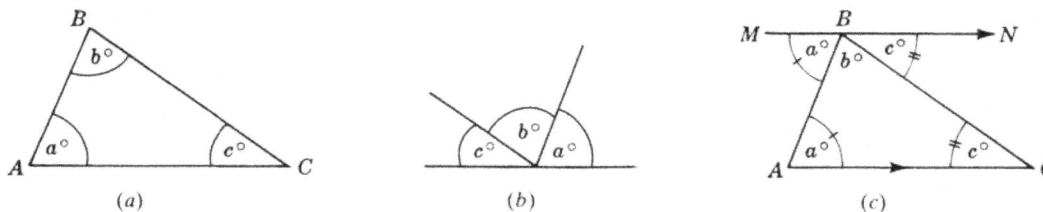

Fig. 4-35

4.3A   Interior and Exterior Angles of a Polygon

An exterior angle of a polygon is formed whenever one of its sides is extended through a vertex. If each of the sides of a polygon is extended, as shown in Fig. 4-36, an exterior angle will be formed at each vertex. Each of these exterior angles is the supplement of its adjacent interior angle.

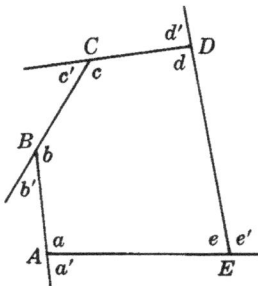

Fig. 4-36

Thus, in the case of pentagon *ABCDE*, there will be five exterior angles, one at each vertex. Note that each exterior angle is the supplement of an adjacent interior angle. For example, $m\angle a + m\angle a' = 180°$.

4.3B   Angle-Measure-Sum Principles

PRINCIPLE 1:   *The sum of the measures of the angles of a triangle equals the measure of a straight angle or* 180°.

Thus in $\triangle ABC$ of Fig. 4-37, $m\angle A + m\angle B + m\angle C = 180°$.

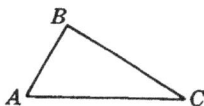

Fig. 4-37

PRINCIPLE 2:   *If two angles of one triangle are congruent respectively to two angles of another triangle, the remaining angles are congruent.*

Thus in $\triangle ABC$ and $\triangle A'B'C'$ in Fig. 4-38, if $\angle A \cong \angle A'$ and $\angle B \cong \angle B'$, then $\angle C \cong \angle C'$.

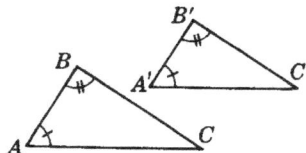

Fig. 4-38

PRINCIPLE 3:   *The sum of the measures of the angles of a quadrilateral equals* 360°.

Thus in quadrilateral *ABCD* (Fig. 4-39), $m\angle A + m\angle B + m\angle C + m\angle D = 360°$.

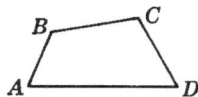

Fig. 4-39

PRINCIPLE 4: *The measure of each exterior angle of a triangle equals the sum of the measures of its two nonadjacent interior angles.*

Thus in $\triangle ABC$ in Fig. 4-40, $m\angle ECB = m\angle A + m\angle B$.

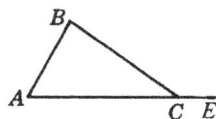

Fig. 4-40

PRINCIPLE 5: *The sum of the measures of the exterior angles of a triangle equals 360°.*

Thus in $\triangle ABC$ in Fig. 4-41, $m\angle a' + m\angle b' + m\angle c' = 360°$.

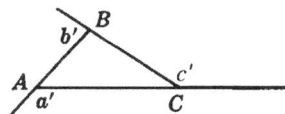

Fig. 4-41

PRINCIPLE 6: *The measure of each angle of an equilateral triangle equals 60°.*

Thus if $\triangle ABC$ in Fig. 4-42 is equilateral, then $m\angle A = 60°$, $m\angle B = 60°$, and $m\angle C = 60°$.

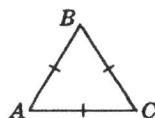

Fig. 4-42

PRINCIPLE 7: *The acute angles of a right triangle are complementary.*

Thus in rt. $\triangle ABC$ in Fig. 4-43, if $m\angle C = 90°$, then $m\angle A + m\angle B = 90°$.

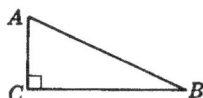

Fig. 4-43

PRINCIPLE 8: *The measure of each acute angle of an isosceles right triangle equals 45°.*

Thus in isos. rt. $\triangle ABC$ in Fig. 4-44, if $m\angle C = 90°$, then $m\angle A = 45°$ and $m\angle B = 45°$.

Fig. 4-44

PRINCIPLE 9: *A triangle can have no more than one right angle.*

Thus in rt. $\triangle ABC$ in Fig. 4-43, if $m\angle C = 90°$, then $\angle A$ and $\angle B$ cannot be rt. $\angle$s

PRINCIPLE 10: *A triangle can have no more than one obtuse angle.*

Thus in obtuse $\triangle ABC$ in Fig. 4-45, if $\angle C$ is obtuse, then $\angle A$ and $\angle B$ cannot be obtuse angles.

Fig. 4-45

PRINCIPLE 11: *Two angles are supplementary if their sides are respectively perpendicular to each other.*

Thus if $l_1 \perp l_3$ and $l_2 \perp l_4$ in Fig. 4-46, then $\angle a \cong \angle b$, and a and $\angle c$ are supplementary.

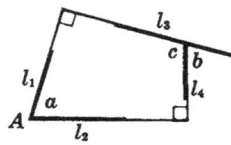

Fig. 4-46

SOLVED PROBLEMS

4.11 Numerical applications of angle-measure-sum principles

In each part of Fig. 4-47, find x and y.

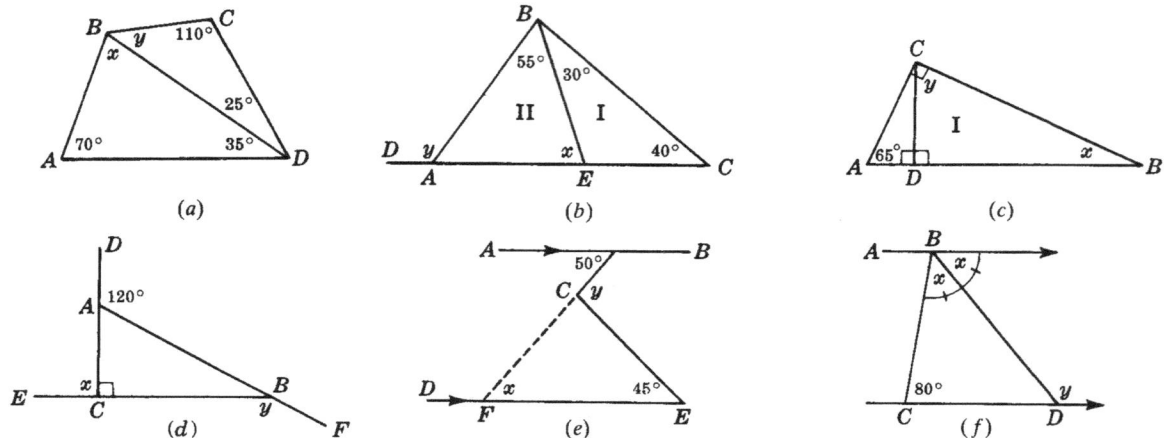

Fig. 4-47

Solutions

(a) $x + 35 + 70 = 180$ (Pr. 1)

$\qquad x = 75°$

$\quad y + 110 + 25 = 180$ (Pr. 1)

$\qquad y = 45°$

Check: The sum of the measures of the angles of quad. *ABCD* should equal 360°.

$70 + 120 + 110 + 60 \overset{?}{=} 360$

$\qquad 360 = 360$

(b) x is ext. $\angle$ of $\triangle$I.

$\quad x = 30 + 40$ (Pr. 4)

$\quad x = 70°$

$\quad y$ is an ext. $\angle$ of $\triangle ABC$.

$\quad y = m\angle B + 40$ (Pr. 4)

$\quad y = 85 + 40 = 125°$

(c) In $\triangle ABC$, $x + 65 = 90$ (Pr. 7)

$\qquad x = 25°$

$\quad$ In $\triangle$I, $x + y = 90$ (Pr. 7)

$\qquad 25 + y = 90$

$\qquad y = 65°$

(d) Since $\overleftrightarrow{DC} \perp \overleftrightarrow{EB}$, $x = 90°$

$\quad x + y + 120 = 360$ (Pr. 7)

$\quad 90 + y + 120 = 360$

$\qquad y = 150°$

(e) Since $\overleftrightarrow{AB} \| \overleftrightarrow{DE}$, $x = 50°$

$\quad y = x + 45$ (Pr. 4)

$\quad y = 50 + 45 = 95°$

(f) Since $\overleftrightarrow{AB} \| \overleftrightarrow{CD}$, $2x + 80 = 180$

$\qquad\qquad\qquad 2x = 100$

$\qquad x = 50°$

$\quad y = x + 80°$ (Pr. 4)

$\quad y = 50 + 80 = 130°$

4.12 Applying angle-measure-sum principles to isosceles and equilateral triangles

Find x and y in each part of Fig. 4-48.

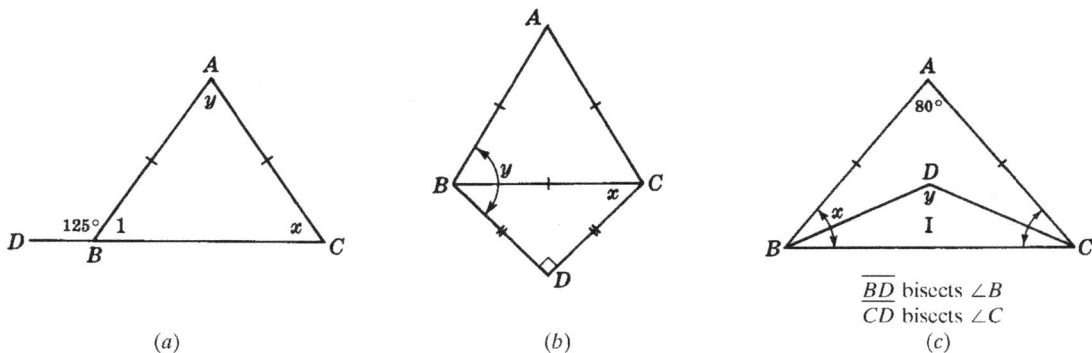

BD bisects $\angle B$
CD bisects $\angle C$

(a)   (b)   (c)

Fig. 4-48

Solutions

(a) Since $\overline{AB} \cong \overline{AC}$, we have $\angle 1 \cong \angle x$

   $x = 180 - 125 = 55°$

   $2x + y = 180$ (Pr. 1)

   $110 + y = 180$

   $y = 70°$

(b) By Pr. 8, $x = 45°$

   Since $m\angle ABC = 60°$ (Pr. 6)

   and $m\angle CBD = 45°$ (Pr. 8)

   $y = 60 + 45 = 105°$

(c) Since $\overline{AB} \cong \overline{AC}$, $\angle ABC \cong \angle ACB$

   $2x + 80 = 180$ (Pr. 1)

   $x = 50°$

   In $\triangle I, \frac{1}{2}x + \frac{1}{2}x + y = 180$ (Pr. 1)

   $x + y = 180$

   $50 + y = 180$

   $y = 130°$

4.13 Applying ratios to angle-measure sums

Find the measure of each angle

(a) Of a triangle if its angle measures are in the ratio of 3:4:5 [Fig. 3-49(a)]

(b) Of a quadrilateral if its angle measures are in the ratio of 3:4:5:6 [(b)]

(c) Of a right triangle if the ratio of the measures of its acute angles is 2:3 [(c)]

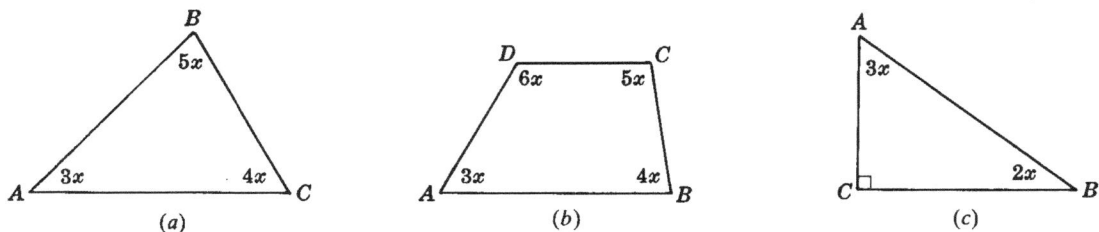

(a)   (b)   (c)

Fig. 4-49

Solutions

(a) Let $3x, 4x,$ and $5x$ represent the measures of the angles. Then $12x = 180$ by Principle 1, so that $x = 15$. Now $3x = 45, 4x = 60,$ and $5x = 75$. *Ans.* 45°, 60°, 75°

(b) Let $3x, 4x, 5x,$ and $6x$ represent the measures of the angles. Then $18x = 360$ by Principle 3, so that $x = 20$. Now $3x = 60, 4x = 80,$ and so forth. *Ans.* 60°, 80°, 100°, 120°

(c) Let $2x$ and $3x$ represent the measures of the acute angles. Then $5x = 90$ by Principle 7 so that $x = 18$. Now $2x = 36$ and $3x = 54$. *Ans.* 36°, 54°, 90°

4.14 Using algebra to prove angle-measure-sum problems

(a) Prove that if the measure of one angle of a triangle equals the sum of the measures of the other two, then the triangle is a right triangle.

(b) Prove that if the opposite angles of a quadrilateral are congruent, then its opposite sides are parallel.

Solutions

(a) **Given:** $\triangle ABC, m\angle C = m\angle A + m\angle B$
To Prove: $\triangle ABC$ is a right triangle.
Plan: Prove $m\angle C = 90°$

ALGEBRAIC PROOF:

Let $\quad a$ = number of degrees in $\angle A$
$\quad\quad\quad b$ = number of degrees in $\angle B$
Then $a + b$ = number of degrees in $\angle C$
$$a + b + (a + b) = 180 \quad \text{(Pr. 1)}$$
$$2a + 2b = 180$$
$$a + b = 90$$
Since $m\angle C = 90°, \triangle ABC$ is a rt. $\triangle$.

(b) **Given:** Quadrilateral $ABCD$
$\quad\quad\quad \angle A \cong \angle C, \angle B \cong \angle D$
To Prove: $\overline{AB} \parallel \overline{CD}, \overline{BC} \parallel \overline{AD}$
Plan: Prove int. $\angle$s on same side of transversal are supplementary.

ALGEBRAIC PROOF:

Let a = number of degrees in $\angle A$ and $\angle C$,
$\quad b$ = number of degrees in $\angle B$ and $\angle D$.
$$2a + 2b = 360 \quad \text{(Pr. 3)}$$
$$a + b = 180$$
Since $\angle A$ and $\angle B$ are supplementary, $\overline{BC} \parallel \overline{AD}$.
Since $\angle A$ and $\angle D$ are supplementary, $\overline{AB} \parallel \overline{CD}$.

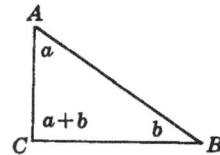

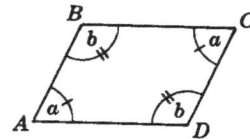

4.4 Sum of the Measures of the Angles of a Polygon

A *polygon* is a closed plane figure bounded by straight line segments as sides. An *n-gon* is a polygon of n sides. Thus, a polygon of 20 sides is a 20-gon.

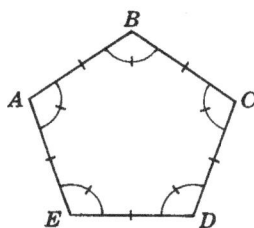

Regular Pentagon

Fig. 4-50

A *regular polygon* is an equilateral and equiangular polygon. Thus, a regular pentagon is a polygon having 5 congruent angles and 5 congruent sides (Fig. 4-50). A square is a regular polygon of 4 sides.

Names of Polygons According to the Number of Sides

Number of Sides	Polygon	Number of Sides	Polygon
3	Triangle	8	Octagon
4	Quadrilateral	9	Nonagon
5	Pentagon	10	Decagon
6	Hexagon	12	Dodecagon
7	Heptagon	n	n-gon

4.4A Sum of the Measures of the Interior Angles of a Polygon

By drawing diagonals from any vertex to each of the other vertices, as in Fig. 4-51, a polygon of 7 sides is divisible into 5 triangles. Note that each triangle has one side of the polygon, except the first and last triangles which have two such sides.

In general, this process will divide a polygon of n sides into $n - 2$ triangles; that is, the number of such triangles is always two less than the number of sides of the polygon.

The sum of the measures of the interior angles of the polygon equals the sum of the measures of the interior angles of the triangles. Hence:

Sum of measures of interior angles of a polygon of n sides $= (n - 2)180°$

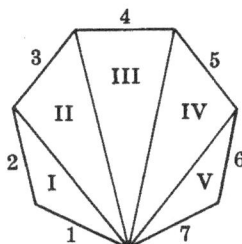

Fig. 4-51

4.4B   Sum of the Measures of the Exterior Angles of a Polygon

The exterior angles of a polygon can be reproduced together, so that they have the same vertex. To do this, draw lines parallel to the sides of the polygon from a point, as shown in Fig. 4-52. If this is done, it can be seen that regardless of the number of sides, the sum of the measures of the exterior angle equals 360°. Hence:

Sum of measures of exterior angles of a polygon of *n* sides = 360°

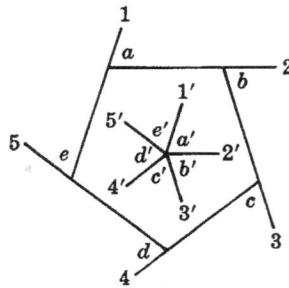

Fig. 4-52

4.4C   Polygon-Angle Principles

For any polygon

PRINCIPLE 1:   *If S is the sum of the measures of the interior angles of a polygon of n sides, then*

$$S = n - 2 \text{ straight angles} = (n - 2)180°$$

The sum of the measures of the interior angles of a polygon of 10 sides (decagon) equals 1440°, since $S = 8(180) = 1440$.

PRINCIPLE 2:   *The sum of the measures of the exterior angles of any polygon equals 360°.*

Thus, the sum of the measures of the exterior angles of a polygon of 23 sides equals 360°.

For a regular polygon

PRINCIPLE 3:   *If a regular polygon of n sides (Fig. 4-53) has an interior angle of measure i and an exterior angle of measure e (in degrees), then*

$$i = \frac{180(n - 2)}{n}, \; e = \frac{360}{n}, \text{ and } i + e = 180$$

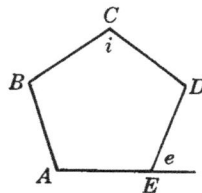

Regular Polygon

Fig. 4-53

Thus for a regular polygon of 20 sides,

$$i = \frac{180(20 - 2)}{20} = 162 \qquad e = \frac{360}{20} = 18 \qquad i + e = 162 + 18 = 180$$

SOLVED PROBLEMS

4.15 Applying angle-measure formulas to a polygon

(a) Find the sum of the measures of the interior angles of a polygon of 9 sides (express your answer in straight angles and in degrees).

(b) Find the number of sides a polygon has if the sum of the measures of the interior angles is 3600°.

(c) Is it possible to have a polygon the sum of whose angle measures is 1890°?

Solutions

(a) S (in straight angles) $= n - 2 = 9 - 2 = 7$ straight angles; $m\angle S = (n - 2)180 = 7(180) = 1260°$.

(b) S (in degrees) $= (n - 2)180$. Then $3600 = (n - 2)180$, from which $n = 22$.

(c) Since $1890 = (n - 2)180$, then $n = 12\frac{1}{2}$. A polygon cannot have $12\frac{1}{2}$ sides.

4.16 Applying angle-measure formulas to a regular polygon

(a) Find each exterior angle measure of a regular polygon having 9 sides.

(b) Find each interior angle measure of a regular polygon having 9 sides.

(c) Find the number of sides a regular polygon has if each exterior angle measure is 5°.

(d) Find the number of sides a regular polygon has if each interior angle measure is 165°.

Solutions

(a) Since $n = 9, m\angle e = \dfrac{360}{n} = \dfrac{360}{9} = 40$. *Ans.* 40°

(b) Since $n = 9, m\angle i = \dfrac{(n - 2)180}{n} = \dfrac{(9 - 2)180}{9} = 140$. *Ans.* 140°

 Another method: Since $i + e = 180, i = 180 - e = 180 - 40 = 140$.

(c) Substituting $e = 5$ in $e = \dfrac{360}{n}$, we have $5 = \dfrac{360}{n}$. Then $5n = 360$, so $n = 72$. *Ans.* 72 sides

(d) Substituting $i = 165$ in $i + e = 180$, we have $165 + e = 180$ or $e = 15$.

 Then, using $e = \dfrac{360}{n}$ with $e = 15$, we have $15 = \dfrac{360}{n}$, or $n = 24$. *Ans.* 24 sides

4.17 Applying algebra to the angle-measure sums of a polygon

Find each interior angle measure of a quadrilateral (a) if its interior angles are represented by $x + 10$, $2x + 20, 3x - 50$, and $2x - 20$; (b) if its exterior angles are in the ratio 2:3:4:6.

Solutions

(a) Since the sum of the measures of the interior $\angle$ is $360°$, we add

$$(x + 10) + (2x + 20) + (3x - 50) + (2x - 20) = 360$$
$$8x - 40 = 360$$
$$x = 50$$

Then $x + 10 = 60$; $2x + 20 = 120$; $3x - 50 = 100$; $2x - 20 = 80$.   *Ans.* $60°, 120°, 100°, 80°$

(b) Let the exterior angles be represented respectively by $2x$, $3x$, $4x$, and $6x$. Then $2x + 3x + 4x + 6x = 360$. Solving gives us $15x = 360$ and $x = 24$. Hence, the exterior angles measure $48°, 72°, 96°$, and $144°$. The interior angles are their supplements.   *Ans.* $132°, 108°, 84°, 36°$

4.5   Two New Congruency Theorems

Three methods of proving triangles congruent have already been introduced here. These are

1. Side-Angle-Side (SAS)
2. Angle-Side-Angle (ASA)
3. Side-Side-Side (SSS)

Two additional methods of proving that triangles are congruent are

4. Side-Angle-Angle (SAA)
5. Hypotenuse-Leg (hy-leg)

4.5A   Two New Congruency Principles

PRINCIPLE 1:   (SAA) *If two angles and a side opposite one of them of one triangle are congruent to the corresponding parts of another, the triangles are congruent.*

Thus if $\angle A \cong \angle A'$, $\angle B \cong \angle B'$, and $\overline{BC} \cong \overline{B'C'}$ in Fig. 4-54, then $\triangle ABC \cong \triangle A'B'C'$.

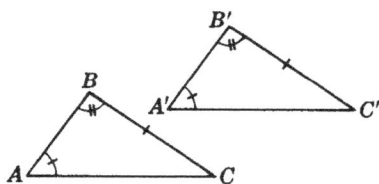
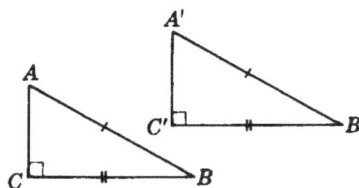

Fig. 4-54   Fig. 4-55

PRINCIPLE 2:   (hy-leg) *If the hypotenuse and a leg of one right triangle are congruent to the corresponding parts of another right triangle, the triangles are congruent.*

Thus if hy$\overline{AB} \cong$ hy$\overline{A'B'}$ and leg $\overline{BC} \cong$ leg $\overline{B'C'}$ in Fig. 4-55, then rt. $\triangle ABC \cong$ rt. $\triangle A'B'C'$.

A proof of this principle is given in Chapter 16.

SOLVED PROBLEMS

4.18 Selecting congruent triangles using hy-leg or SAA

In (a) Fig. 4-56 and (b) Fig. 4-57, select congruent triangles and state the reason for the congruency.

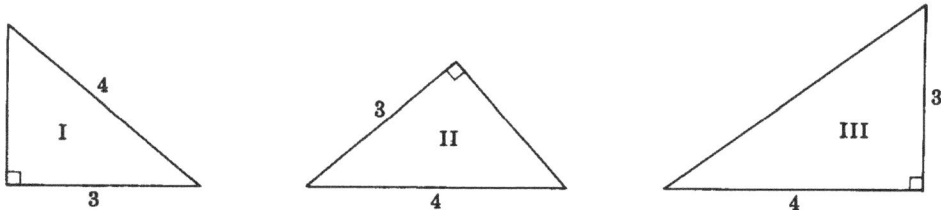

Fig. 4-56

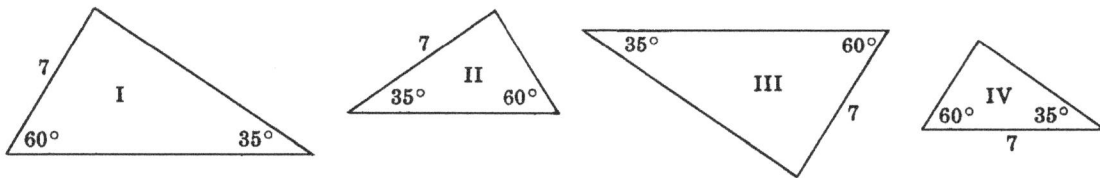

Fig. 4-57

Solutions

(a) $\triangle I \cong \triangle II$ by hy-leg. In $\triangle III$, 4 is not a hypotenuse.

(b) $\triangle I \cong \triangle III$ by SAA. In $\triangle II$, 7 is opposite 60° instead of 35°. In $\triangle IV$, 7 is included between 60° and 35°.

4.19 Determining the reason for congruency of triangles

In each part of Fig. 4-58, $\triangle I$ can be proved congruent to $\triangle II$. Make a diagram showing the congruent parts of both triangles and state the reason for the congruency.

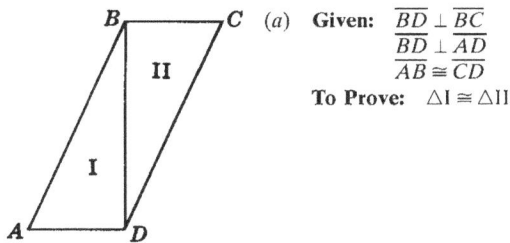
(*a*) **Given:** $\overline{BD} \perp \overline{BC}$
                   $\overline{BD} \perp \overline{AD}$
                   $\overline{AB} \cong \overline{CD}$
        To Prove: $\triangle I \cong \triangle II$

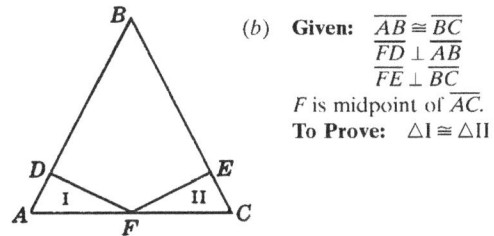
(*b*) **Given:** $\overline{AB} \cong \overline{BC}$
                   $\overline{FD} \perp \overline{AB}$
                   $\overline{FE} \perp \overline{BC}$
        F is midpoint of $\overline{AC}$.
        To Prove: $\triangle I \cong \triangle II$

Fig. 4-58

Solutions

(a) See Fig. 4-59(a). $\triangle I \cong \triangle II$ by hy-leg.

(b) See Fig. 4-59(b). $\triangle I \cong \triangle II$ by SAA.

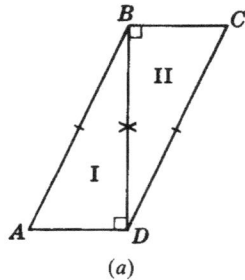 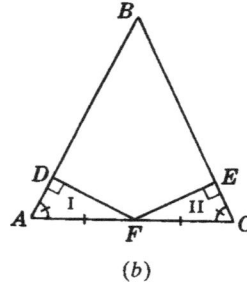

(a) (b)

Fig. 4-59

4.20 Proving a congruency problem

Given: Quadrilateral *ABCD*

$\overline{DF} \perp \overline{AC}, \overline{BE} \perp \overline{AC}$

$\overline{AE} \cong \overline{FC}, \overline{BC} \cong \overline{AD}$

To Prove: $\overline{BE} \cong \overline{FD}$

Plan: Prove $\triangle I \cong \triangle II$

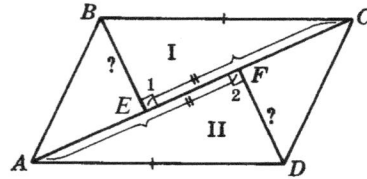

PROOF:

Statements	Reasons
1. $\overline{BC} \cong \overline{AD}$	1. Given
2. $\overline{DF} \perp \overline{AC}, \ \overline{BE} \perp \overline{AC}$	2. Given
3. $\angle 1 \cong \angle 2$	3. Perpendiculars form rt. $\angle$s, and all rt. $\angle$s are congruent.
4. $\overline{AE} \cong \overline{FC}$	4. Given
5. $\overline{EF} \cong \overline{EF}$	5. Identity
6. $\overline{AF} \cong \overline{EC}$	6. If equals are added to equals, the sums are equal. Definition of congruent segments.
7. $\triangle I \cong \triangle II$	7. Hy-leg
8. $\overline{BE} \cong \overline{FD}$	8. Corresponding parts of congruent $\triangle$s are congruent.

4.21 Proving a congruency problem stated in words

Prove that in an isosceles triangle, altitudes to the congruent sides are congruent.

Given: Isosceles $\triangle ABC$ ($\overline{AB} \cong \overline{BC}$)

$\overline{AD}$ is altitude to $\overline{BC}$

$\overline{CE}$ is altitude to $\overline{AB}$

To Prove: $\overline{AD} \cong \overline{CE}$

Plan: Prove $\triangle ACE \cong \triangle CAD$

or $\triangle I \cong \triangle II$

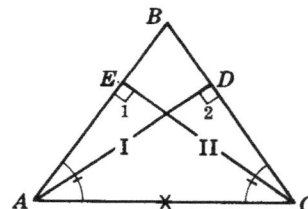

PROOF:

Statements	Reasons
1. $\overline{AB} \cong \overline{BC}$	1. Given
2. $\angle A \cong \angle C$	2. In a $\triangle$, angles opposite equal sides are equal.
3. $\overline{AD}$ is altitude to $\overline{BC}$, $\overline{CE}$ is altitude to $\overline{AB}$.	3. Given
4. $\angle 1 \cong \angle 2$	4. Altitudes form rt. $\angle$ and rt. $\angle$ are congruent.
5. $\overline{AC} \cong \overline{AC}$	5. Identity
6. $\triangle I \cong \triangle II$	6. SAA
7. $\overline{AD} \cong \overline{CE}$	7. Corresponding parts of congruent $\triangle$ are congruent.

SUPPLEMENTARY PROBLEMS

4.1.   In each part of Fig. 4-60, find x and y.   (4.1)

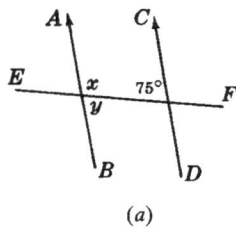

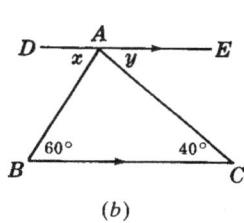

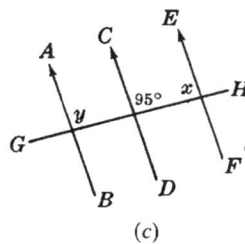

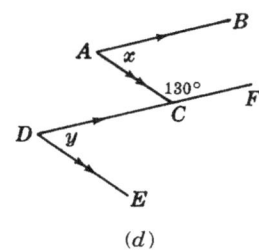

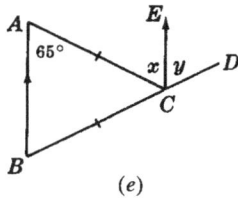

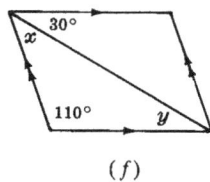

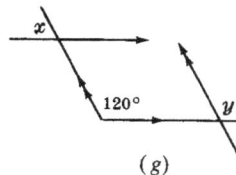

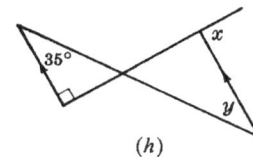

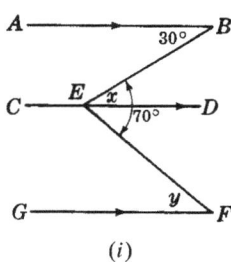

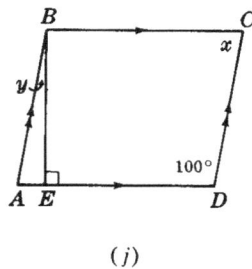

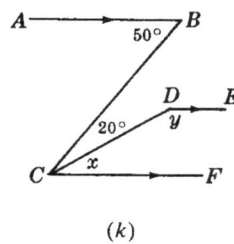

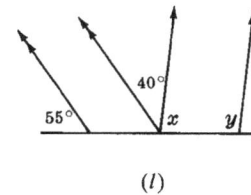

Fig. 4-60

4.2. In each part of Fig. 4-61, find x and y. (4.3)

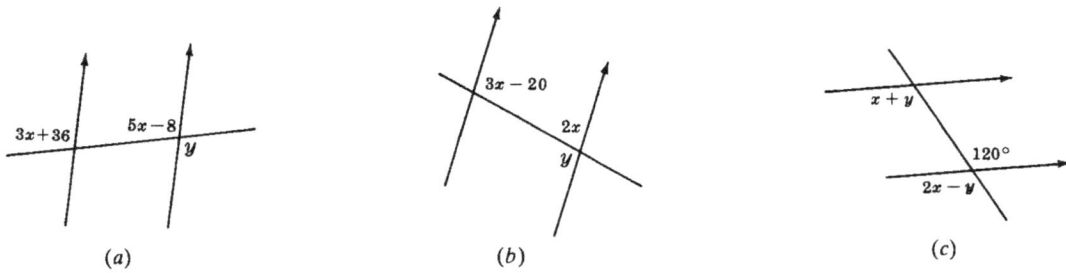

Fig. 4-61

4.3. If two parallel lines are cut by a transversal, find (4.3)

(a) Two alternate interior angles represented by $3x$ and $5x - 70$

(b) Two corresponding angles represented by $2x + 10$ and $4x - 50$

(c) Two interior angles on the same side of the transversal represented by $2x$ and $3x$

4.4. Provide the proofs requested in Fig. 4-62 (4.4)

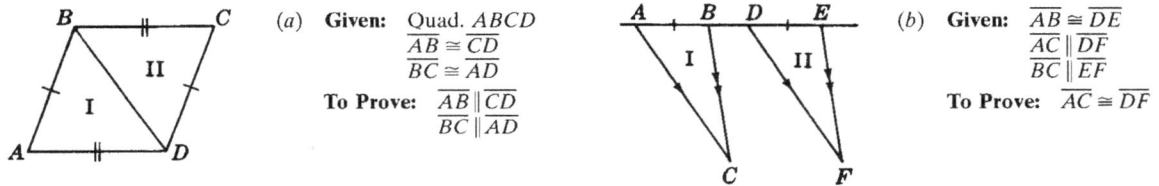

(a) **Given:** Quad. $ABCD$
$\overline{AB} \cong \overline{CD}$
$\overline{BC} \cong \overline{AD}$
To Prove: $\overline{AB} \parallel \overline{CD}$
$\overline{BC} \parallel \overline{AD}$

(b) **Given:** $\overline{AB} \cong \overline{DE}$
$\overline{AC} \parallel \overline{DF}$
$\overline{BC} \parallel \overline{EF}$
To Prove: $\overline{AC} \cong \overline{DF}$

Fig. 4-62

4.5. Provide the proofs requested in Fig. 4-63 (4.4)

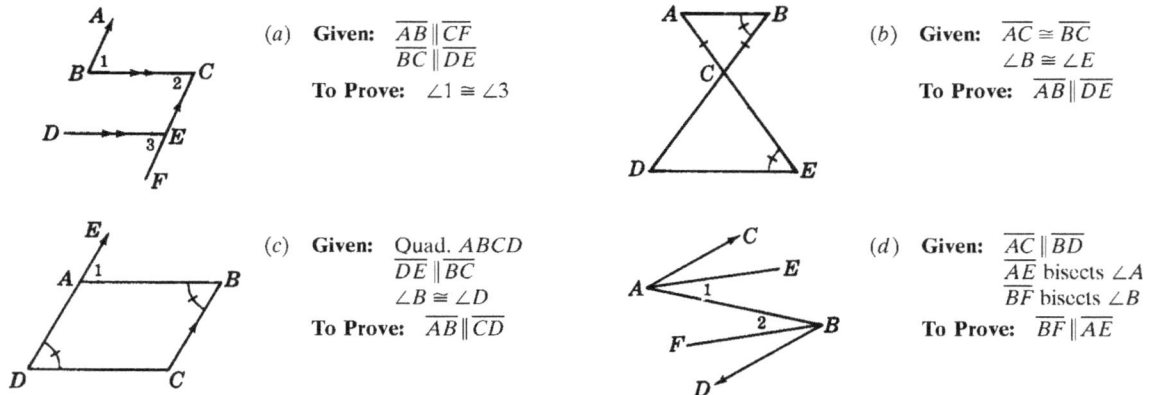

(a) **Given:** $\overline{AB} \parallel \overline{CF}$
$\overline{BC} \parallel \overline{DE}$
To Prove: $\angle 1 \cong \angle 3$

(b) **Given:** $\overline{AC} \cong \overline{BC}$
$\angle B \cong \angle E$
To Prove: $\overline{AB} \parallel \overline{DE}$

(c) **Given:** Quad. $ABCD$
$\overline{DE} \parallel \overline{BC}$
$\angle B \cong \angle D$
To Prove: $\overline{AB} \parallel \overline{CD}$

(d) **Given:** $\overline{AC} \parallel \overline{BD}$
$\overline{AE}$ bisects $\angle A$
$\overline{BF}$ bisects $\angle B$
To Prove: $\overline{BF} \parallel \overline{AE}$

Fig. 4-63

4.6. Prove each of the following: (4.5)

(a) If the opposite sides of a quadrilateral are parallel, then they are also congruent.

(b) If $\overline{AB}$ and $\overline{CD}$ bisect each other at E, then $\overline{AC} \parallel \overline{BD}$.

(c) In quadrilateral $ABCD$, let $\overline{BC} \parallel \overline{AD}$. If the diagonals $\overline{AC}$ and $\overline{BD}$ intersect at E and $\overline{AE} \parallel \overline{DE}$, then $\overline{BE} \cong \overline{CE}$.

(d) $\overleftrightarrow{AB}$ and $\overleftrightarrow{CD}$ are parallel lines cut by a transversal at E and F. If $\overrightarrow{EG}$ and $\overrightarrow{FH}$ bisect a pair of corresponding angles, then $\overrightarrow{EG} \parallel \overrightarrow{FH}$.

(e) If a line through vertex B of $\triangle ABC$ *is parallel to $\overline{AC}$ and bisects the angle formed by extending $\overline{AB}$ through B, then* $\triangle ABC$ is isosceles.

4.7. In Fig. 4-64, find the distance from (a) A to B; (b) E to $\overline{AC}$; (c) A to $\overline{BC}$; (d) $\overline{ED}$ to $\overline{BC}$. (4.6)

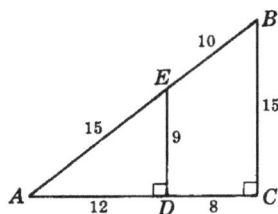
Fig. 4-64

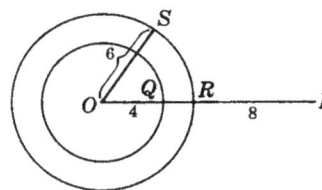
Fig. 4-65

4.8. In Fig. 4-65, find the distance (a) from P to the outer circle; (b) from P to the inner circle; (c) between the concentric circles; (d) from P to O. (4.6)

4.9. In Fig. 4-66 (4.6)

(a) Locate P, a point on $\overline{AD}$, equidistant from B and C. Then locate Q, a point on $\overline{AD}$, equidistant from $\overline{AB}$ and $\overline{BC}$.

(b) Locate R, a point equidistant from A, B, and C. Then locate S, a point equidistant from B, C, and D.

(c) Locate T, a point equidistant from $\overline{BC}, \overline{CD}$, and $\overline{AD}$. Then locate U, a point equidistant from $\overline{AB}, \overline{BC}$, and $\overline{CD}$.

4.10. In each part of Fig. 4-67, describe P, Q, and R as equidistant points and locate them on a bisector.

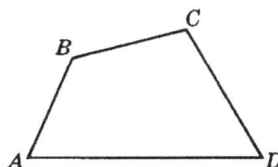
Fig. 4-66

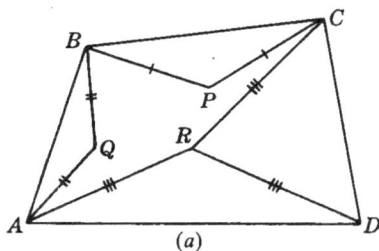
(a)

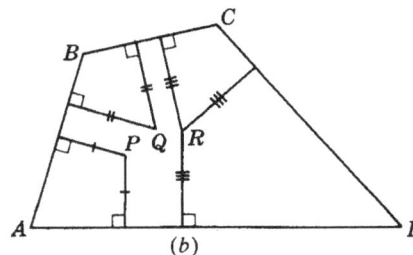
(b)

Fig. 4-67

4.11.   In each part of Fig. 4-68, describe P, Q, and R as equidistant points.   (4.9)

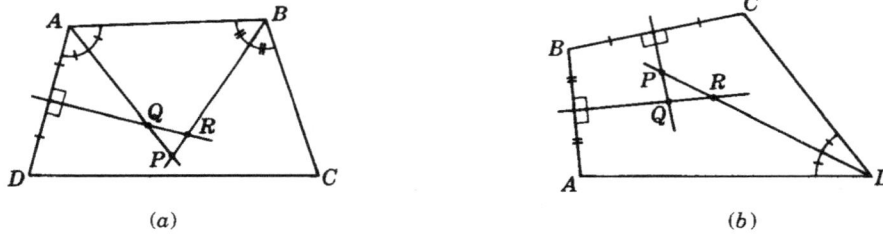

(a)   (b)

Fig. 4-68

4.12.   Find x and y in each part of Fig. 4-69.   (4.11)

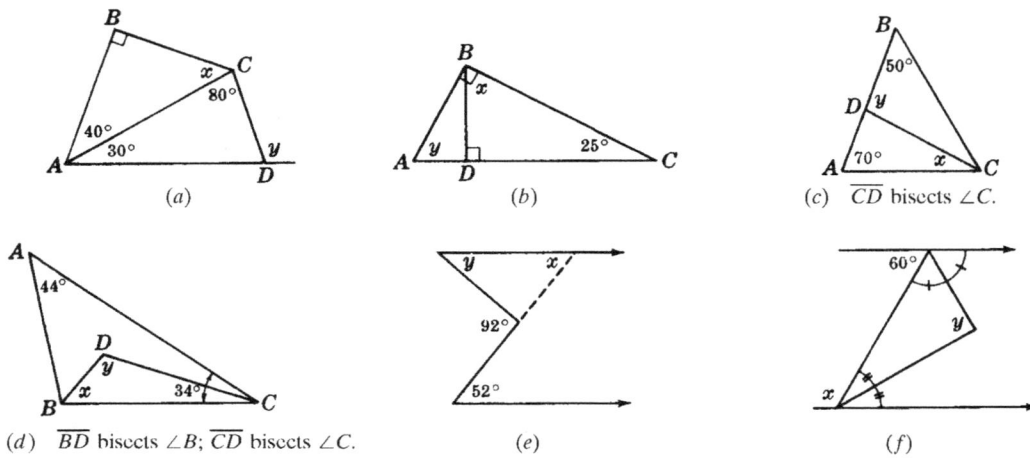

(a)   (b)   (c) $\overline{CD}$ bisects $\angle C$.

(d) $\overline{BD}$ bisects $\angle B$; $\overline{CD}$ bisects $\angle C$.   (e)   (f)

Fig. 4-69

4.13.   Find x and y in each part of Fig. 4-70.   (4.12)

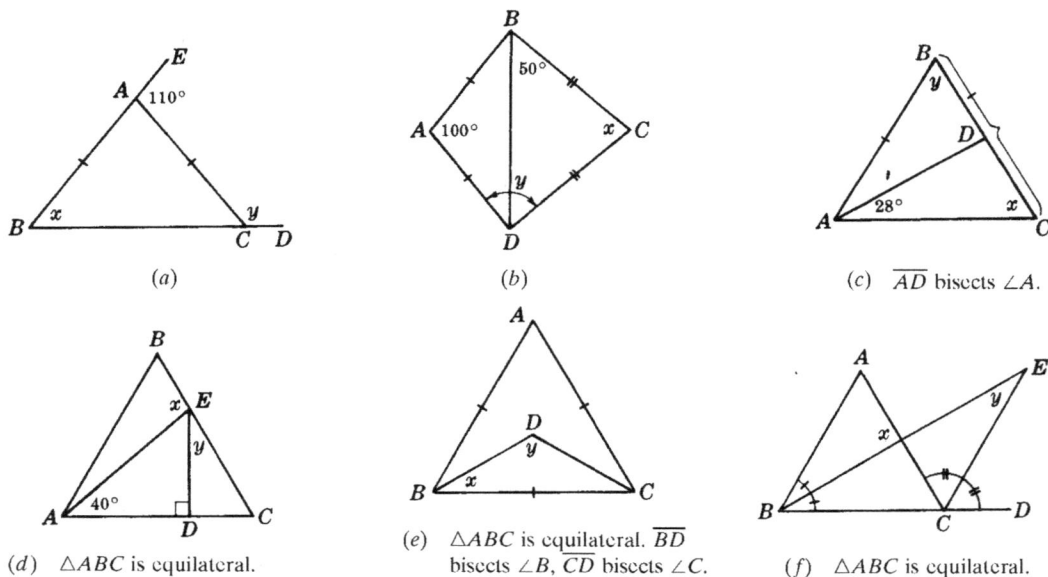

(a)   (b)   (c) $\overline{AD}$ bisects $\angle A$.

(d) $\triangle ABC$ is equilateral.   (e) $\triangle ABC$ is equilateral. $\overline{BD}$ bisects $\angle B$, $\overline{CD}$ bisects $\angle C$.   (f) $\triangle ABC$ is equilateral.

Fig. 4-70

4.14.   Find the measure of each angle   (4.13)

    (a) Of a triangle if its angle measures are in the ratio 1:3:6

    (b) Of a right triangle if its acute angle measures are in the ratio 4:5

    (c) Of an isosceles triangle if the ratio of the measures of its base angle to a vertex angle is 1:3

    (d) Of a quadrilateral if its angle measures are in the ratio 1:2:3:4

    (e) Of a triangle, one of whose angles measures 55° and whose other two angle measures are in the ratio 2:3

    (f) Of a triangle if the ratio of the measures of its exterior angles is 2:3:4

4.15.   Prove each of the following: (4.14)

    (a) In quadrilateral $ABCD$ if $\angle A \cong \angle D$ and $\angle B \cong \angle C$, then $\overline{BC} \parallel \overline{AD}$.

    (b) Two parallel lines are cut by a transversal. Prove that the bisectors of two interior angles on the same side of the transversal are perpendicular to each other.

4.16.   Show that a triangle is   (4.14)

    (a) Equilateral if its angles are represented by $x + 15$, $3x - 75$, and $2x - 30$

    (b) Isosceles if its angles are represented by $x + 15$, $3x - 35$, and $4x$

    (c) A right triangle if its angle measures are in the ratio 2:3:5

    (d) An obtuse triangle if one angle measures 64° and the larger of the other two measures 10° less than five times the measure of the smaller

4.17.   (a) Find the sum of the measures of the interior angles (in straight angles) of a polygon of 9 sides; of 32 sides.   (4.15)

    (b) Find the sum of the measures of the interior angles (in degrees) of a polygon of 11 sides; of 32 sides; of 1002 sides.

    (c) Find the number of sides a polygon has if the sum of the measures of the interior angles is 28 straight angles; 20 right angles; 4500°; 36,000°.

4.18.   (a) Find the measure of each exterior angle of a regular polygon having 18 sides; 20 sides; 40 sides.   (4.16)

    (b) Find the measure of each interior angle of a regular polygon having 18 sides; 20 sides; 40 sides.

    (c) Find the number of sides a regular polygon has if each exterior angle measures 120°; 40°; 18°; 2°.

    (d) Find the number of sides a regular polygon has if each interior angle measures 60°, 150°; 170°, 175°; 179°.

4.19.   (a) Find each interior angle of a quadrilateral if its interior angles are represented by $x - 5$, $x + 20$, $2x - 45$, and $2x - 30$.   (4.17)

    (b) Find the measure of each interior angle of a quadrilateral if the measures of its exterior angles are in the ratio of 1:2:3:3.

4.20.   In each part of Fig. 4-71, select congruent triangles and state the reason for the congruency.   (4.18)

4.21.   In each part of Fig. 4-72, two triangles can be proved congruent. Make a diagram showing the congruent parts of both triangles, and state the reason for the congruency.   (4.19)

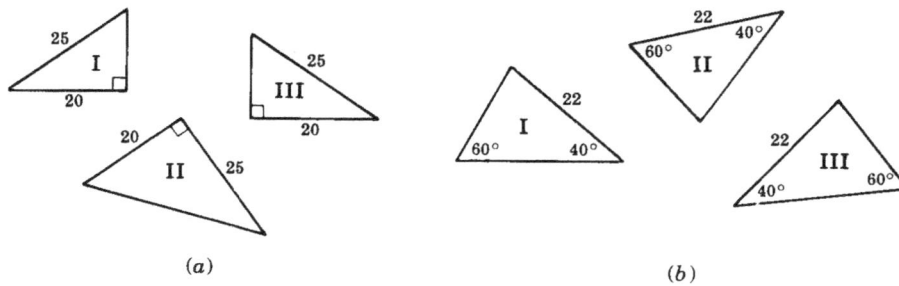

Fig. 4-71

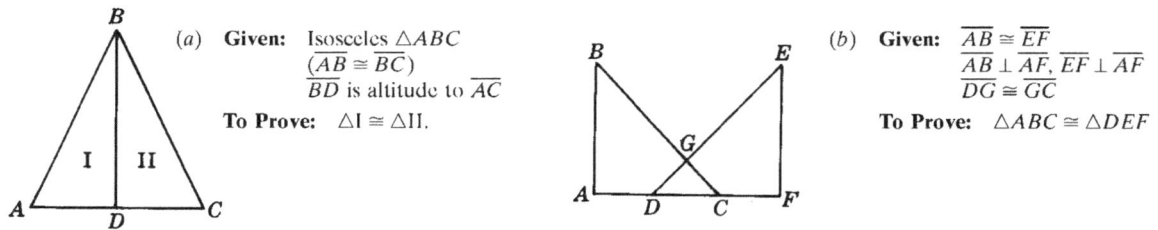

(a) **Given:** Isosceles $\triangle ABC$
$(\overline{AB} \cong \overline{BC})$
$\overline{BD}$ is altitude to $\overline{AC}$
To Prove: $\triangle I \cong \triangle II$.

(b) **Given:** $\overline{AB} \cong \overline{EF}$
$\overline{AB} \perp \overline{AF}, \overline{EF} \perp \overline{AF}$
$\overline{DG} \cong \overline{GC}$
To Prove: $\triangle ABC \cong \triangle DEF$

Fig. 4-72

4.22. Provide the proofs requested in Fig. 4-73.   (4.20)

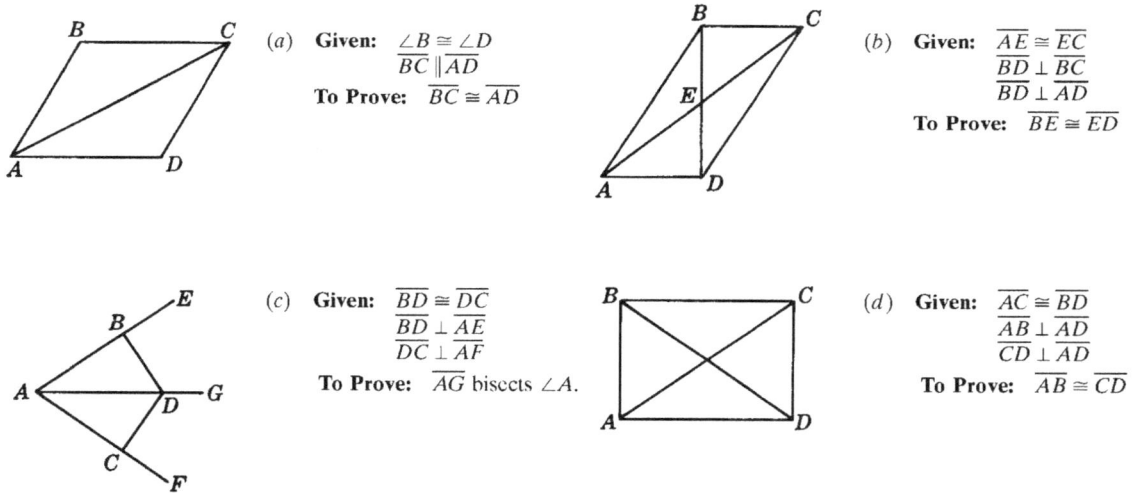

(a) **Given:** $\angle B \cong \angle D$
$\overline{BC} \parallel \overline{AD}$
To Prove: $\overline{BC} \cong \overline{AD}$

(b) **Given:** $\overline{AE} \cong \overline{EC}$
$\overline{BD} \perp \overline{BC}$
$\overline{BD} \perp \overline{AD}$
To Prove: $\overline{BE} \cong \overline{ED}$

(c) **Given:** $\overline{BD} \cong \overline{DC}$
$\overline{BD} \perp \overline{AE}$
$\overline{DC} \perp \overline{AF}$
To Prove: $\overline{AG}$ bisects $\angle A$.

(d) **Given:** $\overline{AC} \cong \overline{BD}$
$\overline{AB} \perp \overline{AD}$
$\overline{CD} \perp \overline{AD}$
To Prove: $\overline{AB} \cong \overline{CD}$

Fig. 4-73

4.23. Prove each of the following: (4.21)

(a) If the perpendiculars to two sides of a triangle from the midpoint of the third side are congruent, then the triangle is isosceles.

(b) Perpendiculars from a point in the bisector of an angle to the sides of the angle are congruent.

(c) If the altitudes to two sides of a triangle are congruent, then the triangle is isosceles.

(d) Two right triangles are congruent if the hypotenuse and an acute angle of one are congruent to the corresponding parts of the other.

CHAPTER 5

Parallelograms, Trapezoids, Medians, and Midpoints

5.1 Trapezoids

A *trapezoid* is a quadrilateral having two, and only two, parallel sides. The *bases* of the trapezoid are its parallel sides; the *legs* are its nonparallel sides. The *median* of the trapezoid is the segment joining the midpoints of its legs.

Thus in trapezoid *ABCD* in Fig. 5-1, the bases are $\overline{AD}$ and $\overline{BC}$, and the legs are $\overline{AB}$ and $\overline{CD}$. If *M* and *N* are midpoints, then $\overline{MN}$ is the median of the trapezoid.

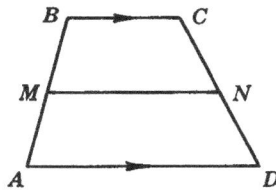

Fig. 5-1

An *isosceles trapezoid* is a trapezoid whose legs are congruent. Thus in isosceles trapezoid *ABCD* in Fig. 5-2 $\overline{AB} \cong \overline{CD}$.

The *base angles* of a trapezoid are the angles at the ends of its longer base: $\angle A$ *and* $\angle D$ are the base angles of isosceles trapezoid *ABCD*.

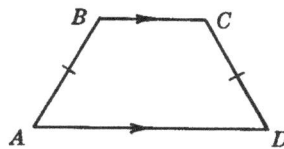

Isosceles Trapezoid

Fig. 5-2

5.1A Trapezoid Principles

PRINCIPLE 1: *The base angles of an isosceles trapezoid are congruent.*

Thus in trapezoid *ABCD* of Fig. 5-3 if $\overline{AB} \cong \overline{CD}$, then $\angle A \cong \angle D$.

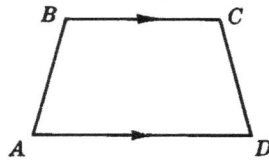

Fig. 5-3

PRINCIPLE 2: *If the base angles of a trapezoid are congruent, the trapezoid is isosceles.*

Thus in Fig. 5-3 if $\angle A \cong \angle D$, then $\overline{AB} \cong \overline{CD}$.

SOLVED PROBLEMS

5.1 Applying algebra to the trapezoid

In each trapezoid in Fig. 5-4, find x and y.

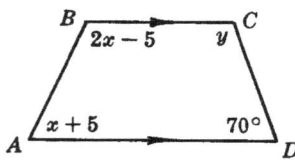

(*a*) *ABCD* is a trapezoid.   (*b*) *ABCD* is an isosceles trapezoid.   (*c*) *ABCD* is an isosceles trapezoid.

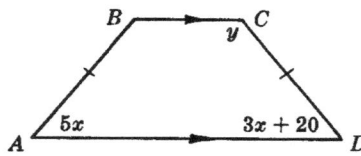

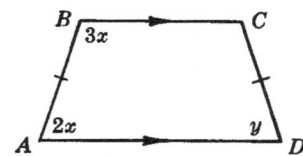

Fig. 5-4

Solutions

(a) Since $\overline{AD} \parallel \overline{BC}, (2x - 5) + (x + 5) = 180$; then $3x = 180$ and $x = 60$.
Also, $y + 70 = 180$ or $y = 110$.

(b) Since $\angle A \cong \angle D, 5x = 3x + 20$, so that $2x = 20$ or $x = 10$.
Since $\overline{BC} \parallel \overline{AD}, y + (3x + 20) = 180$, so $y + 50 = 180$ or $y = 130$.

(c) Since $\overline{BC} \parallel \overline{AD}, 3x + 2x = 180$ or $x = 36$.
Since $\angle D \cong \angle A, y = 2x$ or $y = 72$.

5.2 Proof of a trapezoid principle stated in words

Prove that the base angles of an isosceles trapezoid are congruent.

Given: Isosceles trapezoid *ABCD*
($\overline{BC} \parallel \overline{AD}, \overline{AB} \cong \overline{CD}$)
To Prove: $\angle A \cong \angle D$
Plan: Draw $\perp$s to base from *B* and *C*.
Prove $\triangle \text{I} \cong \text{II}$.

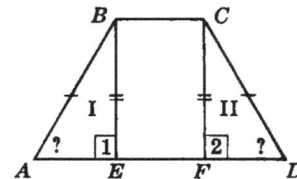

PROOF:

Statements	Reasons
1. Draw $\overline{BE} \perp \overline{AD}$ and $\overline{CF} \perp \overline{AD}$.	1. A $\perp$ may be drawn to a line from an outside point.
2. $\overline{BC} \parallel \overline{AD}, \overline{AB} \cong \overline{CD}$	2. Given
3. $\overline{BE} \cong \overline{CF}$	3. Parallel lines are everywhere equidistant. Definition of congruent segments.
4. $\angle 1 \cong \angle 2$	4. Perpendiculars form rt. $\angle$s. All rt. $\angle$s are congruent.
5. $\triangle \text{I} \cong \triangle \text{II}$	5. Hy-leg
6. $\angle A \cong \angle D$	6. Corresponding parts of congruent $\triangle$ are congruent.

5.2 Parallelograms

A parallelogram is a quadrilateral whose opposite sides are parallel. The symbol for parallelogram is $\square$.
Thus in $\square ABCD$ in Fig. 5-5, $\overline{AB} \parallel \overline{CD}$ in $\overline{AD} \parallel \overline{BC}$.

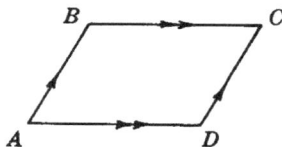

Fig. 5-5

If the opposite sides of a quadrilateral are parallel, then it is a parallelogram. (This is the converse of the above definition.) Thus if $\overline{AB} \parallel \overline{CD}$ and $\overline{AD} \parallel \overline{BC}$, then $ABCD$ is a $\square$.

5.2A Principles Involving Properties of Parallelograms

PRINCIPLE 1: *The opposite sides of a parallelogram are parallel.* (This is the definition.)

PRINCIPLE 2: *A diagonal of a parallelogram divides it into two congruent triangles.*

$\overline{BD}$ is a diagonal of $\square ABCD$ in Fig. 5-6, so $\triangle \text{I} \cong \triangle \text{II}$.

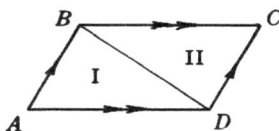

Fig. 5-6

PRINCIPLE 3: *The opposite sides of a parallelogram are congruent.*

Thus in $\square ABCD$ in Fig. 5-5, $\overline{AB} \cong \overline{CD}$ and $\overline{AD} \cong \overline{BC}$.

PRINCIPLE 4: *The opposite angles of a parallelogram are congruent.*

Thus in $\square ABCD$, $\angle A \cong \angle C$ and $\angle B \cong \angle D$.

PRINCIPLE 5: *The consecutive angles of a parallelogram are supplementary.*

Thus in $\square ABCD$, $\angle A$ is the supplement of both $\angle B$ and $\angle D$.

PRINCIPLE 6: *The diagonals of a parallelogram bisect each other.*

Thus in $\square ABCD$ in Fig. 5-7, $\overline{AE} \cong \overline{EC}$ and $\overline{BE} \cong \overline{ED}$.

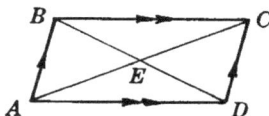

Fig. 5-7

5.2B Proving a Quadrilateral is a Parallelogram

PRINCIPLE 7: *A quadrilateral is a parallelogram if its opposite sides are parallel.*

Thus if $\overline{AB} \parallel \overline{CD}$ and $\overline{AD} \parallel \overline{BC}$, then $ABCD$ is a $\square$.

PRINCIPLE 8: *A quadrilateral is a parallelogram if its opposite sides are congruent.*

Thus if $\overline{AB} \cong \overline{CD}$ and $\overline{AD} \cong \overline{BC}$ in Fig. 5-8, then $ABCD$ is a $\square$.

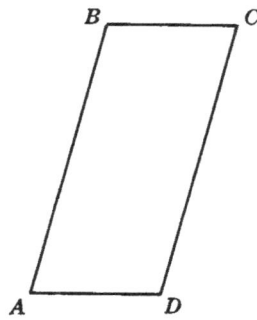

Fig. 5-8

PRINCIPLE 9:   *A quadrilateral is a parallelogram if two sides are congruent and parallel.*

Thus if $\overline{BC} \cong \overline{AD}$ and $\overline{BC} \parallel \overline{AD}$ in Fig. 5-8, then $ABCD$ is a $\square$.

PRINCIPLE 10:   *A quadrilateral is a parallelogram if its opposite angles are congruent.*

Thus if $\angle A \cong \angle C$ and $\angle B \cong \angle D$ in Fig. 5-8, then $ABCD$ is a $\square$.

PRINCIPLE 11:   *A quadrilateral is a parallelogram if its diagonals bisect each other.*

Thus if $\overline{AE} \cong \overline{EC}$ and $\overline{BE} \cong \overline{ED}$ in Fig. 5-9, then $ABCD$ is a $\square$.

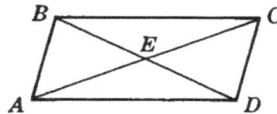

Fig. 5-9

SOLVED PROBLEMS

5.3   Applying properties of parallelograms

Assuming $ABCD$ is a parallelogram, find x and y in each part of Fig. 5-10.

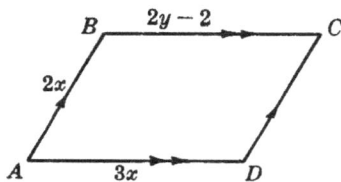
(a) Perimeter = 40.

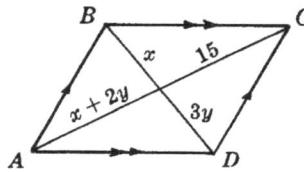
(b)

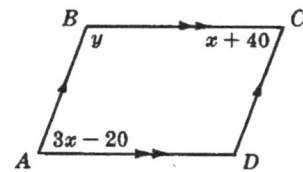
(c) .

Fig. 5-10

Solutions

(a) By Principle 3, $BC = AD = 3x$ and $CD = AB = 2x$; then $2(2x + 3x) = 40$, so that $10x = 40$ or $x = 4$.
By Principle 3, $2y - 2 = 3x$; then $2y - 2 = 3(4)$, so $2y = 14$ or $y = 7$.

(b) By Principle 6, $x + 2y = 15$ and $x = 3y$.
Substituting $3y$ for x in the first equation yields $3y + 2y = 15$ or $y = 3$. Then $x = 3y = 9$.

(c) By Principle 4, $3x - 20 = x + 40$, so $2x = 60$ for $x = 30$.
By Principle 5, $y + (x + 40) = 180$. Then $y + (30 + 40) = 180$ or $y = 110$.

5.4   Applying principle 7 to determine parallelograms

Name the parallelograms in each part of Fig. 5-11.

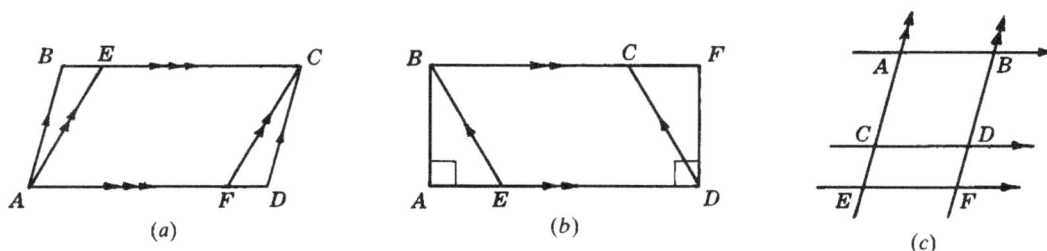

Fig. 5-11

Solutions

(a) *ABCD*, *AECF*; (b) *ABFD*, *BCDE*; (c) *ABDC*, *CDFE*, *ABFE*.

5.5 Applying principles 9, 10, and 11

State why *ABCD* is a parallelogram in each part of Fig. 5-12.

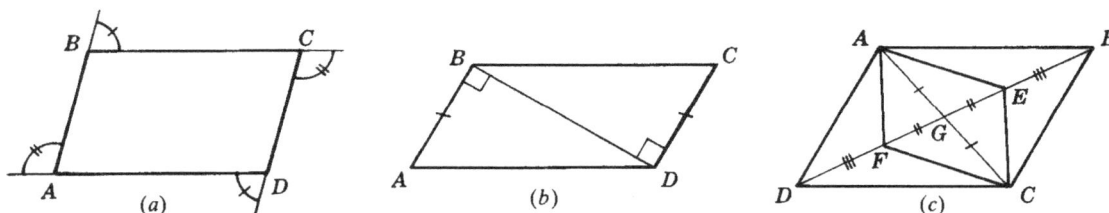

Fig. 5-12

Solutions

(a) Since supplements of congruent angles are congruent, opposite angles of *ABCD* are congruent. Thus by Principle 10, *ABCD* is a parallelogram.

(b) Since perpendiculars to the same line are parallel, $\overline{AB} \parallel \overline{CD}$. Hence by Principle 9, *ABCD* is a parallelogram.

(c) By the addition axiom, $\overline{DG} \cong \overline{GB}$. Hence by Principle 11, *ABCD* is a parallelogram.

5.6 Proving a parallelogram problem

Given: $\square ABCD$
   E is midpoint of $\overline{BC}$.
   G is midpoint of $\overline{AD}$.
   $\overline{EF} \perp \overline{BD}$, $\overline{GH} \perp \overline{BD}$
To Prove: $\overline{EF} \cong \overline{GH}$
Plan: Prove $\triangle BFE \cong \triangle GHD$

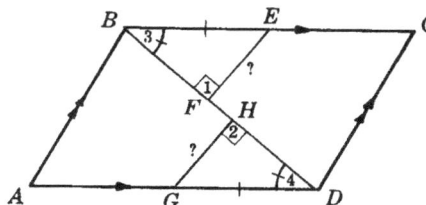

PROOF:

Statements	Reasons
1. *E* is midpoint of $\overline{BC}$. *G* is midpoint of $\overline{AD}$.	1. Given
2. $BE = \frac{1}{2}BC$, $GC = \frac{1}{2}AD$	2. A midpoint cuts a segment in half.
3. *ABCD* is a $\square$.	3. Given
4. $\overline{BC} \cong \overline{AD}$	4. Opposite sides of a $\square$ are congruent.
5. $\overline{BE} \cong \overline{GD}$	5. Halves of equals are equal.
6. $\overline{EF} \perp \overline{BD}$, $\overline{GH} \perp \overline{BD}$	6. Given
7. $\angle 1 \cong \angle 2$	7. Perpendiculars form rt. $\angle$. Rt. $\angle$ are $\cong$.
8. $\overline{BC} \parallel \overline{AD}$	8. Opposite sides of a $\square$ are $\parallel$.
9. $\angle 3 \cong \angle 4$	9. Alternate interior $\angle$ of $\parallel$ lines are $\cong$.
10. $\triangle BFE \cong \triangle GHD$	10. SAA
11. $\overline{EF} \cong \overline{GH}$	11. Corresponding parts of congruent $\triangle$ are $\cong$.

5.3　Special Parallelograms: Rectangle, Rhombus, and Square

5.3A　Definitions and Relationships among the Special Parallelograms

Rectangles, rhombuses, and squares belong to the set of parallelograms. Each of these may be defined as a parallelogram, as follows:

1. A *rectangle* is an equiangular parallelogram.
2. A *rhombus* is an equilateral parallelogram.
3. A *square* is an equilateral and equiangular parallelogram. Thus, a square is both a rectangle and a rhombus.

The relations among the special parallelograms can be pictured by using a circle to represent each set. Note the following in Fig. 5-13:

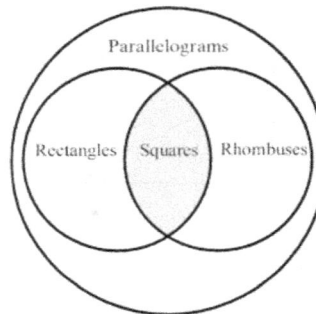

Fig. 5-13

1. Since every rectangle and every rhombus must be a parallelogram, the circle for the set of rectangles and the circle for the set of rhombuses must be inside the circle for the set of parallelograms.
2. Since every square is both a rectangle and a rhombus, the overlapping shaded section must represent the set of squares.

5.3B　Principles Involving Properties of the Special Parallelograms

PRINCIPLE 1:　*A rectangle, rhombus, or square has all the properties of a parallelogram.*

PRINCIPLE 2:　*Each angle of a rectangle is a right angle.*

PRINCIPLE 3:　*The diagonals of a rectangle are congruent.*

Thus in rectangle *ABCD* in Fig. 5-14, $\overline{AC} \cong \overline{BD}$.

PRINCIPLE 4:　*All sides of a rhombus are congruent.*

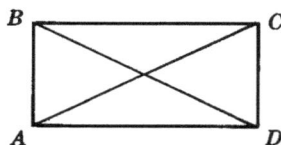

Rectangle

Fig. 5-14

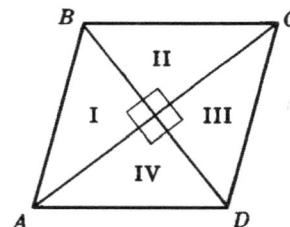

Rhombus

Fig. 5-15

PRINCIPLE 5: *The diagonals of a rhombus are perpendicular bisectors of each other.*

Thus in rhombus *ABCD* in Fig. 5-15, $\overline{AC}$ and $\overline{BD}$ are ⊥ bisectors of each other.

PRINCIPLE 6: *The diagonals of a rhombus bisect the vertex angles.*

Thus in rhombus *ABCD*, $\overline{AC}$ bisects ∠*A* and ∠*C*.

PRINCIPLE 7: *The diagonals of a rhombus form four congruent triangles.*

Thus in rhombus *ABCD*, △I ≅ △II ≅ △III ≅ △IV.

PRINCIPLE 8: *A square has all the properties of both the rhombus and the rectangle.*

By definition, a square is both a rectangle and a rhombus.

5.3C Diagonal Properties of Parallelograms, Rectangles, Rhombuses, and Squares

Each check in the following table indicates a diagonal property of the figure.

Diagonal Properties	Parallelogram	Rectangle	Rhombus	Square
Diagonals bisect each other.	✓	✓	✓	✓
Diagonals are congruent.		✓		✓
Diagonals are perpendicular.			✓	✓
Diagonals bisect vertex angles.			✓	✓
Diagonals form 2 pairs of congruent triangles.	✓	✓	✓	✓
Diagonals form 4 congruent triangles.			✓	✓

5.3D Proving that a Parallelogram is a Rectangle, Rhombus, or Square

Proving that a Parallelogram is a Rectangle

The basic or minimum definition of a rectangle is this: *A rectangle is a parallelogram having one right angle.* Since the consecutive angles of a parallelogram are supplementary, if one angle is a right angle, the remaining angles must be right angles.

The converse of this basic definition provides a useful method of proving that a parallelogram is a rectangle, as follows:

PRINCIPLE 9: *If a parallelogram has one right angle, then it is a rectangle.*

Thus if *ABCD* in Fig. 5-16 is a ▱ and $m\angle A = 90°$, then *ABCD* is a rectangle.

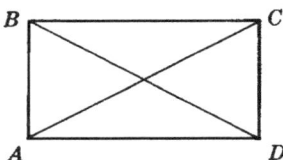

Fig. 5-16

PRINCIPLE 10: *If a parallelogram has congruent diagonals, then it is a rectangle.*

Thus if *ABCD* is a ▱ and $\overline{AC} \cong \overline{BD}$, then *ABCD* is a rectangle.

Proving that a Parallelogram is a Rhombus

The basic or minimum definition of a rhombus is this: *A rhombus is a parallelogram having two congruent adjacent sides.*

The converse of this basic definition provides a useful method of proving that a parallelogram is a rhombus, as follows:

PRINCIPLE 11: *If a parallelogram has congruent adjacent sides, then it is a rhombus.*

Thus if *ABCD* in Fig. 5-17 is a ▱ and $\overline{AB} \cong \overline{BC}$, then *ABCD* is a rhombus.

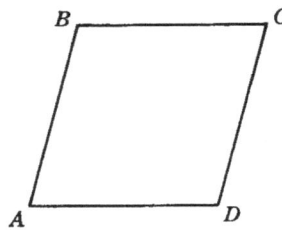

Fig. 5-17

Proving that a Parallelogram is a Square

PRINCIPLE 12: *If a parallelogram has a right angle and two congruent adjacent sides, then it is a square.*

This follows from the fact that a square is both a rectangle and a rhombus.

SOLVED PROBLEMS

5.7   Applying algebra to the rhombus

Assuming *ABCD* is a rhombus, find *x* and *y* in each part of Fig. 5-18.

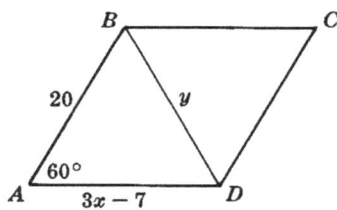

(*a*)

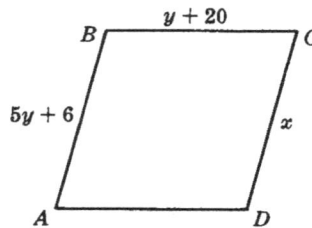

(*b*)

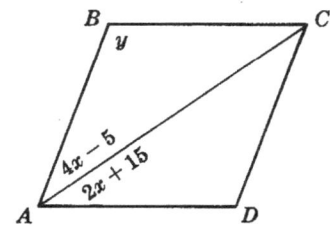

(*c*)

Fig. 5-18

Solutions

(a) Since $\overline{AB} \cong \overline{AD}$, $3x - 7 = 20$ or $x = 9$. Since $\triangle ABD$ is equiangular it is equilateral, and so $y = 20$.

(b) Since $\overline{BC} \cong \overline{AB}$, $5y + 6 = y + 20$ or $y = 3\frac{1}{2}$. Since $\overline{CD} \cong \overline{BC}$, $x = y + 20$ or $x = 23\frac{1}{2}$.

(c) Since $\overline{AC}$ bisects $\angle A$, $4x - 5 = 2x + 15$ or $x = 10$. Hence, $2x + 15 = 35$ and $m\angle A = 2(35°) = 70°$. Since $\angle B$ and $\angle A$ are supplementary, $y + 70 = 180$ or $y = 110$.

5.8   Proving a special parallelogram problem

Given: Rectangle *ABCD*
          E is midpoint of $\overline{BC}$.
To Prove: $\overline{AE} \cong \overline{ED}$
Plan: Prove $\triangle AEB \cong \triangle DEC$

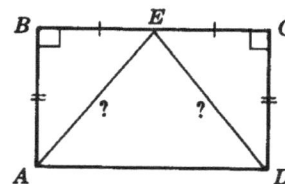

PROOF:

Statements	Reasons
1. *ABCD* is a rectangle.	1. Given
2. *E* is midpoint of $\overline{BC}$.	2. Given
3. $\overline{BE} \cong \overline{EC}$	3. A midpoint divides a line into two congruent parts.
4. $\angle B \cong \angle C$	4. A rectangle is equiangular.
5. $\overline{AB} \cong \overline{CD}$	5. Opposite sides of a ▱ are congruent.
6. $\triangle AEB \cong \triangle DEC$	6. SAS
7. $\overline{AE} \cong \overline{ED}$	7. Corresponding parts of congruent △ are congruent.

5.9 Proving a special parallelogram problem stated in words

Prove that a diagonal of a rhombus bisects each vertex angle through which it passes.

Solution

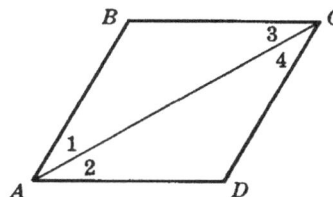

Given: Rhombus *ABCD*
         $\overline{AC}$ is a diagonal.
To Prove: $\overline{AC}$ bisects ∠*A and* ∠*C*.
Plan: Prove (1) ∠1 and ∠2 are congruent to ∠3.
            (2) ∠3 and ∠4 are congruent to ∠1.

PROOF:

Statements	Reasons
1. *ABCD* is a rhombus.	1. Given
2. $\overline{AB} \cong \overline{BC}$	2. A rhombus is equilateral.
3. ∠1 ≅ ∠3	3. In a △, angles opposite congruent sides are congruent.
4. $\overline{BC} \parallel \overline{AD}$, $\overline{AB} \parallel \overline{CD}$	4. Opposite sides of a ▱ are ∥.
5. ∠2 ≅ ∠3, ∠1 ≅ ∠4	5. Alternate interior ⦤ of ∥ lines are congruent.
6. ∠1 ≅ ∠2, ∠3 ≅ ∠4	6. Things congruent to the same thing are congruent to each other.
7. $\overline{AC}$ bisects ∠*A* and ∠*C*.	7. To divide into two congruent parts is to bisect.

5.4 Three or More Parallels; Medians and Midpoints

5.4A Three or More Parallels

PRINCIPLE 1: *If three or more parallels cut off congruent segments on one transversal, then they cut off congruent segments on any other transversal.*

Thus if $l_1 \parallel l_2 \parallel l_3$ in Fig. 5-19 and segments *a* and *b* of transversal $\overleftrightarrow{AB}$ are congruent, then segments *c* and *d* of transversal $\overleftrightarrow{CD}$ are congruent.

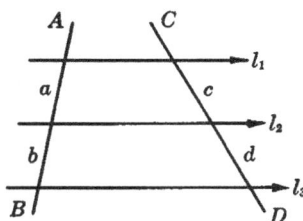

Fig. 5-19

5.4B Midpoint and Median Principles of Triangles and Trapezoids

PRINCIPLE 2: *If a line is drawn from the midpoint of one side of a triangle and parallel to a second side, then it passes through the midpoint of the third side.*

Thus in △*ABC* in Fig. 5-20 if *M* is the midpoint of $\overline{AB}$ and $\overline{MN} \parallel \overline{AC}$, then *N* is the midpoint of $\overline{BC}$.

PRINCIPLE 3: *If a line joins the midpoints of two sides of a triangle, then it is parallel to the third side and its length is one-half the length of the third side.*

Thus in △*ABC*, if *M* and *N* are the midpoints of $\overline{AB}$ and $\overline{BC}$, *then* $\overline{MN} \parallel \overline{AC}$ and $MN = \frac{1}{2} AC$.

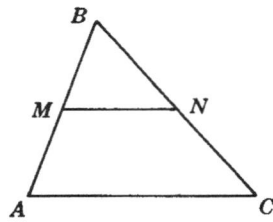

Fig. 5-20

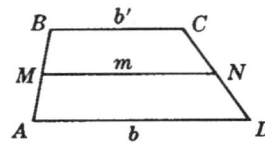

Fig. 5-21

PRINCIPLE 4:   *The median of a trapezoid is parallel to its bases, and its length is equal to one-half of the sum of their lengths.*

Thus if $\overline{MN}$ is the median of trapezoid $ABCD$ in Fig. 5-21 then $\overline{MN} \parallel \overline{AD}$, $\overline{MN} \parallel \overline{BC}$, and $m = \frac{1}{2}(b + b')$.

PRINCIPLE 5:   *The length of the median to the hypotenuse of a right triangle equals one-half the length of the hypotenuse.*

Thus in rt. $\triangle ABC$ in Fig. 5-22, if $\overline{CM}$ is the median to hypotenuse $\overline{AB}$, then $CM = \frac{1}{2}AB$; that is, $\overline{CM} \cong \overline{AM} \cong \overline{MB}$.

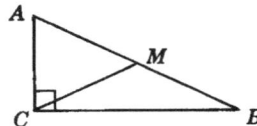

Fig. 5-22

PRINCIPLE 6:   *The medians of a triangle meet in a point which is two-thirds of the distance from any vertex to the midpoint of the opposite side.*

Thus if $\overline{AN}$, $\overline{BP}$, and $\overline{CM}$ are medians of $\triangle ABC$ in Fig. 5-23, then they meet in a point G which is two-thirds of the distance from A to N, B to P, and C to M.

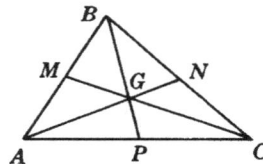

Fig. 5-23

SOLVED PROBLEMS

5.10   Applying principle 1 to three or more parallels

Find x and y in each part of Fig. 5-24.

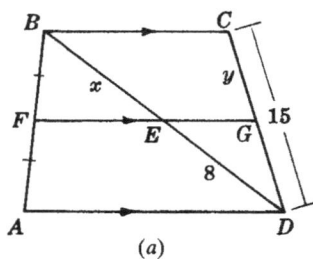

(a)

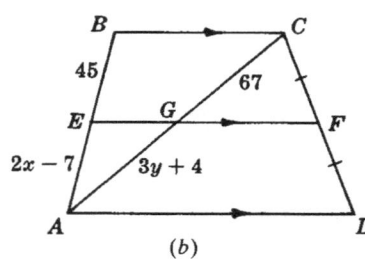

(b)

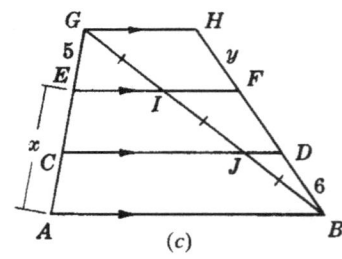

(c)

Fig. 5-24

Solutions

(a) Since $BE = ED$ and $GC = \frac{1}{2}CD$, $x = 8$ and $y = 7\frac{1}{2}$.

(b) Since $BE = EA$ and $CG = AG$, $2x - 7 = 45$ and $3y + 4 = 67$. Hence $x = 26$ and $y = 21$.

(c) Since $AC = CE = EG$ and $HF = FD = DB$, $x = 10$ and $y = 6$.

5.11 Applying principles 2 and 3

Find x and y in each part of Fig. 5-25.

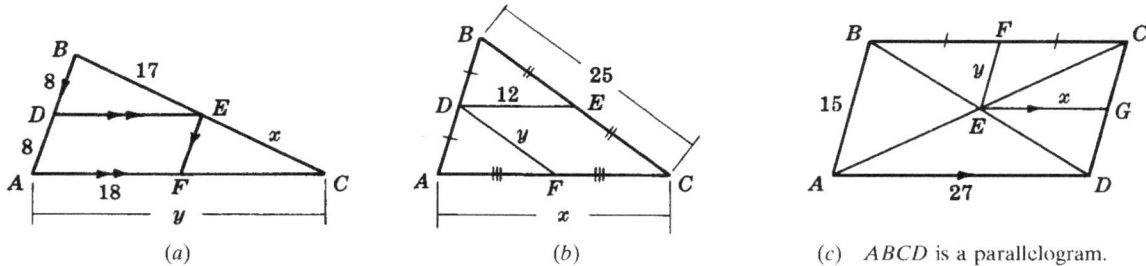

(a)　　　　　　　　　　(b)　　　　　　(c) *ABCD* is a parallelogram.

Fig. 5-25

Solutions

(a) By Principle 2, E is the midpoint of $\overline{BC}$ and F is the midpoint of $\overline{AC}$. Hence $x = 17$ and $y = 36$.

(b) By Principle 3, $DE = \frac{1}{2}AC$ and $DF = \frac{1}{2}BC$. Hence $x = 24$ and $y = 12\frac{1}{2}$.

(c) Since $ABCD$ is a parallelogram, E is the midpoint of $\overline{AC}$. Then by Principle 2, G is the midpoint of $\overline{CD}$. By Principle 3, $x = \frac{1}{2}(27) = 13\frac{1}{2}$ and $y = \frac{1}{2}(15) = 7\frac{1}{2}$.

5.12 Applying principle 4 to the median of a trapezoid

If $\overline{MP}$ is the median of trapezoid $ABCD$ in Fig. 5-26,

(a) Find m if $b = 20$ and $b' = 28$.

(b) Find b' if $b = 30$ and $m = 26$.

(c) Find b if $b' = 35$ and $m = 40$.

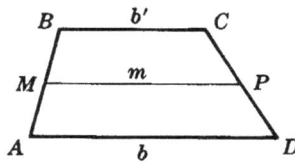

Fig. 5-26

Solutions

In each case, we apply the formula $m = \frac{1}{2}(b + b')$. The results are

(a) $m = \frac{1}{2}(20 + 28)$ or $m = 24$

(b) $26 = \frac{1}{2}(30 + b')$ or $b' = 22$

(c) $40 = \frac{1}{2}(b + 35)$ or $b = 45$

5.13 Applying principles 5 and 6 to the medians of a triangle

Find x and y in each part of Fig. 5-27.

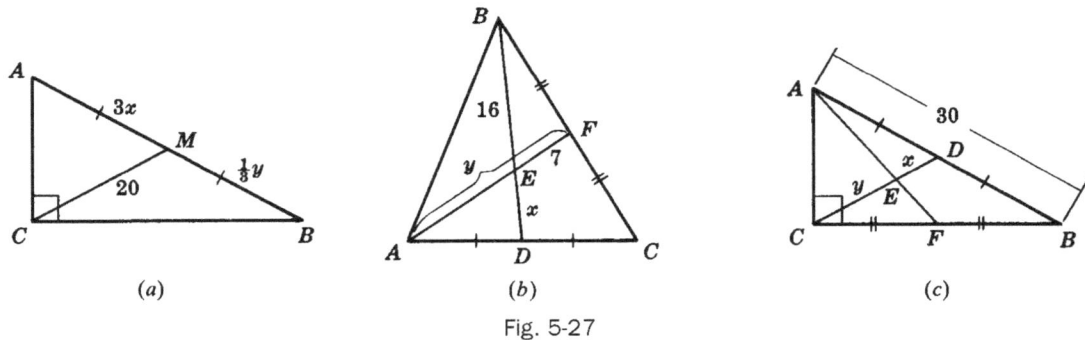

(a) (b) (c)

Fig. 5-27

Solutions

(a) Since $AM = MB$, $\overline{CM}$ is the median to hypotenuse $\overline{AB}$. Hence by Principle 5, $3x = 20$ and $\frac{1}{3}y = 20$. Thus, $x = 6\frac{2}{3}$ and $y = 60$.

(b) $\overline{BD}$ and $\overline{AF}$ are medians of $\triangle ABC$. Hence by Principle 6, $x = \frac{1}{2}(16) = 8$ and $y = 3(7) = 21$.

(c) $\overline{CD}$ is the median to hypotenuse $\overline{AB}$; hence by Principle 5, $CD = 15$.

$\overline{CD}$ and $\overline{AF}$ are medians of $\triangle ABC$; hence by Principle 6, $x = \frac{1}{3}(15) = 5$ and $y = \frac{2}{3}(15) = 10$.

5.14 Proving a midpoint problem

Given: Quadrilateral $ABCD$
$E, F, G,$ and H are midpoints of $\overline{AD}, \overline{AB}, \overline{BC},$ and $\overline{CD}$, respectively.
To Prove: $EFGH$ is a $\square$.
Plan: Prove $\overline{EF}$ and $\overline{GH}$ are congruent and parallel.

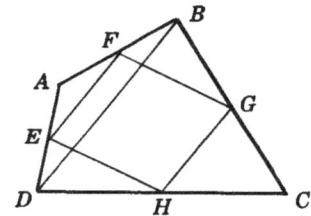

PROOF:

Statements	Reasons
1. Draw $\overline{BD}$.	1. A segment may be drawn between any two points.
2. $E, F, G,$ and H are midpoints.	2. Given
3. $\overline{EF} \| \overline{BD}$ and $\overline{GH} \| \overline{BD}$ $EF = \frac{1}{2}BD$ and $GH = \frac{1}{2}BD$	3. A line segment joining the midpoints of two sides of a $\triangle$ is parallel to the third side and equal in length to half the third side.
4. $\overline{EF} \| \overline{GH}$	4. Two lines parallel to a third line are parallel to each other.
5. $\overline{EF} \cong \overline{GH}$	5. Segments of the same length are congruent.
6. $EFGH$ is a $\square$.	6. If two sides of a quadrilateral are $\cong$ and $\|$, the quadrilateral is a $\square$.

SUPPLEMENTARY PROBLEMS

5.1. Find x and y in each part of Fig. 5-28. (5.1)

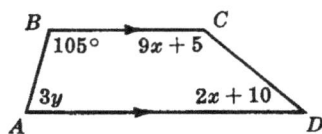

(a) $ABCD$ is a trapezoid. (b) $ABCD$ is an isosceles trapezoid. (c) $ABCD$ is an isosceles trapezoid.

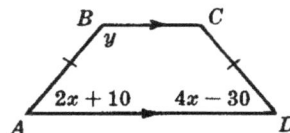

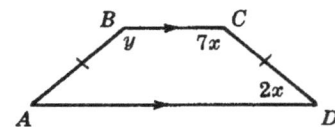

Fig. 5-28

5.2. Prove that if the base angles of a trapezoid are congruent, the trapezoid is isosceles. (5.2)

5.3. Prove that (a) the diagonals of an isosceles trapezoid are congruent; (b) if the nonparallel sides $\overline{AB}$ and $\overline{CD}$ of an isosceles trapezoid are extended until they meet at E, triangle ADE thus formed is isosceles. (5.2)

5.4. Name the parallelograms in each part of Fig. 5-29. (5.4)

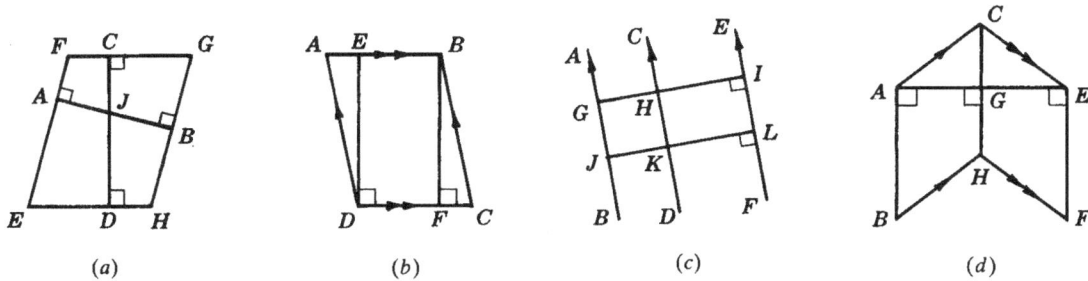

(a) (b) (c) (d)

Fig. 5-29

5.5. State why *ABCD* in each part of Fig. 5-30 is a parallelogram. (5.5)

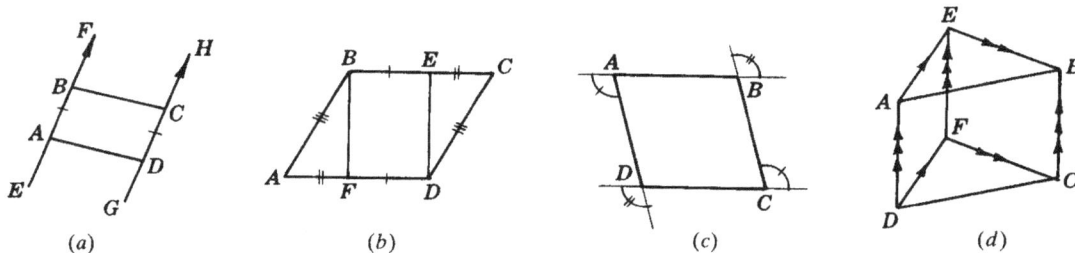

(a) (b) (c) (d)

Fig. 5-30

5.6. Assuming *ABCD* in Fig. 5-31 is a parallelogram, find *x* and *y* if (5.3)

(a) $AD = 5x, AB = 2x, CD = y$, perimeter $= 84$

(b) $AB = 2x, BC = 3y + 8, CD = 7x - 25, AD = 5y - 10$

(c) $m\angle A = 4y - 60, m\angle C = 2y, m\angle D = x$

(d) $m\angle A = 3x, m\angle B = 10x - 15, m\angle C = y$

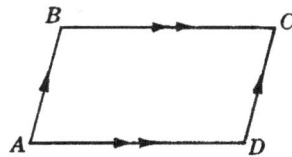 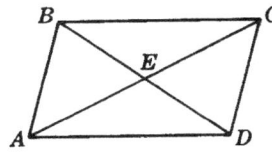

Fig. 5-31   Fig. 5-32

5.7. Assuming *ABCD* in Fig. 5-32 is a parallelogram, find *x* and *y* if (5.3)

(a) $AE = x + y, EC = 20, BE = x - y, ED = 8$

(b) $AE = x, EC = 4y, BE = x - 2y, ED = 9$

(c) $AE = 3x - 4, EC = x + 12, BE = 2y - 7, ED = x - y$

(d) $AE = 2x + y, AC = 30, BE = x + y, BD = 24$

5.8. Provide the proofs requested in Fig. 5-33.   (5.6)

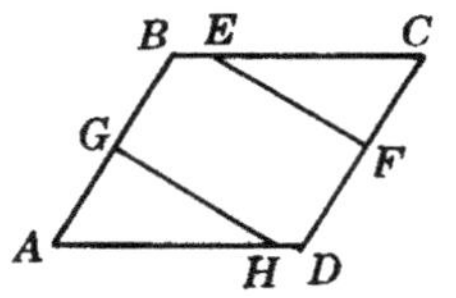

(a) **Given:** $\square ABCD$
  G is midpoint of $\overline{AB}$.
  F is midpoint of $\overline{CD}$.
  $\overline{HG} \perp AB, \overline{EF} \perp CD$
To Prove: $\overline{EF} \cong \overline{GH}$

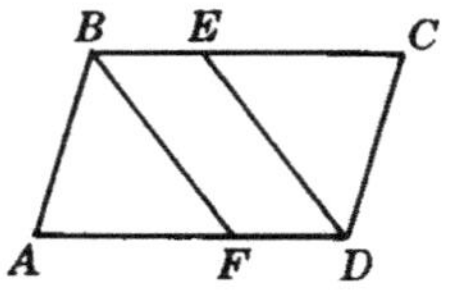

(b) **Given:** $\square ABCD$
  $\overline{BE} \cong \overline{FD}$
To Prove: $\overline{BF} \cong \overline{ED}$

(c) **Given:** $\square ABCD$
  $\overline{BF}$ bisects $\angle B$,
  $\overline{ED}$ bisects $\angle D$.
To Prove: $\overline{BF} \cong \overline{ED}$

Fig. 5-33

5.9. Prove each of the following:

(a) The opposite sides of a parallelogram are congruent (Principle 3).

(b) If the opposite sides of a quadrilateral are congruent, then the quadrilateral is a parallelogram (Principle 8).

(c) If two sides of a quadrilateral are congruent and parallel, the quadrilateral is a parallelogram (Principle 9).

(d) The diagonals of a parallelogram bisect each other (Principle 6).

(e) If the diagonals of a quadrilateral bisect each other, then the quadrilateral is a parallelogram (Principle 11).

5.10. Assuming $ABCD$ in Fig. 5-34 is a rhombus, find x and y if   (5.7)

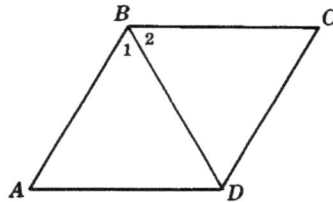

Fig. 5-34

(a) $BC = 35, CD = 8x - 5, BD = 5y, m\angle C = 60°$

(b) $AB = 43, AD = 4x + 3, BD = y + 8, m\angle B = 120°$

(c) $AB = 7x, AD = 3x + 10, BC = y$

(d) $AB = x + y, AD = 2x - y, BC = 12$

(e) $m\angle B = 130°, m\angle 1 = 3x - 10, m\angle A = 2y$

(f) $m\angle 1 = 8x - 29, m\angle 2 = 5x + 4, m\angle D - y$

5.11. Provide the proofs requested in Fig. 5-35.   (5.8)

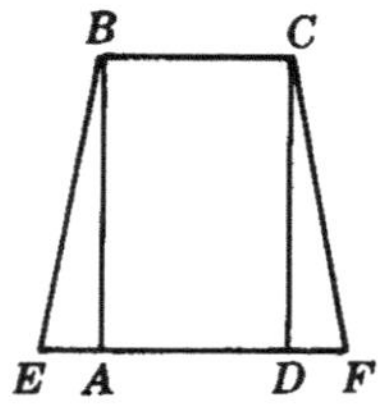

(a) **Given:** Rectangle $ABCD$
  $\overline{EA} \cong \overline{DF}$
To Prove: $\overline{BE} \cong \overline{CF}$

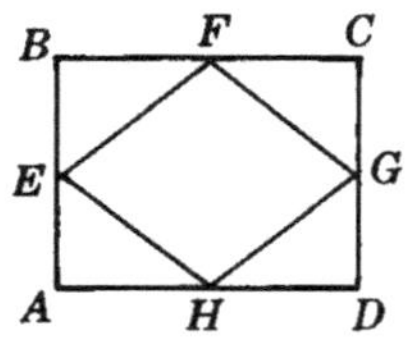

(b) **Given:** Rectangle $ABCD$
  $E, F, G,$ and H are the midpoints of the sides of the rectangle.
To Prove: $EFGH$ is a rhombus.

Fig. 5-35

5.12. Prove each of the following: (5.9)

(a) If the diagonals of a parallelogram are congruent, the parallelogram is a rectangle.

(b) If the diagonals of a parallelogram are perpendicular to each other, the parallelogram is a rhombus.

(c) If a diagonal of a parallelogram bisects a vertex angle, then the parallelogram is a rhombus.

(d) The diagonals of a rhombus divide it into four congruent triangles.

(e) The diagonals of a rectangle are congruent.

5.13. Find x and y in each part of Fig. 5-36. (5.10)

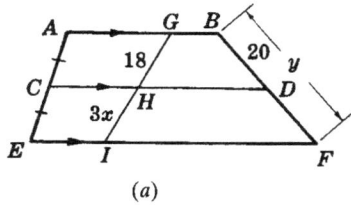

(a)

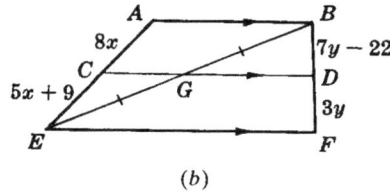

(b)

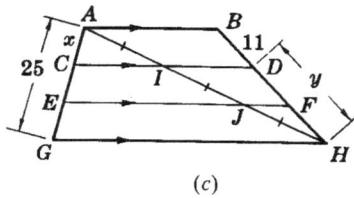

(c)

Fig. 5-36

5.14. Find x and y in each part of Fig. 5-37. (5.11)

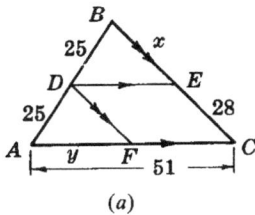

(a)

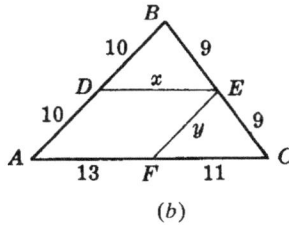

(b)

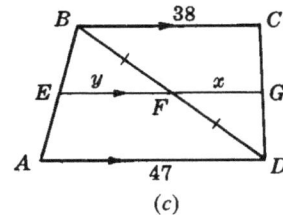

(c)

Fig. 5-37

5.15. If $\overline{MP}$ is the median of trapezoid $ABCD$ in Fig. 5-38 (5.12)

(a) Find m if $b = 23$ and $b' = 15$.

(b) Find b' if $b = 46$ and $m = 41$.

(c) Find b if $b' = 51$ and $m = 62$.

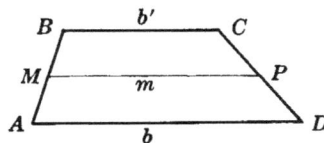

Fig. 5-38

5.16. Find x and y in each part of Fig. 5-39.   (5.11 and 5.12)

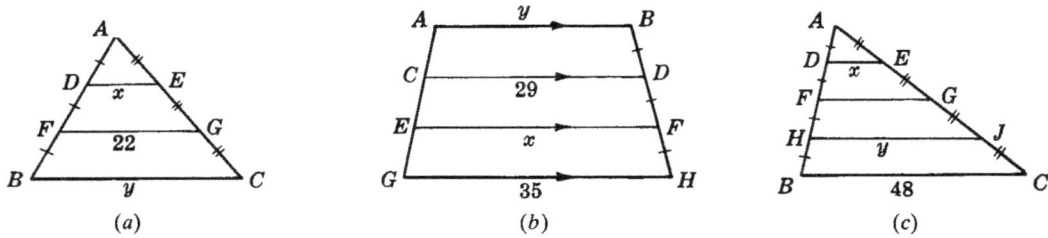

(a)   *(b)*   *(c)*

Fig. 5-39

5.17. In a right triangle   (5.13)

     (a) Find the length of the median to a hypotenuse whose length is 45.

     (b) Find the length of the hypotenuse if the length of its median is 35.

5.18. If the medians of $\triangle ABC$ meet in D (5.13)

     (a) Find the length of the median whose shorter segment is 7.

     (b) Find the length of the median whose longer segment is 20.

     (c) Find the length of the shorter segment of the median of length 42.

     (d) Find the length of the longer segment of the median of length 39.

5.19. Prove each of the following: (5.14)

     (a) If the midpoints of the sides of a rhombus are joined in order, the quadrilateral formed is a rectangle.

     (b) If the midpoints of the sides of a square are joined in order, the quadrilateral formed is a square.

     (c) In $\triangle ABC$, let M, P, and Q be the midpoints of $\overline{AB}$, $\overline{BC}$, and $\overline{AC}$, respectively. Prove that $QMPC$ is a parallelogram.

     (d) In right $\triangle ABC$, $m\angle C = 90°$. If Q, M, and P are the midpoints of $\overline{AC}$, $\overline{AB}$, and $\overline{BC}$, respectively, prove that $QMPC$ is a rectangle.

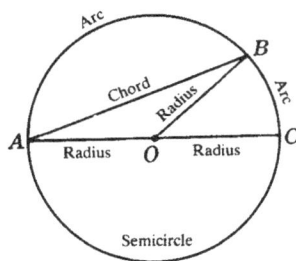

The oval at the top right contains:

CHAPTER 6

Circles

6.1 The Circle; Circle Relationships

The following terms are associated with the circle. Although some have been defined previously, they are repeated here for ready reference.

A *circle* is the set of all points in a plane that are at the same distance from a fixed point called the center. The symbol for circle is ⊙; for circles ⊚.

The *circumference* of a circle is the distance around the circle. It contains 360°.

A *radius* of a circle is a line segment joining the center to a point on the circle.

Note: Since all radii of a given circle have the same length, we may at times use the word *radius* to mean the number that is "the length of the radius."

A *central angle* is an angle formed by two radii.

An *arc* is a continuous part of a circle. The symbol for arc is ⌒. A *semicircle* is an arc measuring one-half the circumference of a circle.

A *minor arc* is an arc that is less than a semicircle. A *major arc* is an arc that is greater than a semicircle.

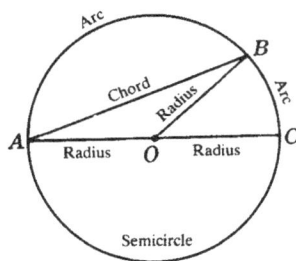

Fig. 6-1

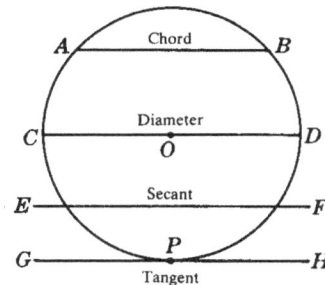

Fig. 6-2

Thus in Fig. 6-1, $\overset{\frown}{BC}$ is a minor arc and $\overset{\frown}{BAC}$ is a major arc. Three letters are needed to indicate a major arc.

To intercept an arc is to cut off the arc.

Thus in Fig. 6-1, ∠BAC and ∠BOC intercept $\overset{\frown}{BC}$.

A *chord* of a circle is a line segment joining two points of the circumference.

Thus in Fig. 6-2, $\overline{AB}$ is a chord.

A *diameter* of a circle is a chord through the center. A *secant* of a circle is a line that intersects the circle at two points. A *tangent* of a circle is a line that touches the circle at one and only one point no matter how far produced.

Thus, $\overline{CD}$ is a diameter of circle O in Fig. 6-2, $\overleftrightarrow{EF}$ is a secant, and $\overleftrightarrow{GH}$ is a tangent to the circle at P. P is the point of contact or the point of tangency.

An *inscribed polygon* is a polygon all of whose sides are chords of a circle. A *circumscribed circle* is a circle passing through each vertex of a polygon.

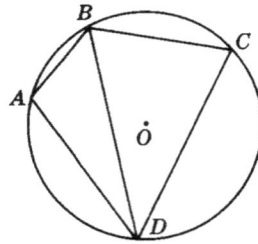

Inscribed Polygons
Circumscribed Circle
Fig. 6-3

Thus △*ABD*, △*BCD*, and quadrilateral *ABCD* are inscribed polygons of circle O in Fig. 6-3. Circle O is a circumscribed circle of quadrilateral *ABCD*.

A *circumscribed polygon* is a polygon all of whose sides are tangents to a circle. An *inscribed circle* is a circle to which all the sides of a polygon are tangents.

Thus, △*ABC* is a circumscribed polygon of circle O in Fig. 6-4. Circle O is an inscribed circle of △*ABC*.

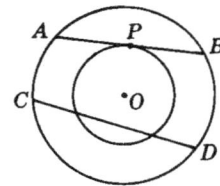

Concentric circles are circles that have the same center.

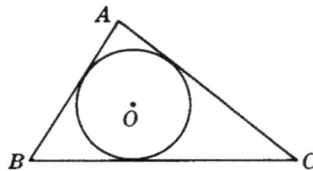

Circumscribed Polygon
Inscribed Circle
Fig. 6-4

Concentric Circles
Fig. 6-5

Thus, the two circles shown in Fig. 6-5 are concentric circles. $\overline{AB}$ is a tangent of the inner circle and a chord of the outer one. $\overline{CD}$ is a secant of the inner circle and a chord of the outer one.

Two circles are *equal* if their radii are equal in length; two circles are *congruent* if their radii are congruent.

Two arcs are congruent if they have equal degree measure and length. We use the notation $m\overarc{AC}$ to denote "measure of arc *AC*."

6.1A Circle Principles

PRINCIPLE 1: *A diameter divides a circle into two equal parts.*

Thus, diameter $\overline{AB}$ divides circle O of Fig. 6-6 into two congruent semicircles, $\overarc{ACB}$ and $\overarc{ADB}$.

PRINCIPLE 2:   *If a chord divides a circle into two equal parts, then it is a diameter.* (This is the converse of Principle 1.)

Thus if $\overset{\frown}{ACB} \cong \overset{\frown}{ADB}$ in Fig. 6-6, then $\overline{AB}$ is a diameter.

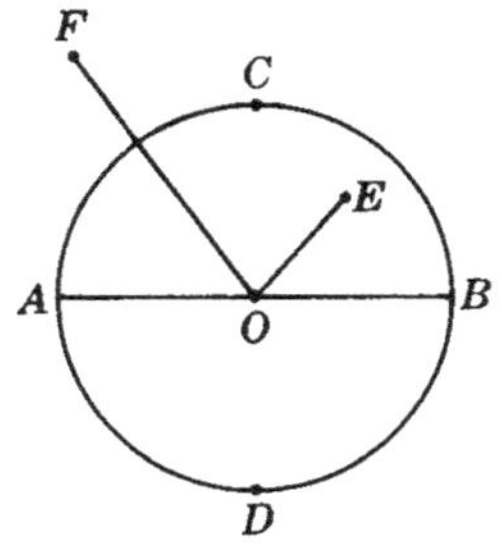

Fig. 6-6

PRINCIPLE 3:   *A point is outside, on, or inside a circle according to whether its distance from the center is greater than, equal to, or smaller than the radius.*

F is outside circle O in Fig. 6-6, since $\overline{FO}$ is greater in length than a radius. E is inside circle O since $\overline{EO}$ is smaller in length than a radius. A is on circle O since $\overline{AO}$ is a radius.

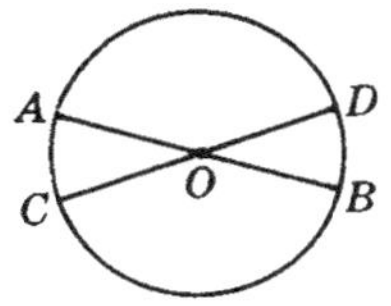

Fig. 6-7

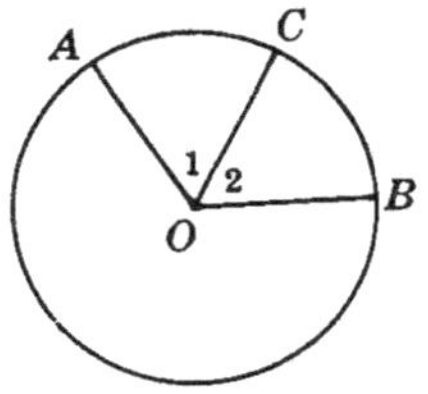

Fig. 6-8

PRINCIPLE 4:   *Radii of the same or congruent circles are congruent.*

Thus in circle O of Fig. 6-7, $\overline{OA} \cong \overline{OC}$.

PRINCIPLE 5:   *Diameters of the same or congruent circles are congruent.*

Thus in circle O of Fig. 6-7, $\overline{AB} \cong \overline{CD}$.

PRINCIPLE 6:   *In the same or congruent circles, congruent central angles have congruent arcs.*

Thus in circle O of Fig. 6-8, if $\angle 1 \cong \angle 2$, then $\overset{\frown}{AC} \cong \overset{\frown}{CB}$.

PRINCIPLE 7:   *In the same or congruent circles, congruent arcs have congruent central angles.*

Thus in circle O of Fig. 6-8, if $\overset{\frown}{AC} \cong \overset{\frown}{CB}$, then $\angle 1 \cong \angle 2$.

(Principles 6 and 7 are converses of each other.)

PRINCIPLE 8:   *In the same or congruent circles, congruent chords have congruent arcs.*

Thus in circle O of Fig. 6-9, if $\overline{AB} \cong \overline{AC}$, then $\overset{\frown}{AB} \cong \overset{\frown}{AC}$.

PRINCIPLE 9: *In the same or congruent circles, congruent arcs have congruent chords.*

Thus in circle O of Fig. 6-9, if $\overarc{AB} \cong \overarc{AC}$, then $\overline{AB} \cong \overline{AC}$.

(Principles 8 and 9 are converses of each other.)

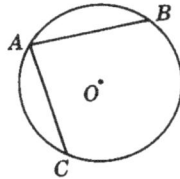

Fig. 6-9

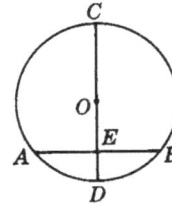

Fig. 6-10

PRINCIPLE 10: *A diameter perpendicular to a chord bisects the chord and its arcs.*

Thus in circle O of Fig. 6-10, if $\overline{CD} \perp \overline{AB}$, then $\overline{CD}$ bisects $\overline{AB}$, $\overarc{AB}$, and $\overarc{ACB}$.

A proof of this principle is given in Chapter 16.

PRINCIPLE 11: *A perpendicular bisector of a chord passes through the center of the circle.*

Thus in circle O of Fig. 6-11, if $\overline{PD}$ is the perpendicular bisector of $\overline{AB}$, then $\overline{PD}$ passes through center O.

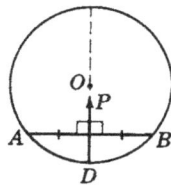

Fig. 6-11

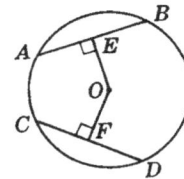

Fig. 6-12

PRINCIPLE 12: *In the same or congruent circles, congruent chords are equally distant from the center.*

Thus in circle O of Fig. 6-12, if $\overline{AB} \cong \overline{CD}$, if $\overline{OE} \perp \overline{AB}$, and if $\overline{OF} \perp \overline{CD}$, then $\overline{OE} \cong \overline{OF}$.

PRINCIPLE 13: *In the same or congruent circles, chords that are equally distant from the center are congruent.*

Thus in circle O of Fig. 6-12, if $\overline{OE} \cong \overline{OF}$, $\overline{OE} \perp \overline{AB}$, and $\overline{OF} \perp \overline{CD}$, then $\overline{AB} \cong \overline{CD}$.

(Principles 12 and 13 are converses of each other.)

SOLVED PROBLEMS

6.1 Matching test of circle vocabulary

Match each part of Fig. 6-13 on the left with one of the names on the right:

(a) $\overline{OE}$ 1. Radius

(b) $\overline{FG}$ 2. Central angle

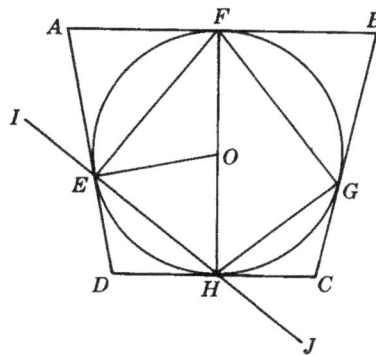

Fig. 6-13

(c) $\overline{FH}$	3. Semicircle
(d) $\overline{CD}$	4. Minor arc
(e) $\overline{IJ}$	5. Major arc
(f) $\overset{\frown}{EF}$	6. Chord
(g) $\overset{\frown}{FGH}$	7. Diameter
(h) $\overset{\frown}{FEG}$	8. Secant
(i) $\angle EOF$	9. Tangent
(j) Circle O about $EFGH$	10. Inscribed polygon
(k) Circle O in $ABCD$	11. Circumscribed polygon
(l) Quadrilateral $EFGH$	12. Inscribed circle
(m) Quadrilateral $ABCD$	13. Circumscribed circle

Solutions

(a) 1 (c) 7 (e) 8 (g) 3 (i) 2 (k) 12 (m) 11

(b) 6 (d) 9 (f) 4 (h) 5 (j) 13 (l) 10

6.2 Applying principles 4 and 5

In Fig. 6-14, (a) what kind of triangle is OCD; (b) what kind of quadrilateral is $ABCD$? (c) In Fig. 6-15 if circle O = circle Q, what kind of quadrilateral is $OAQB$?

Solutions

Radii or diameters of the same or equal circles have equal lengths.

(a) Since $\overline{OC} \cong \overline{OD}$, $\triangle OCD$ is isosceles.

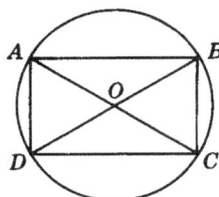

Fig. 6-14

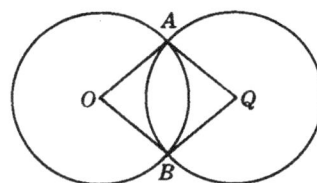

Fig. 6-15

(b) Since diagonals $\overline{AC}$ and $\overline{BD}$ are equal in length and bisect each other, $ABCD$ is a rectangle.

(c) Since the circles are equal, $\overline{OA} \cong \overline{AQ} \cong \overline{QB} \cong \overline{BO}$ and $OAQB$ is a rhombus.

6.3   Proving a circle problem

Given: $\overline{AB} \cong \overline{DE}$
  $\overline{BC} \cong \overline{EF}$
To Prove: $\angle B \cong \angle E$
Plan: Prove $\triangle I \cong \triangle II$.

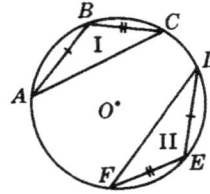

PROOF:

Statements	Reasons
1. $\overline{AB} \cong \overline{DE}, \overline{BC} \cong \overline{EF}$	1. Given
2. $\overarc{AB} \cong \overarc{DE}, \overarc{BC} \cong \overarc{EF}$	2. In a circle, $\cong$ chords have $\cong$ arcs.
3. $\overarc{ABC} \cong \overarc{DEF}$	3. If equals are added to equals, the sums are equal. Definition of $\cong$ arcs.
4. $\overline{AC} \cong \overline{DF}$	4. In a circle, $\cong$ arcs have $\cong$ chords.
5. $\triangle I \cong \triangle II$	5. SSS
6. $\angle B \cong \angle E$	6. Corresponding parts of congruent $\triangle$ are $\cong$.

6.4   Proving a circle problem stated in words

Prove that if a radius bisects a chord, then it is perpendicular to the chord.

Solution

Given: Circle O
  $\overline{OC}$ bisects $\overline{AB}$.
To Prove: $\overline{OC} \perp \overline{AB}$
Plan: Prove $\triangle AOD \cong \triangle BOD$ to show $\angle 1 \cong \angle 2$.
  Also, $\angle 1$ and $\angle 2$ are supplementary.

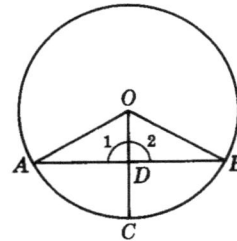

PROOF:

Statements	Reasons
1. Draw $\overline{OA}$ and $\overline{OB}$.	1. A straight line segment may be drawn between any two points.
2. $\overline{OA} \cong \overline{OB}$	2. Radii of a circle are congruent.
3. $\overline{OC}$ bisects $\overline{AB}$.	3. Given
4. $\overline{AD} \cong \overline{DB}$	4. To bisect is to divide into two $\cong$ parts.
5. $\overline{OD} \cong \overline{OD}$	5. Reflexive property
6. $\triangle AOD \cong \triangle BOD$	6. SSS
7. $\angle 1 \cong \angle 2$	7. Corresponding parts of congruent $\triangle$ are $\cong$.
8. $\angle 1$ is the supplement of $\angle 2$.	8. Adjacent $\angle$ are supplementary if exterior sides lie in a straight line.
9. $\angle 1$ and $\angle 2$ are right angles.	9. Congruent supplementary angles are right angles.
10. $\overline{OC} \perp \overline{AB}$.	10. Rt. $\angle$ are formed by perpendiculars.

6.2   Tangents

The *length of a tangent* from a point to a circle is the length of the segment of the tangent from the given point to the point of tangency. Thus, *PA* is the length of the tangent from *P* to circle *O* in Fig. 6-16.

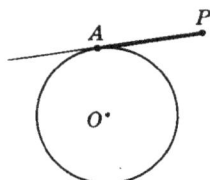

Fig. 6-16

6.2A   Tangent Principles

PRINCIPLE 1:   *A tangent is perpendicular to the radius drawn to the point of contact.*

Thus if $\overleftrightarrow{AB}$ is a tangent to circle *O* at *P* in Fig. 6-17, and $\overline{OP}$ is drawn, then $\overline{AB} \perp \overline{OP}$.

PRINCIPLE 2:   *A line is tangent to a circle if it is perpendicular to a radius at its outer end.*

Thus if $\overleftrightarrow{AB} \perp$ radius $\overline{OP}$ at *P* of Fig. 6-17, then $\overleftrightarrow{AB}$ is tangent to circle *O*.

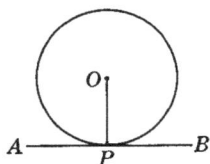   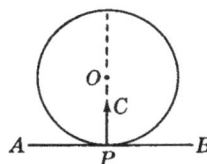

Fig. 6-17   Fig. 6-18

PRINCIPLE 3:   *A line passes through the center of a circle if it is perpendicular to a tangent at its point of contact.*

Thus if $\overleftrightarrow{AB}$ is tangent to circle *O* at *P* in Fig. 6-18, and $\overline{CP} \perp \overleftrightarrow{AB}$ at *P*, then $\overline{CP}$ extended will pass through the center *O*.

PRINCIPLE 4:   *Tangents to a circle from an outside point are congruent.*

Thus if $\overline{AP}$ and $\overline{AQ}$ are tangent to circle *O* at *P* and *Q* (Fig. 6-19), then $\overline{AP} \cong \overline{AQ}$.

PRINCIPLE 5:   *The segment from the center of a circle to an outside point bisects the angle between the tangents from the point to the circle.*

Thus $\overline{OA}$ bisects $\angle PAQ$ in Fig. 6-19 if $\overline{AP}$ and $\overline{AQ}$ are tangents to circle *O*.

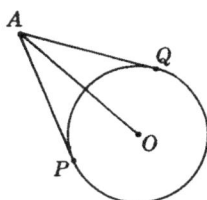

Fig. 6-19

6.2B   Two Circles in Varying Relative Positions

The *line of centers of two circles* is the line joining their centers. Thus, $\overline{OO'}$ is the line of centers of circles O and O' in Fig. 6-20.

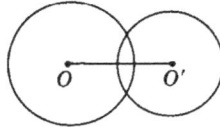

Fig. 6-20

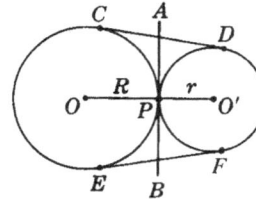

Fig. 6-21

Circles Tangent Externally

Circles O and O' in Fig. 6-21 are tangent externally at P. $\overleftrightarrow{AB}$ is the common internal tangent of both circles. The line of centers $\overline{OO'}$ passes through P, is perpendicular to $\overleftrightarrow{AB}$, and is equal in length to the sum of the radii, $R + r$. Also $\overleftrightarrow{AB}$ bisects each of the common external tangents, $\overline{CD}$ and $\overline{EF}$.

Circles Tangent Internally

Circles O and O' in Fig. 6-22 are tangent internally at P. $\overleftrightarrow{AB}$ is the common external tangent of both circles. The line of centers $\overline{OO'}$ if extended passes through P, is perpendicular to $\overleftrightarrow{AB}$, and is equal in length to the difference of the radii, $R - r$.

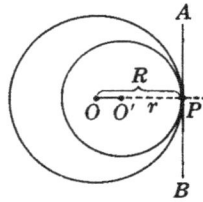

Fig. 6-22

Overlapping Circles

Circles O and O' in Fig. 6-23 overlap. Their common chord is $\overline{AB}$. If the circles are unequal, their (equal) common external tangents $\overrightarrow{CD}$ and $\overrightarrow{EF}$ meet at P. The line of centers $\overline{OO'}$ is the perpendicular bisector of $\overline{AB}$ and, if extended, passes through P.

Circles Outside Each Other

Circles O and O' in Fig. 6-24 are entirely outside each other. The common internal tangents, $\overline{AB}$ and $\overline{CD}$ meet at P. If the circles are unequal, their common external tangents, $\overline{EF}$ and $\overline{GH}$ if extended, meet at P'. The line of centers $\overline{OO'}$ passes through P and P'. Also, $AB = CD$ and $EF = GH$.

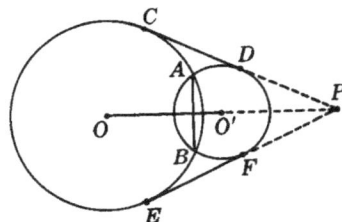

Fig. 6-23

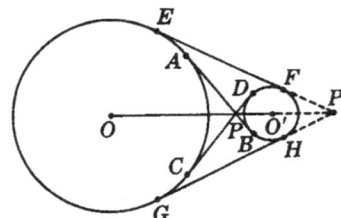

Fig. 6-24

SOLVED PROBLEMS

6.5 Triangles and quadrilaterals having tangent sides

Points *P*, *Q*, and R in Fig. 6-25 are points of tangency.

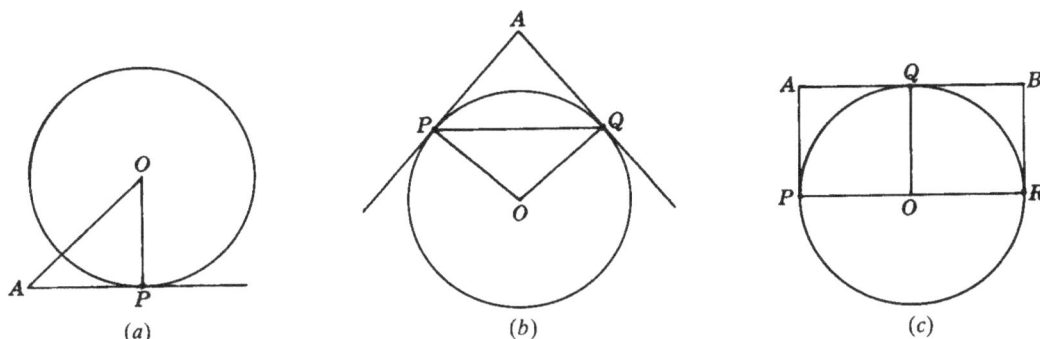

Fig. 6-25

(a) In Fig. 6-25(a), if $AP = OP$, what kind of triangle is OPA?

(b) In Fig. 6-25(b), if $AP = PQ$, what kind of triangle is APQ?

(c) In Fig. 6-25(b), if $AP = OP$, what kind of quadrilateral is $OPAQ$?

(d) In Fig. 6-25(c), if $\overline{OQ} \perp \overline{PR}$, what kind of quadrilateral is $PABR$?

Solutions

(a) $\overline{AP}$ is tangent to the circle at *P*; then by Principle 1, $\angle OPA$ is a right angle. Also, $AP = OP$. Hence, $\triangle OAP$ is an isosceles right triangle.

(b) $\overline{AP}$ and $\overline{AQ}$ are tangents from a point to the circle; hence by Principle 4, $AP = AQ$. Also, $AP = PQ$. Then $\triangle APQ$ is an equilateral triangle.

(c) By Principle 4, $AP = AQ$. Also, $\overline{OP}$ and $\overline{OQ}$ are $\cong$ radii. And $AP = OP$. By Principle 1, $\angle APO$ is a rt. $\angle$. Then $AP = AQ = OP = OQ$; hence, $OPAQ$ is a rhombus with a right angle, or a square.

(d) By Principle 1, $\overline{AP} \perp \overline{PR}$ and $\overline{BR} \perp \overline{PR}$. Then $\overline{AP} \parallel \overline{BR}$, since both are $\perp$ to $\overline{PR}$. By Principle 1, $\overline{AB} \perp \overline{OQ}$; also, $\overline{PR} \perp \overline{OQ}$ (Given). Then $\overline{AB} \parallel \overline{PR}$, since both are $\perp$ to $\overline{OQ}$. Hence, $PABR$ is a parallelogram with a right angle, or a rectangle.

6.6 Applying principle 1

(a) In Fig. 6-26(a), $\overline{AP}$ is a tangent. Find $\angle A$ if $m\angle A : m\angle O = 2{:}3$.

(b) In Fig. 6-26(b), $\overline{AP}$ and $\overline{AQ}$ are tangents. Find $m\angle 1$ if $m\angle O = 140°$.

(c) In Fig. 6-26(c), $\overline{DP}$ and $\overline{CQ}$ are tangents. Find $m\angle 2$ and $m\angle 3$ if $\angle OPD$ is trisected and $\overline{PQ}$ is a diameter.

Solutions

(a) By Principle 1, $m\angle P = 90°$. Then $m\angle A + m\angle O = 90°$. If $m\angle A = 2x$ and $m\angle O = 3x$, then $5x = 90$ and $x = 18$. Hence, $m \angle A = 36°$.

CHAPTER 6 Circles

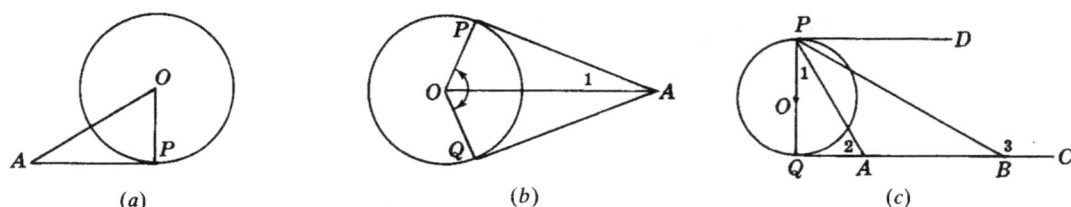

Fig. 6-26

(b) By Principle 1, $m\angle P = m\angle Q = 90°$. Since $m\angle P + m\angle Q + m\angle O + m\angle A = 360°$, $m\angle A + m\angle O = 180°$. Since $m\angle O = 140°$, $m\angle A = 40°$. By Principle 5, $m\angle 1 = \frac{1}{2} m\angle A = 20°$.

(c) By Principle 1, $m\angle DPQ = m\angle PQC = 90°$. Since $m\angle 1 = 30°$, $m\angle 2 = 60°$. Since $\angle 3$ is an exterior angle of $\triangle PBQ$, $m\angle 3 = 90° + 60° = 150°$.

6.7 Applying principle 4

(a) $\overline{AP}$, $\overline{BQ}$, and $\overline{AB}$ in Fig. 6-27(a) are tangents. Find y.

(b) $\triangle ABC$ in Fig. 6-27(b) is circumscribed. Find x.

(c) Quadrilateral $ABCD$ in Fig. 6-27(c) is circumscribed. Find x.

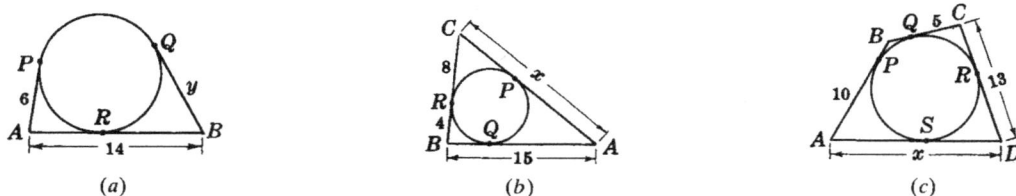

Fig. 6-27

Solutions

(a) By Principle 4, $AR = 6$, and $RB = y$. Then $RB = AB - AR = 14 - 6 = 8$. Hence, $y = RB = 8$.

(b) By Principle 4, $PC = 8$, $QB = 4$, and $AP = AQ$. Then $AQ = AB - QB = 11$. Hence, $x = AP + PC = 11 + 8 = 19$.

(c) By Principle 4, $AS = 10$, $CR = 5$, and $RD = SD$. Then $RD = CD - CR = 8$. Hence, $x = AS + SD = 10 + 8 = 18$.

6.8 Finding the line of centers

Two circles have radii of 9 and 4, respectively. Find the length of their line of centers (a) if the circles are tangent externally, (b) if the circles are tangent internally, (c) if the circles are concentric, (d) if the circles are 5 units apart. (See Fig. 6-28.)

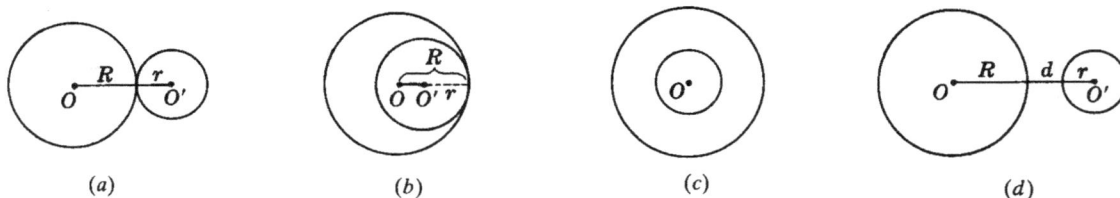

Fig. 6-28

Solutions

Let R = radius of larger circle, r = radius of smaller circle.

(a) Since $R = 9$ and $r = 4$, $OO' = R + r = 9 + 4 = 13$.

(b) Since $R = 9$ and $r = 4$, $OO' = R - r = 9 - 4 = 5$.

(c) Since the circles have the same center, their line of centers has zero length.

(d) Since $R = 9, r = 4$, and $d = 5$, $OO' = R + d + r = 9 + 5 + 4 = 18$.

6.9 Proving a tangent problem stated in words

Prove: Tangents to a circle from an outside point are congruent (Principle 4).

Given: Circle O
$\overline{AP}$ is tangent at P.
$\overline{AQ}$ is tangent at Q.

To Prove: $\overline{AP} \cong \overline{AQ}$

Plan: Draw $\overline{OP}, \overline{OQ}$, and $\overline{OA}$ and prove $\triangle AOP \cong \triangle AOQ$.

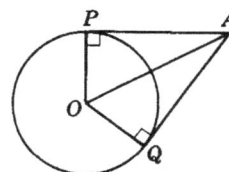

PROOF:

Statements	Reasons
1. Draw $\overline{OP}, \overline{OQ}$, and $\overline{OA}$.	1. A straight line may be drawn between any two points.
2. $\overline{OP} \cong \overline{OQ}$	2. Radii of a circle are congruent.
3. $\angle P$ and $\angle Q$ are right angles.	3. A tangent is $\perp$ to radius drawn to point of contact.
4. $\overline{OA} \cong \overline{OA}$	4. Reflexive property
5. $\triangle AOP \cong \triangle AOQ$	5. Hy-leg
6. $\overline{AP} \cong \overline{AQ}$	6. Corresponding parts of congruent $\triangle$ are congruent.

6.3 Measurement of Angles and Arcs in a Circle

A *central angle* has the same number of degrees as the arc it intercepts. Thus, as shown in Fig. 6-29, a central angle which is a right angle intercepts a 90° arc; a 40° central angle intercepts a 40° arc, and a central angle which is a straight angle intercepts a semicircle of 180°.

Since the numerical measures in degrees of both the central angle and its intercepted arc are the same, we may restate the above principle as follows: A central angle is measured by its intercepted arc. The symbol $\stackrel{\circ}{=}$ may be used to mean "is measured by." (Do not say that the central angle equals its intercepted arc. An angle cannot *equal* an arc.)

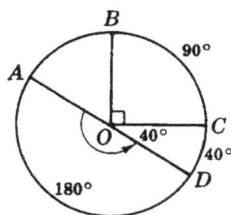

Fig. 6-29

An *inscribed angle* is an angle whose vertex is on the circle and whose sides are chords. *An angle inscribed in an arc* has its vertex on the arc and its sides passing through the ends of the arc. Thus, $\angle A$ in Fig. 6-30 is an inscribed angle whose sides are the chords $\overline{AB}$ and $\overline{AC}$. Note that $\angle A$ intercepts $\overset{\frown}{BC}$ and is inscribed in $\overset{\frown}{BAC}$.

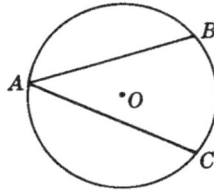

Fig. 6-30

6.3A Angle-Measurement Principles

PRINCIPLE 1:   *A central angle is measured by its intercepted arc.*

PRINCIPLE 2:   *An inscribed angle is measured by one-half its intercepted arc.*

A proof of this principle is given in Chapter 16.

PRINCIPLE 3:   *In the same or congruent circles, congruent inscribed angles have congruent intercepted arcs.*

Thus in Fig. 6-31, if $\angle 1 \cong \angle 2$, then $\overset{\frown}{BC} \cong \overset{\frown}{DE}$.

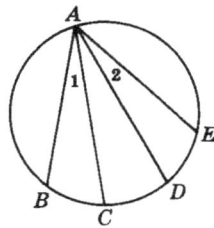

Fig. 6-31

PRINCIPLE 4:   *In the same or congruent circles, inscribed angles having congruent intercepted arcs are congruent.* (This is the converse of Principle 3.)

Thus in Fig. 6-31, if $\overset{\frown}{BC} \cong \overset{\frown}{DE}$, then $\angle 1 \cong \angle 2$.

PRINCIPLE 5:   *Angles inscribed in the same or congruent arcs are congruent.*

Thus in Fig. 6-32, if $\angle C$ and $\angle D$ are inscribed in $\overset{\frown}{ACB}$, then $\angle C \cong \angle D$.

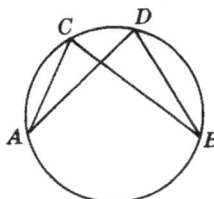

Fig. 6-32

PRINCIPLE 6: *An angle inscribed in a semicircle is a right angle.*

Thus in Fig. 6-33, since $\angle C$ is inscribed in semicircle $\overset{\frown}{ACD}$, $m\angle C = 90°$.

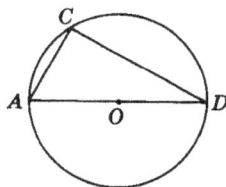

Fig. 6-33

PRINCIPLE 7: *Opposite angles of an inscribed quadrilateral are supplementary.*

Thus in Fig. 6-34, if *ABCD* is an inscribed quadrilateral, $\angle A$ is the supplement of $\angle C$.

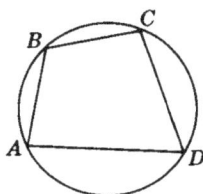

Fig. 6-34

PRINCIPLE 8: *Parallel lines intercept congruent arcs on a circle.*

Thus in Fig. 6-35, if $\overline{AB} \parallel \overline{CD}$, then $\overset{\frown}{AC} \cong \overset{\frown}{BD}$. If tangent $\overleftrightarrow{FG}$ is parallel to $\overline{CD}$, then $\overset{\frown}{PC} \cong \overset{\frown}{PD}$.

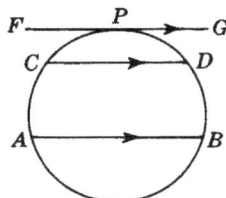

Fig. 6-35

PRINCIPLE 9: *An angle formed by a tangent and a chord is measured by one-half its intercepted arc.*

PRINCIPLE 10: *An angle formed by two intersecting chords is measured by one-half the sum of the intercepted arcs.*

PRINCIPLE 11: *An angle formed by two secants intersecting outside a circle is measured by one-half the difference of the intercepted arcs.*

PRINCIPLE 12: *An angle formed by a tangent and a secant intersecting outside a circle is measured by one-half the difference of the intercepted arcs.*

PRINCIPLE 13: *An angle formed by two tangents intersecting outside a circle is measured by one-half the difference of the intercepted arcs.*

Proofs of Principles 10 to 13 are given in Chapter 16.

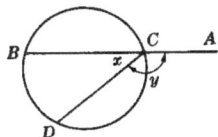

Fig. 6-36

6.3B Table of Angle-Measurement Principles

Position of Vertex	Kind of Angle	Diagram	Measurement Formula	Method of Measurement
Center of circle	Central angle (apply Principle 1)		$\angle O \doteq \overset{\frown}{AB}$ $m\angle O = a°$	By intercepted arc
On the circle	Inscribed angle (apply Principle 2)		$\angle A \doteq \frac{1}{2}\overset{\frown}{BC}$ $m\angle A = \frac{1}{2}a°$	By one-half intercepted arc
	Angle formed by a tangent and a chord (apply Principle 9)			
Inside the circle	Angle formed by two intersecting chords (apply Principle 10)		$\angle 1 \doteq \frac{1}{2}(\overset{\frown}{AC} + \overset{\frown}{BD})$ $m\angle 1 = \frac{1}{2}(a° + b°)$	By one-half sum of intercepted arcs
Outside the circle	Angle formed by two secants (apply Principle 10)		$\angle A \doteq \frac{1}{2}(\overset{\frown}{BC} - \overset{\frown}{DE})$ $m\angle A = \frac{1}{2}(a° - b°)$	By one-half difference of intercepted arcs
	Angle formed by a secant and a tangent (apply Principle 12)		$\angle A \doteq \frac{1}{2}(\overset{\frown}{BC} - \overset{\frown}{BD})$ $m\angle A = \frac{1}{2}(a° - b°)$	
	Angle formed by two tangents (apply Principle 13)		$\angle A \doteq \frac{1}{2}(\overset{\frown}{BDC} - \overset{\frown}{BC})$ $m\angle A = \frac{1}{2}(a° - b°)$ Also, $m\angle A = (180 - b)°$	

Note: To find the angle formed by a secant and a chord meeting on the circle, first find the measure of the inscribed angle adjacent to it and then subtract from 180°. Thus if secant $\overline{AB}$ meets chord $\overline{CD}$ at C on the circle in Fig. 6-36, to find $m\angle y$, first find the measure of inscribed $\angle x$. Obtain $m\angle y$ by subtracting $m\angle x$ from 180°.

SOLVED PROBLEMS

6.10 Applying principles 1 and 2

(a) In Fig. 6-37(a), if $m\angle y = 46°$, find $m\angle x$.

(b) In Fig. 6-37(b), if $m\angle y = 112°$, find $m\angle x$.

(c) In Fig. 6-37(c), if $m\angle x = 75°$, find $m\widehat{y}$.

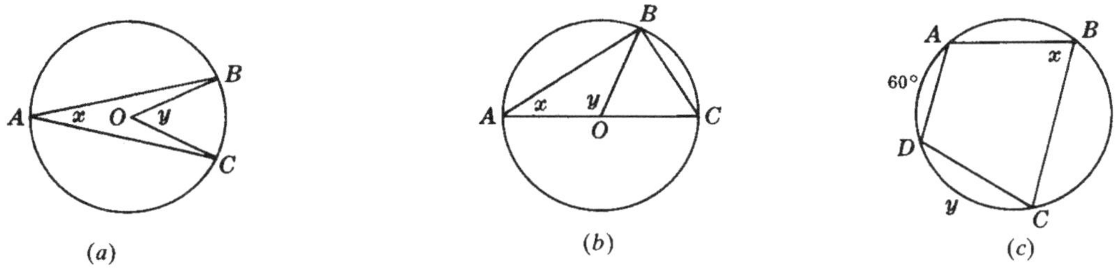

Fig. 6-37

Solutions

(a) $\angle y \stackrel{\circ}{=} \widehat{BC}$, so $m\widehat{BC} = 46°$. Then $\angle x \stackrel{\circ}{=} \frac{1}{2}\widehat{BC} = \frac{1}{2}(46°) = 23°$, so $m\angle x = 23°$.

(b) $\angle y \stackrel{\circ}{=} \widehat{AB}$ so $m\widehat{AB} = 112°$.
$m\widehat{BC} = m(\widehat{ABC} - \widehat{AB}) = 180° - 112° = 68°$. Then $\angle x \stackrel{\circ}{=} \frac{1}{2}\widehat{BC} = \frac{1}{2}(68°) = 34°$, so $m\angle x = 34°$.

(c) $\angle x \stackrel{\circ}{=} \frac{1}{2}\widehat{ADC}$, so $m\widehat{ADC} = 150°$. Then $m\widehat{y} = m(\widehat{ADC} - \widehat{AD}) = 150° - 60° = 90°$.

6.11 Applying principles 3 to 8

Find x and y in each part of Fig. 6-38.

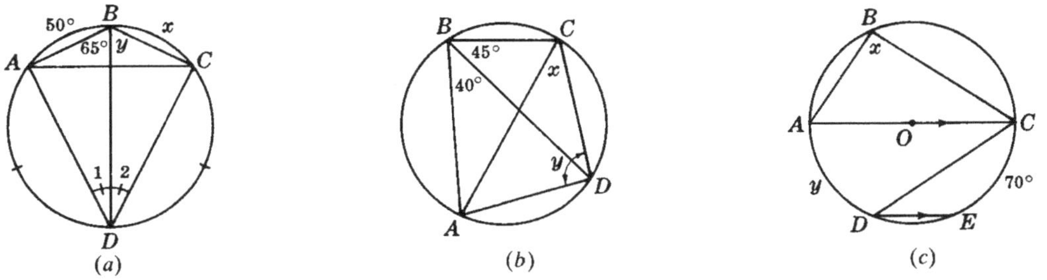

Fig. 6-38

Solutions

(a) Since $m\angle 1 = m\angle 2$, $m\widehat{x} = m\widehat{AB} = 50°$. Since $\widehat{AD} \cong \widehat{CD}$, $m\angle y = m\angle ABD = 65°$.

(b) $\angle ABD$ and $\angle x$ are inscribed in $\widehat{ABD}$; hence, $m\angle x = m\angle ABD = 40°$.
 $ABCD$ is an inscribed quadrilateral; hence, $m\angle y = 180° - m\angle B = 95°$.

(c) Since $\angle x$ is inscribed in a semicircle, $m\angle x = 90°$. Since $\overline{AC} \parallel \overline{DE}$, $m\widehat{y} = m\widehat{CE} = 70°$.

6.12 Applying principle 9

In each part of Fig. 6-39, *CD* is a tangent at *P*.

(a) If $m\widehat{y} = 220°$ in part (a), find $m\angle x$.

(b) If $m\widehat{y} = 140°$ in part (b), find $m\angle x$.

(c) If $m\angle y = 75°$ in part (c), find $m\angle x$.

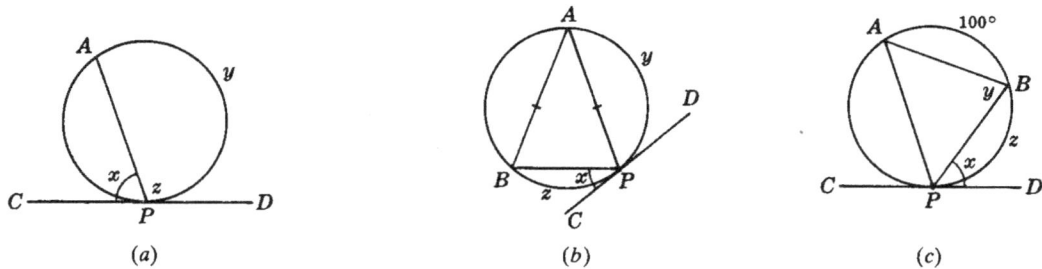

Fig. 6-39

Solutions

(a) $\angle z \triangleq \frac{1}{2}\widehat{y} = \frac{1}{2}(220°) = 110°$. So $m\angle x = 180° - 110° = 70°$.

(b) Since $AB = AP$, $m\widehat{AB} = m\widehat{y} = 140°$. Then $m\widehat{z} = 360° - 140° - 140° = 80°$.
Since $\angle x \triangleq \frac{1}{2}\widehat{z} = 40°$, $m\angle x = 40°$.

(c) $\angle y \triangleq \frac{1}{2}\widehat{AP}$, so $m\widehat{AP} = 150°$. Then $m\widehat{z} = 360° - 100° - 150° = 110°$.
Since $\angle x \triangleq \frac{1}{2}\widehat{z} = 55°$, $m\angle x = 55°$.

6.13 Applying Principle 10

(a) If $m\angle x = 95°$ in Fig. 6-40(a), find $m\widehat{y}$.

(b) If $m\widehat{y} = 80°$ in Fig. 6-40(b), find $m\angle x$.

(c) If $m\widehat{x} = 78°$ in Fig. 6-40(c), find $m\angle y$.

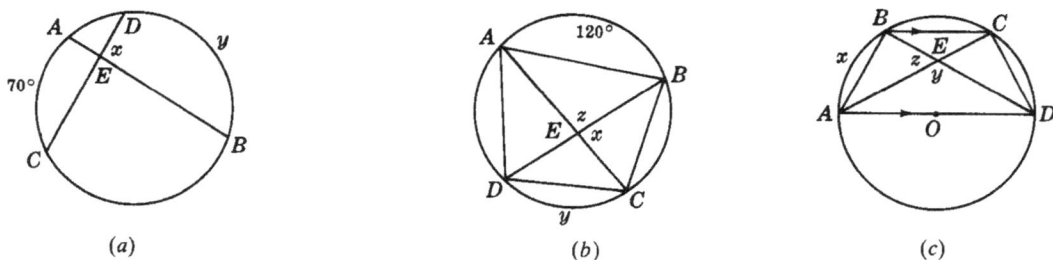

Fig. 6-40

Solutions

(a) $\angle x \triangleq \frac{1}{2}(\widehat{AC} + \widehat{y})$; thus $95° = \frac{1}{2}(70° + m\widehat{y})$, so $m\widehat{y} = 120°$.

(b) $\angle z \triangleq \frac{1}{2}(\widehat{y} + \widehat{AB}) = \frac{1}{2}(80° + 120°) = 100°$. Then $m\angle x = 180° - m\angle z = 80°$.

(c) Because $\overline{BC} \parallel \overline{AD}$, $m\widehat{CD} = m\widehat{x} = 78°$. Also, $\angle z \triangleq \frac{1}{2}(\widehat{x} + \widehat{CD}) = 78°$. Then $m\angle y = 180° - m\angle z = 102°$.

6.14 Applying principles 11 to 13

(a) If $m\angle x = 40°$ in Fig. 6-41(a), find $m\widehat{y}$.

(b) If $m\angle x = 67°$ in Fig. 6-41(b), find $m\widehat{y}$.

(c) If $m\angle x = 61°$ in Fig. 6-41(c), find $m\widehat{y}$.

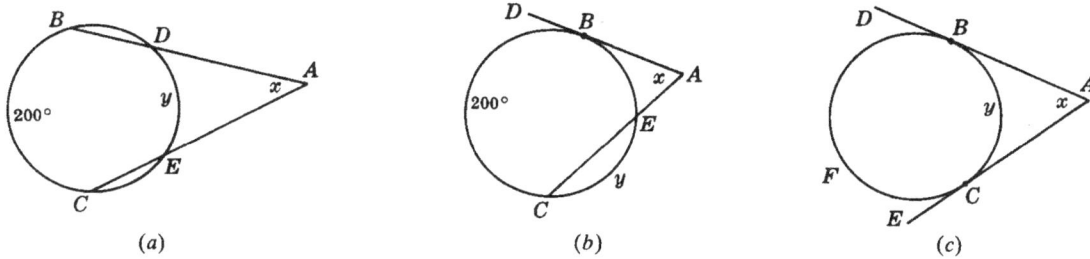

Fig. 6-41

Solutions

(a) $\angle x \overset{\circ}{=} \frac{1}{2}(\widehat{BC} - \widehat{y})$, so $40° = \frac{1}{2}(200° - m\widehat{y})$ or $m\widehat{y} = 120°$.

(b) $\angle x \overset{\circ}{=} \frac{1}{2}(\widehat{BC} - \widehat{BE})$, so $67° = \frac{1}{2}(200° - m\widehat{BE})$ or $m\widehat{BE} = 66°$.
Then $m\widehat{y} = 360° - 200° - 66° = 94°$.

(c) $\angle x \overset{\circ}{=} \frac{1}{2}(\widehat{BFC} - \widehat{y})$ and $m\widehat{BFC} = 360° - m\widehat{y}$. Then $61° = \frac{1}{2}[(360° - m\widehat{y}) - m\widehat{y}] = 180° - m\widehat{y}$. Thus $m\widehat{y} = 119°$.

6.15 Using equations in two unknowns to find arcs

In each part of Fig. 6-42, find x and y using equations in two unknowns.

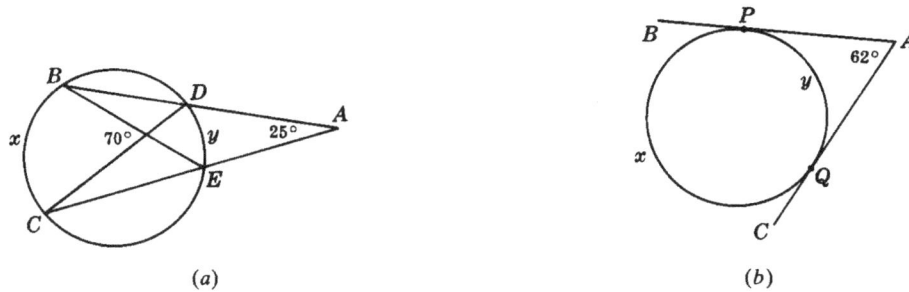

Fig. 6-42

Solutions

(a) By Principle 10, $70° = \frac{1}{2}(m\widehat{x} + m\widehat{y})$
By Principle 11, $25° = \frac{1}{2}(m\widehat{x} - m\widehat{y})$
If we add these two equations, we get $m\widehat{x} = 95°$. If we subtract one from the other, we get $m\widehat{y} = 45°$.

(b) Since $m\widehat{x} + m\widehat{y} = 360°$, $\frac{1}{2}(m\widehat{x} + m\widehat{y}) = 180°$
By Principle 13, $\frac{1}{2}(m\widehat{x} - m\widehat{y}) = 62°$
If we add these two equations, we find that $m\widehat{x} = 242°$. If we subtract one from the other, we get $m\widehat{y} = 118°$.

6.16 Measuring angles and arcs in general

Find x and y in each part of Fig. 6-43.

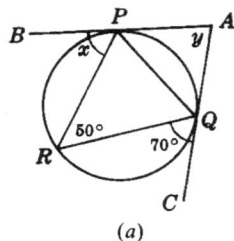

(a)

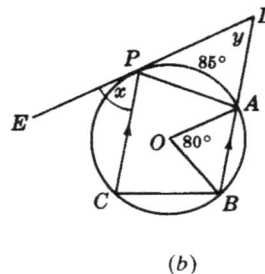

(b)

Fig. 6-43

Solutions

(a) By Principle 2, $50° = \frac{1}{2}m\widehat{PQ}$ or $m\widehat{PQ} = 100°$. Also, by Principle 9, $70° = \frac{1}{2}m\widehat{QR}$ or $m\widehat{QR} = 140°$.
Then $m\widehat{PR} = 360° - m\widehat{PQ} - m\widehat{QR} = 120°$.
By Principle 9, $x = \frac{1}{2}m\widehat{PR} = 60°$.
By Principle 13, $y = \frac{1}{2}(m\widehat{PRQ} - m\widehat{PQ}) = \frac{1}{2}(260° - 100°) = 80°$.

(b) By Principle 1, $m\widehat{AB} = 80°$. Also, by Principle 8, $m\widehat{BC} = m\widehat{PA} = 85°$. Then $m\widehat{PC} = 360° - m\widehat{PA} - m\widehat{AB} - m\widehat{BC} = 110°$.
By Principle 9, $x = \frac{1}{2}m\widehat{PC} = 55°$.
By Principle 12, $y = \frac{1}{2}(m\widehat{PCB} - m\widehat{PA}) = \frac{1}{2}(195° - 85°) = 55°$.

6.17 Proving an angle measurement problem

Given: $\widehat{BD} = \widehat{CE}$

To Prove: $AB = AC$

Plan: First prove $\widehat{CD} = \widehat{BE}$.
Use this to show that $\angle B \cong \angle C$.

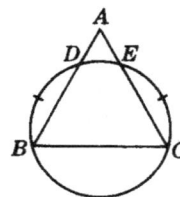

PROOF:

Statements	Reasons
1. $\widehat{BD} = \widehat{CE}$	1. Given
2. $\widehat{DE} = \widehat{DE}$	2. Reflexive property
3. $\widehat{BE} = \widehat{CD}$	3. If =s are added to =s, the sums are =.
4. $\angle B \cong \angle C$	4. In a circle, inscribed angles having equal intercepted arcs are $\cong$.
5. $AB = AC$	5. In a triangle, sides opposite $\cong$ angles are equal in length.

6.18 Proving an angle measurement problem stated in words

Prove that parallel chords drawn at the ends of a diameter are equal in length.

Solution

Given: Circle O

$\overline{AB}$ is a diameter.

$\overline{AC} \parallel \overline{BD}$

To Prove: $AC = BD$

Plan: Prove $\overset{\frown}{AC} \cong \overset{\frown}{BD}$

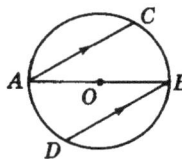

PROOF:

Statements	Reasons
1. $\overline{AB}$ is a diameter	1. Given
2. $\overset{\frown}{ACB} \cong \overset{\frown}{ADB}$	2. A diameter cuts a circle into two equal semicircles.
3. $\overline{AC} \parallel \overline{BD}$	3. Given
4. $\overset{\frown}{AD} \cong \overset{\frown}{BC}$	4. Parallel lines intercept $\cong$ arcs on a circle.
5. $\overset{\frown}{AC} \cong \overset{\frown}{BD}$	5. If equals are subtracted from equals, the differences are equal. Definition of $\cong$ arcs.
6. $AC = BD$	6. In a circle, equal arcs have chords which are equal in length.

SUPPLEMENTARY PROBLEMS

6.1. Provide the proofs requested in Fig. 6-44. (6.3)

(a) **Given:** $AB = DE$
$AC = DF$
To Prove: $\angle B \cong \angle E$

(b) **Given:** Circle O, $AB = BC$
Diameter $\overline{BD}$
To Prove: $\overline{BD}$ bisects $\angle AOC$.

(c) **Given:** Circle O
$\overset{\frown}{AB} = \overset{\frown}{CD}$
To Prove: $\angle AOC \cong \angle BOD$

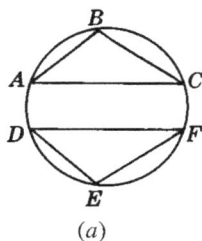

(a)

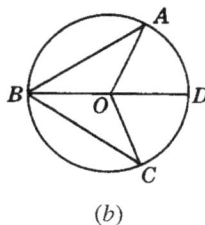

(b)

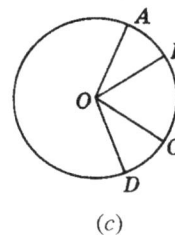

(c)

Fig. 6-44

6.2. Provide the proofs requested in Fig. 6-45. Please refer to figure 6-45(a) for problems 6.2(a) and (b); to figure 6-45(b) for problems 6.2(c) and (d); and figure 6-45(c) for problems 6.2(e) and (f). (6.3)

(a) **Given:** $AB = AC$
To Prove: $\overset{\frown}{ABC} \cong \overset{\frown}{ACB}$

(b) **Given:** $\overset{\frown}{ABC} \cong \overset{\frown}{ACB}$
To Prove: $AB = AC$

(c) **Given:** Circle O, $AB = AD$
Diameter $\overline{AC}$
To Prove: $BC = CD$

(d) **Given:** Circle O
$AB = AD$, $BC = CD$
To Prove: $\overline{AC}$ is a diameter.

(e) **Given:** $AD = BC$
To Prove: $AC = BD$

(f) **Given:** $AC = BD$
To Prove: $AD = BC$

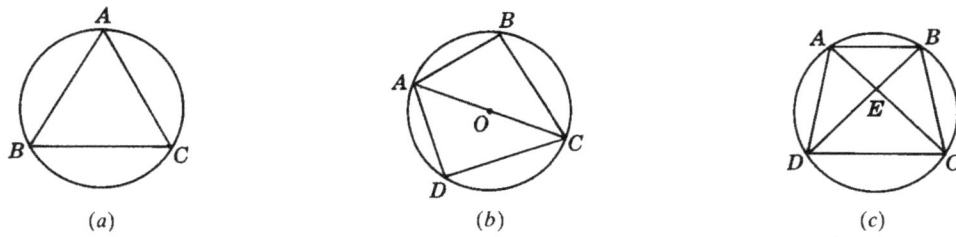

(a)   (b)   (c)

Fig. 6-45

6.3. Prove each of the following: (6.4)

(a) If a radius bisects a chord, then it bisects its arcs.

(b) If a diameter bisects the major arc of a chord, then it is perpendicular to the chord.

(c) If a diameter is perpendicular to a chord, it bisects the chord and its arcs.

6.4. Prove each of the following: (6.4)

(a) A radius through the point of intersection of two congruent chords bisects an angle formed by them.

(b) If chords drawn from the ends of a diameter make congruent angles with the diameter, the chords are congruent.

(c) In a circle, congruent chords are equally distant from the center of the circle.

(d) In a circle, chords that are equally distant from the center are congruent.

6.5. Determine each of the following, assuming t, t', and t'' in Fig. 6-46 are tangents. (6.5)

(a) If $m\angle A = 90°$ in Fig. 6-46(a), what kind of quadrilateral is $PAQO$?

(b) If $BR = RC$ in Fig. 6-46(b), what kind of triangle is ABC?

(c) What kind of quadrilateral is $PABQ$ in Fig. 6-46(c) if $\overline{PQ}$ is a diameter?

(d) What kind of triangle is AOB in Fig. 6-46(c)?

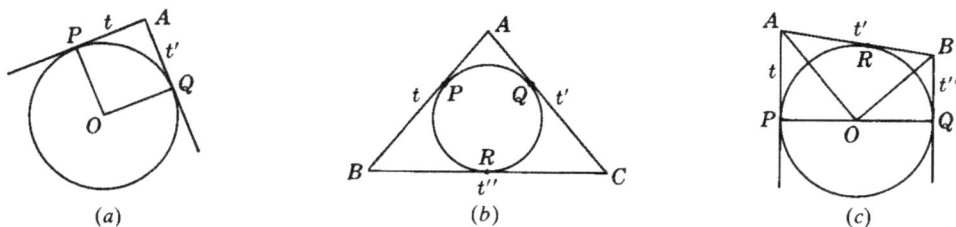

(a)   (b)   (c)

Fig. 6-46

6.6. In circle O, radii $\overline{OA}$ and $\overline{OB}$ are drawn to the points of tangency of $\overline{PA}$ and $\overline{PB}$. Find $m\angle AOB$ if $m\angle APB$ equals (a) 40°; (b) 120°; (c) 90°; (d) $x°$; (e) $(180 - x)°$; (f) $(90 - x)°$. (6.6)

6.7. Find each of the following (t and t' in Fig. 6-47 are tangents). (6.6)

In Fig. 6-47(a)

(a) If $m\angle POQ = 80°$, find $m\angle PAQ$.

(b) If $m\angle PBO = 25°$, find $m\angle 1$ and $m\angle PAQ$.

(c) If $m\angle PAQ = 72°$, find $m\angle 1$ and $m\angle PBO$.

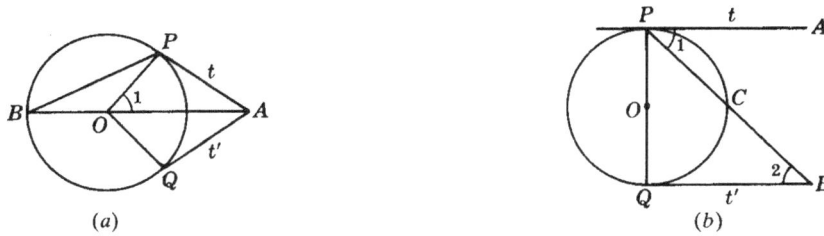

Fig. 6-47

In Fig. 6-47(b)

(d) If $\overline{PB}$ bisects $\angle APQ$, find $m\angle 2$.

(e) If $m\angle 1 = 35°$, find $m\angle 2$.

(f) If $PQ = QB$, find $m\angle 1$.

6.8. In Fig. 6-48(a), $\triangle ABC$ is circumscribed. (a) If $y = 9$, find x. (b) If $x = 25$, find y.   (6.7)
In Fig. 6-48(b), quadrilateral $ABCD$ is circumscribed. (c) Find $AB + CD$. (d) Find perimeter of $ABCD$.
In Fig. 6-48(c), quadrilateral $ABCD$ is circumscribed. (e) If $r = 10$, find x. (f) If $x = 25$, find r.

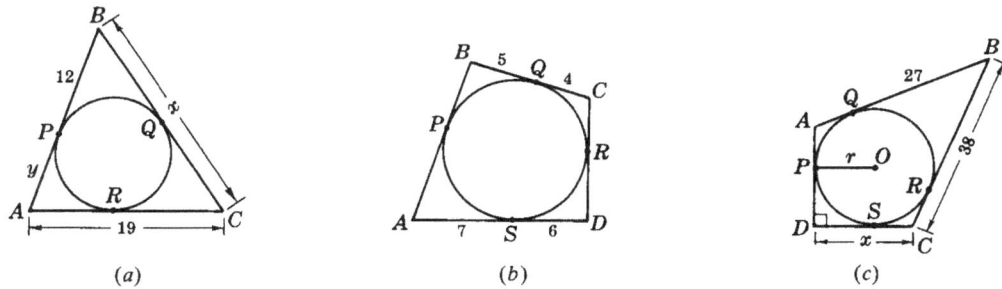

Fig. 6-48

6.9. If two circles have radii of 20 and 13, respectively, find their line of centers: (6.8)

(a) If the circles are concentric

(b) If the circles are 7 units apart

(c) If the circles are tangent externally

(d) If the circles are tangent internally

6.10. If the line of centers of two circles measures 30, what is the relation between the two circles: (6.8)

(a) If their radii are 25 and 5?

(b) If their radii are 35 and 5?

(c) If their radii are 20 and 5?

(d) If their radii are 25 and 10?

6.11. What is the relation between two circles if the length of their line of centers is (a) 0; (b) equal to the difference of their radii; (c) equal to the sum of their radii; (d) greater than the sum of their radii, (e) less than the difference of their radii and greater than 0; (f) greater than the difference and less than the sum of their radii?   (6.8)

6.12. Prove each of the following: (6.9)

    (a) The line from the center of a circle to an outside point bisects the angle between the tangents from the point to the circle.

    (b) If two circles are tangent externally, their common internal tangent bisects a common external tangent.

    (c) If two circles are outside each other, their common internal tangents are congruent.

    (d) In a circumscribed quadrilateral, the sum of the lengths of the two opposite sides equals the sum of the lengths of the other two.

6.13. Find the number of degrees in a central angle which intercepts an arc of (a) 40°; (b) 90°; (c) 170°; (d) 180°; (e) $2x°$; (f) $(180 - x)°$; (g) $(2x - 2y)°$.   (6.10)

6.14. Find the number of degrees in an inscribed angle which intercepts an arc of (a) 40°; (b) 90°; (c) 170°; (d) 180°; (e) 260°; (f) 348°; (g) $2x°$; (h) $(180 - x)°$; (i) $(2x - 2y)°$.   (6.10)

6.15. Find the number of degrees in the arc intercepted by   (6.10)

    (a) A central angle of 85°

    (b) An inscribed angle of 85°

    (c) A central angle of $c°$

    (d) An inscribed angle of $i°$

    (e) The central angle of a triangle formed by two radii and a chord equal to a radius

    (f) The smallest angle of an inscribed triangle whose angles intercept arcs in the ratio of 1:2:3

6.16. Find the number of degrees in each of the arcs intercepted by the angles of an inscribed triangle if the measures of these angles are in the ratio of (a) 1:2:3; (b) 2:3:4; (c) 5:6:7; (d) 1:4:5.   (6.10)

6.17. (a) If $m\widehat{y} = 40°$ in Fig. 6-49(a), find $m\angle x$.   (d) If $m\angle x = 108°$ in Fig. 6-49(b), find $m\angle y$.

    (b) If $m\angle x = 165°$ in Fig. 6-49(a), find $m\widehat{y}$.   (e) If $m\widehat{y} = 105°$ in Fig. 6-49(c), find $m\angle x$.

    (c) If $m\angle y = 115°$ in Fig. 6-49(b), find $m\angle x$.   (f) If $m\angle x = 96°$ in Fig. 6-49(c), find $m\widehat{y}$.

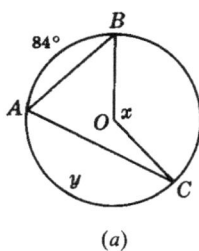

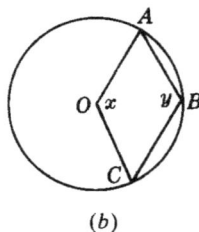

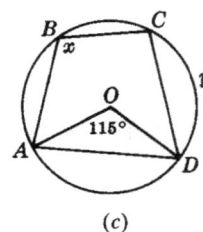

                (*a*)   (*b*)   (*c*)

Fig. 6-49

6.18. If quadrilateral *ABCD* is inscribed in a circle in Fig. 6-50, find (6.11)

(a) $m\angle A$ if $m\angle C = 45°$

(e) $m\angle A$ if $m\widehat{BAD} = 160°$

(b) $m\angle B$ if $m\angle D = 90°$

(f) $m\angle B$ if $m\widehat{ABC} = 200°$

(c) $m\angle C$ if $m\angle A = x°$

(g) $m\angle C$ if $m\widehat{BC} = 140°$ and $m\widehat{CD} = 110°$

(d) $m\angle D$ if $m\angle B = (90 - x)°$

(h) $m\angle D$ if $m\angle D : m\angle B = 2:3$

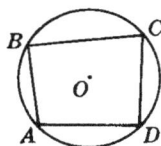

Fig. 6-50

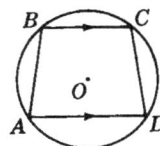

Fig. 6-51

6.19. If *BC* and *AD* are the parallel sides of inscribed trapezoid *ABCD* in Fig. 6-51, find (6.11)

(a) $m\widehat{AB}$ if $m\widehat{CD} = 85°$

(e) $m\angle A$ if $m\angle D = 72°$

(b) $m\widehat{CD}$ if $m\widehat{AB} = y°$

(f) $m\angle A$ if $m\angle C = 130°$

(c) $m\widehat{AB}$ if $m\widehat{BC} = 60°$ and $m\widehat{AD} = 80°$

(g) $m\angle B$ if $m\angle C = 145°$

(d) $m\widehat{CD}$ if $m\widehat{AD} + m\widehat{BC} = 170°$

(h) $m\angle B$ if $m\widehat{AD} = 90°$ and $m\widehat{AB} = 84°$

6.20. A diameter is parallel to a chord. Find the number of degrees in an arc between the diameter and chord if the chord intercepts (a) a minor arc of 80°; (b) a major arc of 300°. (6.11)

6.21. Find *x* and *y* in each part of Fig. 6-52. (6.11)

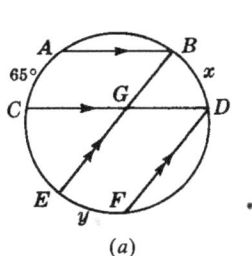

(a)

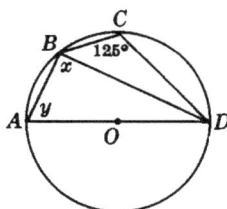

(b)

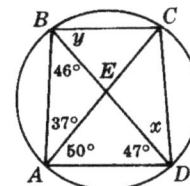

(c)

Fig. 6-52

6.22. Find the number of degrees in the angle formed by a tangent and a chord drawn to the point of tangency if the intercepted arc has measure (a) 38°; (b) 90°; (c) 138°; (d) 180°; (e) 250°; (f) 334°; (g) $x°$; (h) $(360 - x)°$; (i) $(2x + 2y)°$. (6.12)

6.23. Find the number of degrees in the arc intercepted by an angle formed by a tangent and a chord drawn to the point of tangency if the angle measures (a) 55°; (b) $67\frac{1}{2}°$; (c) 90°; (d) 135°; (e) $(90 - x)°$; (f) $(180 - x)°$; (g) $(x - y)°$; (h) $3\frac{1}{2}x°$. (6.12)

6.24. Find the number of degrees in the acute angle formed by a tangent through one vertex and an adjacent side of an inscribed (a) square; (b) equilateral triangle; (c) regular hexagon; (d) regular decagon. (6.12)

6.25. Find x and y in each part of Fig. 6-53 (t and t' are tangents). (6.12)

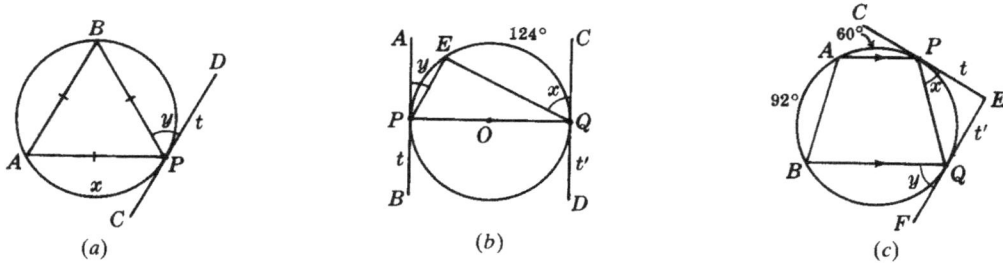

Fig. 6-53

6.26. If $\overline{AC}$ and $\overline{BD}$ are chords intersecting in a circle as shown in Fig. 6-54, find (6.13)

(a) $m\angle x$ if $m\overset{\frown}{AB} = 90°$ and $m\overset{\frown}{CD} = 60°$

(b) $m\angle x$ if $m\overset{\frown}{AB}$ and $m\overset{\frown}{CD}$ each equals 75°

(c) $m\angle x$ if $m\overset{\frown}{AB} + m\overset{\frown}{CD} = 230°$

(d) $m\angle x$ if $m\overset{\frown}{BC} + m\overset{\frown}{AD} = 160°$

(e) $m\overset{\frown}{AB} + m\overset{\frown}{CD}$ if $m\angle x = 70°$

(f) $m\overset{\frown}{BC} + m\overset{\frown}{AD}$ if $m\angle x = 65°$

(g) $m\overset{\frown}{BC}$ if $m\angle x = 60°$ and $m\overset{\frown}{AD} = 160°$

(h) $m\overset{\frown}{BC}$ if $m\angle y = 72°$ and $m\overset{\frown}{AD} = 2m\overset{\frown}{BC}$

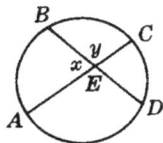

Fig. 6-54

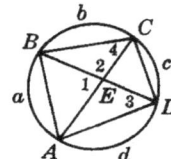

Fig. 6-55

6.27. If $\overline{AC}$ and $\overline{BD}$ are diagonals of an inscribed quadrilateral $ABCD$ as shown in Fig. 6-55, find (6.13)

(a) $m\angle 1$ if $m\overset{\frown}{a} = 95°$ and $m\overset{\frown}{c} = 75°$

(b) $m\angle 1$ if $m\overset{\frown}{b} = 88°$ and $m\overset{\frown}{d} = 66°$

(c) $m\angle 1$ if $m\overset{\frown}{b}$ and $m\overset{\frown}{d}$ each equals 100°

(d) $m\angle 1$ if $m\overset{\frown}{a}:m\overset{\frown}{b}:m\overset{\frown}{c}:m\overset{\frown}{d} = 1:2:3:4$

(e) $m\angle 2$ if $m\overset{\frown}{b} + m\overset{\frown}{d} = m\overset{\frown}{a} + m\overset{\frown}{c}$

(f) $m\angle 2$ if $\overset{\frown}{BC} \parallel \overset{\frown}{AD}$ and $m\overset{\frown}{a} = 70°$

(g) $m\angle 2$ if $\overline{AD}$ is a diameter and $m\overset{\frown}{b} = 80°$

(h) $m\angle 2$ if $ABCD$ is a rectangle and $m\overset{\frown}{a} = 70°$

6.28. Find x and y in each part of Fig. 6-56. (6.13)

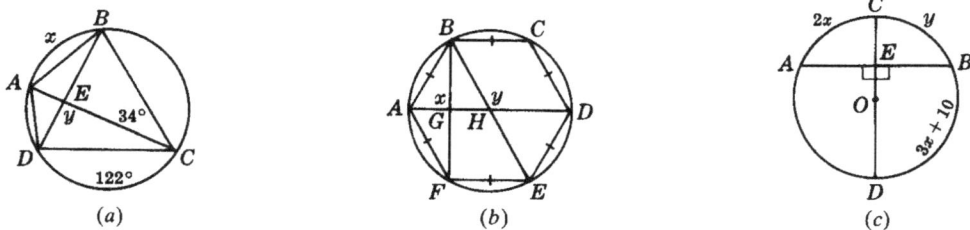

Fig. 6-56

6.29. If $\overline{AB}$ and $\overline{AC}$ are intersecting secants as shown in Fig. 6-57, find (6.14)

(a) $m\angle A$ if $m\widehat{c} = 100°$ and $m\widehat{a} = 40°$

(e) $m\widehat{a}$ if $m\widehat{c} = 160°$ and $m\angle A = 20°$

(b) $m\angle A$ if $m\widehat{c} - m\widehat{a} = 74°$

(f) $m\widehat{c}$ if $m\widehat{a} = 60°$ and $m\angle A = 35°$

(c) $m\angle A$ if $m\widehat{c} = m\widehat{a} + 40°$

(g) $m\widehat{c} - m\widehat{a}$ if $m\angle A = 47°$

(d) $m\angle A$ if $m\widehat{a}:m\widehat{b}:m\widehat{c}:m\widehat{d} = 1:4:3:2$

(h) $m\widehat{a}$ if $m\widehat{c} = 3\widehat{a}$ and $m\angle A = 25°$

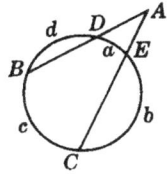

Fig. 6-57

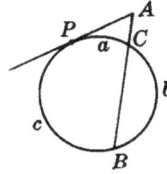

Fig. 6-58

6.30. If tangent $\overline{AP}$ and secant $\overline{AB}$ intersect as shown in Fig. 6-58, find (6.14)

(a) $m\angle A$ if $m\widehat{c} = 150°$ and $m\widehat{a} = 60°$

(f) $m\widehat{a}$ if $m\widehat{c} = 220°$ and $m\angle A = 40°$

(b) $m\angle A$ if $m\widehat{c} = 200°$ and $m\widehat{b} = 110°$

(g) $m\widehat{c}$ if $m\widehat{a} = 55°$ and $m\angle A = 30°$

(c) $m\angle A$ if $m\widehat{b} = 120°$ and $m\widehat{a} = 70°$

(h) $m\widehat{a}$ if $m\widehat{c} = 3\,m\widehat{a}$ and $m\angle A = 45°$

(d) $m\angle A$ if $m\widehat{c} - m\widehat{a} = 73°$

(i) $m\widehat{a}$ if $m\widehat{b} = 100°$ and $m\angle A = 50°$

(e) $m\angle A$ if $m\widehat{a}:m\widehat{b}:m\widehat{c} = 1:4:7$

6.31. If $\vec{AP}$ and $\vec{AQ}$ are intersecting tangents as shown in Fig. 6-59, find (6.14)

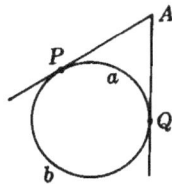

Fig. 6-59

(a) $m\angle A$ if $m\widehat{b} = 200°$

(i) $m\angle A$ if $m\widehat{b}:m\widehat{a} = 7:3$

(b) $m\angle A$ if $m\widehat{a} = 95°$

(j) $m\angle A$ if $m\widehat{b} = 5m\widehat{a} - 60°$

(c) $m\angle A$ if $m\widehat{a} = x°$

(k) $m\widehat{a}$ if $m\angle A = 35°$

(d) $m\angle A$ if $m\widehat{a} = (90 - x)°$

(l) $m\widehat{a}$ if $m\angle A = y°$

(e) $m\angle A$ if $m\widehat{b} = 3\,m\widehat{a}$

(m) $m\widehat{b}$ if $m\angle A = 60°$

(f) $m\angle A$ if $m\widehat{b} = m\widehat{a} + 50°$

(n) $m\widehat{b}$ if $m\angle A = x°$

(g) $m\angle A$ if $m\widehat{b} - m\widehat{a} = 84°$

(o) $m\widehat{b}$ if $\vec{AP} \perp \vec{AQ}$

(h) $m\angle A$ if $m\widehat{b}:m\widehat{a} = 5:1$

6.32. Find x and y in each part of Fig. 6-60 (t and t' are tangents). (6.14)

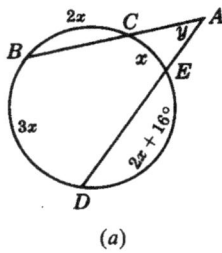

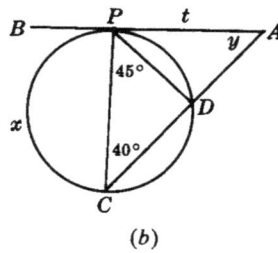

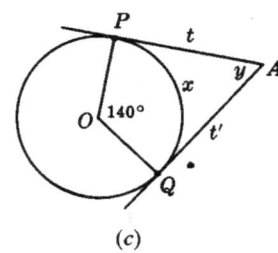

Fig. 6-60

6.33. If $\overline{AB}$ and $\overline{AC}$ are intersecting secants as shown in Fig. 6-61, find (6.15)

(a) $m\widehat{x}$ if $m\angle 1 = 80°$ and $m\angle A = 40°$

(b) $m\widehat{x}$ if $m\angle 1 + m\angle A = 150°$

(c) $m\widehat{x}$ if $\angle 1$ and $\angle A$ are supplementary

(d) $m\widehat{y}$ if $m\angle 1 = 95°$ and $m\angle A = 45°$

(e) $m\widehat{y}$ if $m\angle 1 - m\angle A = 22\frac{1}{2}°$

(f) $m\widehat{y}$ if $m\widehat{x} + m\widehat{y} = 190°$ and $m\angle A = 50°$

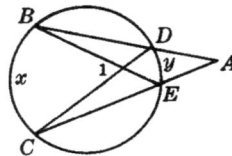

Fig. 6-61

6.34. Find x and y in each part of Fig. 6-62 (t and t' are tangents). (6.15)

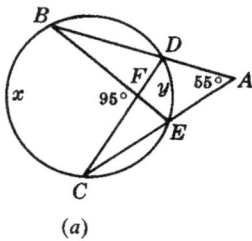

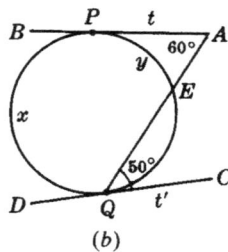

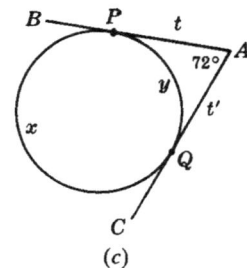

Fig. 6-62

6.35. If ABC is an inscribed triangle as shown in Fig. 6-63, find (6.16)

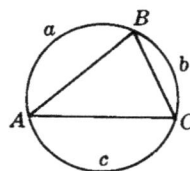

Fig. 6-63

(a) $m\angle A$ if $m\widehat{a} = 110°$ and $m\widehat{c} = 200°$

(f) $m\angle B$ if $m\widehat{ABC} = 208°$

(b) $m\angle A$ if $\overline{AC} \perp \overline{BC}$ and $m\widehat{a} = 102°$

(g) $m\angle B$ if $m\widehat{a} + m\widehat{b} = 3m\widehat{c}$

(c) $m\angle A$ if $\overline{AC}$ is a diameter and $m\widehat{a} = 80°$

(h) $m\angle B$ if $m\widehat{a} = 75°$ and $m\widehat{c} = 2m\widehat{b}$

(d) $m\angle A$ if $m\widehat{a}:m\widehat{b}:m\widehat{c} = 3:1:2$

(i) $m\angle C$ if $\overline{AB} \perp \overline{BC}$ and $m\widehat{a} = 5m\widehat{b}$

(e) $m\angle A$ in $\overline{AC}$ is a diameter and $m\widehat{a}:m\widehat{b} = 5:4$

(j) $m\widehat{c}$ if $m\angle A:m\angle B:m\angle C = 5:4:3$

6.36. If $ABCP$ is an inscribed quadrilateral, $\overleftrightarrow{PD}$ a tangent, and $\overrightarrow{AF}$ a secant in Fig. 6-64, find (6.16)

(a) $m\angle 1$ if $m\widehat{a} = 94°$ and $m\widehat{c} = 54°$

(g) $m\widehat{a}$ if $\overline{BC} \parallel \overline{AP}$ and $m\angle 6 = 42°$

(b) $m\angle 2$ if $\overline{AP}$ is a diameter

(h) $m\widehat{a}$ if $\overline{AC}$ is a diameter and $m\angle 5 = 35°$

(c) $m\angle 3$ if $m\widehat{CPA} = 250°$

(i) $m\widehat{b}$ if $\overline{AC} \perp \overline{BP}$ and $m\angle 2 = 57°$

(d) $m\angle 3$ if $m\angle ABC = 120°$

(j) $m\widehat{c}$ if $\overline{AC}$ and $\overline{BP}$ are diameters and $m\angle 5 = 41°$

(e) $m\angle 4$ if $m\widehat{BCP} = 130°$ and $m\widehat{b} = 50°$

(k) $m\widehat{d}$ if $m\angle 1 = 95°$ and $m\widehat{b} = 95°$

(f) $m\angle 4$ if $\overline{BC} \parallel \overline{AP}$ and $m\widehat{a} = 74°$

(l) $m\angle CPA$ if $m\angle 3 = 79°$

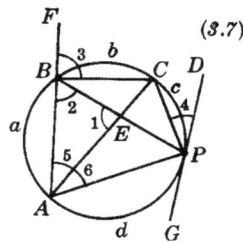

Fig. 6-64

6.37. Find x and y in each part of Fig. 6-65 (t and t' are tangents). (6.16)

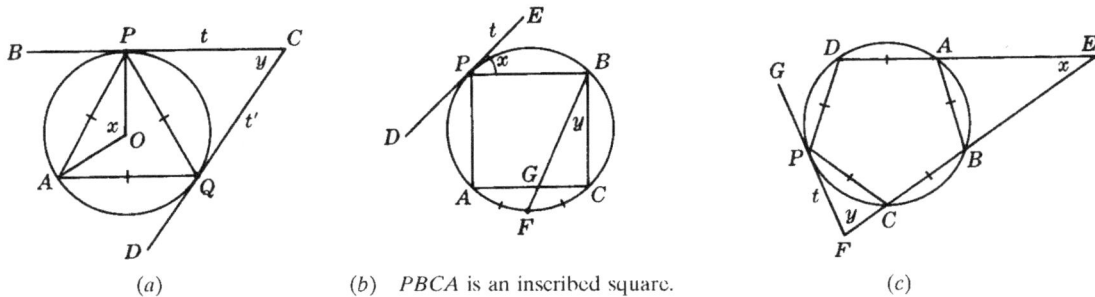

(a)   (b) *PBCA* is an inscribed square.   (c)

Fig. 6-65

6.38. Find x and y in each part of Fig. 6-66. (6.16)

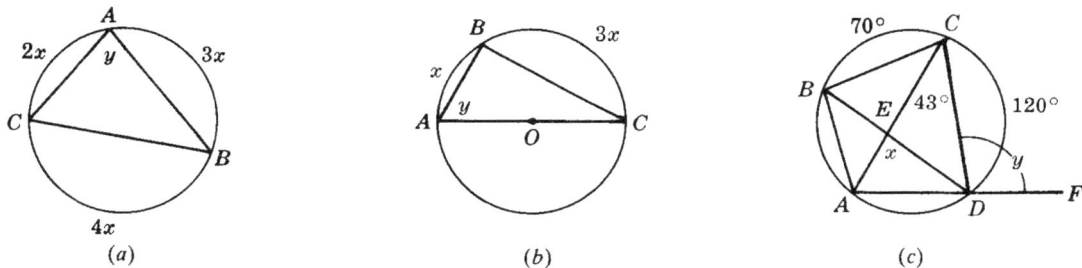

(a)   (b)   (c)

Fig. 6-66

6.39.   Provide the proofs requested in Fig. 6-67.   (6.17)

(a) **Given:** $\overline{AC}$ bisects $\angle A$
        To Prove: $\overline{BC} \cong \overline{CD}$

(b) **Given:** $BC \cong \overline{CD}$
        To Prove: $\overline{AC}$ bisects $\angle A$

(c) **Given:** $\overline{AB} \parallel \overline{CD}$
                 $\overline{AB}$ is a tangent.
        To Prove: $PC = PD$

(d) **Given:** $PC = PD$
                 $\overleftrightarrow{AB}$ is a tangent.
        To Prove: $\overline{AB} \parallel \overline{CD}$

(e) **Given:** $m\widehat{AC} = m\widehat{BD}$
        To Prove: $CE = ED$

(f) **Given:** $CE = EB$
        To Prove: $\widehat{AC} \cong \widehat{BD}$

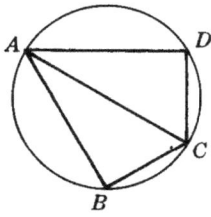 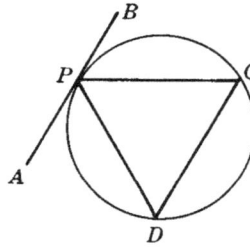 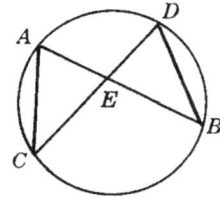

Fig. 6-67

6.40.   Prove each of the following: (6.18)

(a) The base angles of an inscribed trapezoid are congruent.

(b) A parallelogram inscribed in a circle is a rectangle.

(c) In a circle, parallel chords intercept equal arcs.

(d) Diagonals drawn from a vertex of a regular inscribed pentagon trisect the vertex angle.

(e) If a tangent through a vertex of an inscribed triangle is parallel to its opposite side, the triangle is isosceles.

CHAPTER 7

Similarity

7.1 Ratios

Ratios are used to compare quantities by division. The ratio of two quantities is the first divided by the second. A ratio is an abstract number, that is, a number without a unit of measure. Thus, the ratio of 10 ft to 5 ft is 10 ft ÷ 5 ft, which equals 2.

A ratio can be expressed in the following ways: (1) using a colon, as in $3:4$; (2) using "to" as in 3 to 4; (3) as a common fraction, as in $\frac{3}{4}$; (4) as a decimal, 0.75; and (5) as a percent, 75%.

The quantities involved in a ratio must have the same unit. A ratio should be simplified by reducing to lowest terms and eliminating fractions. Thus to find the ratio of 1 ft to 4 in, we first change the foot to 12 in, and then take the ratio of 12 in to 4 in; the result is a ratio of 3 to 1, or 3. Also, the ratio of $2\frac{1}{2}:\frac{1}{2}$ would be restated as $5:1$ or 5.

The ratio of three or more quantities may be expressed as a *continued ratio*. Thus, the ratio of $2 to $3 to $5 is the continued ratio $2:3:5$. This enlarged ratio is a combination of three separate ratios; these are $2:3$, $3:5$, and $2:5$.

Throughout this chapter, readers should use a calculator whenever they choose.

SOLVED PROBLEMS

7.1 Ratio of two quantities with the same unit

Express each of the following ratios in lowest terms: (a) $15°$ to $3°$; (b) $1.25 to $5; (c) $2\frac{1}{2}$ years to 2 years.

Solutions

(a) $\frac{15}{3} = 5$

(b) $\frac{1.25}{5} = \frac{1}{4}$

(c) $\frac{2\frac{1}{2}}{2} = \frac{5}{4}$

7.2 Ratio of two quantities with different units

Express each of the following ratios in lowest terms: (a) 2 years to 3 months; (b) 80 cents to $3.20.

Solutions

(a) 2 years to 3 months = 24 months to 3 months $= \frac{24}{3} = 8$

(b) 80 cents to $3.20 = 80 cents to 320 cents $= \frac{80}{320} = \frac{1}{4}$

7.3 Continued ratio of three quantities

Express each of the following ratios in lowest terms: (a) 1 gal to 2 qt to 2 pt; (b) 1 ton to 1lb to 8 oz.

Solutions

(a) 1 gal to 2 qt to 2 pt = 4 qt to 2 qt to 1 qt = $4:2:1$

(b) 1 ton to 1 lb to 8 oz = 2000 lb to 1 lb to $\frac{1}{2}$lb = $2000:1:\frac{1}{2}$ = $4000:2:1$

7.4 Numerical and algebraic ratios

Express each of the following ratios in lowest terms: (a) 50 to 60; (b) 6.3 to 0.9; (c) 12 to $\frac{3}{8}$; (d) $2x$ to $5x$; (e) $5s^2$ to s^3; (f) x to $5x$ to $7x$.

Solutions

(a) $\dfrac{50}{60} = \dfrac{5}{6}$
(d) $\dfrac{2x}{5x} = \dfrac{2}{5}$

(b) $\dfrac{6.3}{0.9} = 7$
(e) $\dfrac{5s^2}{s^3} = \dfrac{5}{s}$

(c) $12 \div \dfrac{3}{8} = 32$
(f) $x:5x:7x = 1:5:7$

7.5 Using ratios in angle problems

If two angles are in the ratio of $3:2$, find the angles if (a) they are adjacent and form an angle measuring $40°$; (b) they are acute angles of a right triangle; (c) they are two angles of a triangle whose third angle measures $70°$.

Solutions

Let the measures of the angles be $3x$ and $2x$. Then:

(a) $3x + 2x = 40$, so that $5x = 40$ or $x = 8$; hence, the angles measure $24°$ and $16°$.

(b) $3x + 2x = 90$, so $5x = 90$ or $x = 18$; hence, the angles measure $54°$ and $36°$.

(c) $3x + 2x + 70 = 180$, so $5x = 110$ or $x = 22$; hence, the angles measure $66°$ and $44°$.

7.6 Three angles having a fixed ratio

Three angles are in the ratio of $4:3:2$. Find the angles if (a) the first and the third are supplementary; (b) the angles are the three angles of a triangle.

Solutions

Let the measures of the angles be $4x$, $3x$, and $2x$. Then:

(a) $4x + 2x = 180$, so that $6x = 180$ for $x = 30$; hence, the angles measure $120°, 90°$, and $60°$.

(b) $4x + 3x + 2x = 180$, so $9x = 180$ or $x = 20$; hence, the angles measure $80°, 60°$, and $40°$.

7.2 Proportions

A *proportion* is an equality of two ratios. Thus, $2:5 = 4:10$ (or $\frac{2}{5} = \frac{4}{10}$) is a proportion.

The fourth term of a proportion is the *fourth proportional* to the other three taken in order. Thus in $2:3 = 4:x$, x is the fourth proportional to 2, 3, and 4.

The *means* of a proportion are its middle terms, that is, its second and third terms. The *extremes* of a proportion are its outside terms, that is, its first and fourth terms. Thus in $a:b = c:d$, the means are b and c, and the extremes are a and d.

If the two means of a proportion are the same, either mean is the *mean proportional* between the first and fourth terms. Thus in $9:3 = 3:1$, 3 is the mean proportional between 9 and 1.

7.2A Proportion Principles

PRINCIPLE 1: *In any proportion, the product of the means equals the product of the extremes.*

Thus if $a:b = c:d$, then $ad = bc$.

PRINCIPLE 2: *If the product of two numbers equals the product of two other numbers, either pair may be made the means of a proportion and the other pair may be made the extremes.*

Thus if $3x = 5y$, then $x:y = 5:3$ or $y:x = 3:5$ or $3:y = 5:x$ or $5:x = 3:y$.

7.2B Methods of Changing a Proportion into an Equivalent Proportion

PRINCIPLE 3: *(Inversion method) A proportion may be changed into an equivalent proportion by inverting each ratio.*

Thus if $\frac{1}{x} = \frac{4}{5}$, then $\frac{x}{1} = \frac{5}{4}$.

PRINCIPLE 4: *(Alternation method) A proportion may be changed into an equivalent proportion by interchanging the means or by interchanging the extremes.*

Thus if $\frac{x}{3} = \frac{y}{2}$, then $\frac{x}{y} = \frac{3}{2}$ or $\frac{2}{3} = \frac{y}{x}$.

PRINCIPLE 5: *(Addition method) A proportion may be changed into an equivalent proportion by adding terms in each ratio to obtain new first and third terms.*

Thus if $\frac{a}{b} = \frac{c}{d}$, then $\frac{a+b}{b} = \frac{c+d}{d}$. If $\frac{x-2}{2} = \frac{9}{1}$, then $\frac{x}{2} = \frac{10}{1}$.

PRINCIPLE 6: *(Subtraction method) A proportion may be changed into an equivalent proportion by subtracting terms in each ratio to obtain new first and third terms.*

Thus if $\frac{a}{b} = \frac{c}{d}$, then $\frac{a-b}{b} = \frac{c-d}{d}$. If $\frac{x+3}{3} = \frac{9}{1}$, then $\frac{x}{3} = \frac{8}{1}$.

7.2C Other Proportion Principles

PRINCIPLE 7: *If any three terms of one proportion equal the corresponding three terms of another proportion, the remaining terms are equal.*

Thus if $\frac{x}{y} = \frac{3}{5}$ and $\frac{x}{4} = \frac{3}{5}$, then $y = 4$.

PRINCIPLE 8: *In a series of equal ratios, the sum of any of the numerators is to the sum of the corresponding denominators as any numerator is to its denominator.*

Thus if $\frac{a}{b} = \frac{c}{d} = \frac{e}{f}$, then $\frac{a+c+e}{b+d+f} = \frac{a}{b}$. If $\frac{x-y}{4} = \frac{y-3}{5} = \frac{3}{1}$, then $\frac{x-y+y-3+3}{4+5+1} = \frac{3}{1}$ or $\frac{x}{10} = \frac{3}{1}$.

SOLVED PROBLEMS

7.7 Finding unknowns in proportions

Solve the following proportions for x:

(a) $x:4 = 6:8$ (c) $x:5 = 2x:(x+3)$ (e) $\dfrac{x}{2x-3} = \dfrac{3}{5}$

(b) $3:x = x:27$ (d) $\dfrac{3}{x} = \dfrac{2}{5}$ (f) $\dfrac{x-2}{4} = \dfrac{7}{x+2}$

Solutions

(a) Since $4(6) = 8x$, $8x = 24$ or $x = 3$.

(b) Since $x^2 = 3(27)$, $x^2 = 81$ or $x = \pm 9$.

(c) Since $5(2x) = x(x + 3)$, we have $10x = x^2 + 3x$. Then $x^2 - 7x = 0$, so $x = 0$ or 7.

(d) Since $2x = 3(5)$, $2x = 15$ or $x = 7\frac{1}{2}$.

(e) Since $3(2x - 3) = 5x$, we have $6x - 9 = 5x$, so $x = 9$.

(f) Since $4(7) = (x - 2)(x + 2)$, we have $28 = x^2 - 4$. Then $x^2 = 32$, so $x = \pm 4\sqrt{2}$.

7.8 Finding fourth proportionals to three given numbers

Find the fourth proportional to (a) $2, 4, 6$; (b) $4, 2, 6$; (c) $\frac{1}{2}, 3, 4$; (d) b, d, c.

Solutions

(a) We have $2 : 4 = 6 : x$, so $2x = 24$ or $x = 12$.

(b) We have $4 : 2 = 6 : x$, so $4x = 12$ or $x = 3$.

(c) We have $\frac{1}{2} : 3 = 4 : x$, so $\frac{1}{2}x = 12$ or $x = 24$.

(d) We have $b : d = c : x$, so $bx = cd$ or $x = cd/b$.

7.9 Finding the mean proportional to two given numbers

Find the positive mean proportional x between (a) 5 and 20; (b) $\frac{1}{2}$ and $\frac{8}{9}$.

Solutions

(a) We have $5 : x = x : 20$, so $x^2 = 100$ or $x = 10$.

(b) We have $\frac{1}{2} : x = x : \frac{8}{9}$, so $x^2 = \frac{4}{9}$ or $x = \frac{2}{3}$.

7.10 Changing equal products into proportions

(a) Form a proportion whose fourth term is x and such that $2bx = 3s^2$.

(b) Find the ratio x to y if $ay = bx$.

Solutions

(a) $2b : 3s = s : x$ or $2b : 3 = s^2 : x$ or $2b : s^2 = 3 : x$ (b) $x : y = a : b$

7.11 Changing proportions into new proportions

Use each of the following to form a new proportion whose first term is x:

(a) $\dfrac{15}{x} = \dfrac{3}{4}$ (b) $\dfrac{x - 6}{6} = \dfrac{5}{3}$ (c) $\dfrac{x + 8}{8} = \dfrac{4}{3}$ (d) $\dfrac{5}{2} = \dfrac{15}{x}$

Solutions

(a) By Principle 3, $\dfrac{x}{15} = \dfrac{4}{3}$ (c) By Principle 6, $\dfrac{x}{8} = \dfrac{1}{3}$

(b) By Principle 5, $\dfrac{x}{6} = \dfrac{8}{3}$ (d) By Principle 4, $\dfrac{x}{2} = \dfrac{15}{5}$; thus, $\dfrac{x}{2} = \dfrac{3}{1}$

7.12 Combining numerators and denominators of proportions

Use Principle 8 to find x in each of the following proportions:

(a) $\dfrac{x - 2}{9} = \dfrac{2}{3}$ (b) $\dfrac{x + y}{8} = \dfrac{x - y}{4} = \dfrac{2}{3}$ (c) $\dfrac{3x - y}{15} = \dfrac{y - 3}{10} = \dfrac{3}{5}$

Solutions

(a) Adding numerators and denominators yields $\frac{x-2+2}{9+3} = \frac{2}{3}$ or $\frac{x}{12} = \frac{2}{3}$, so $x = 8$.

(b) Here we have $\frac{(x+y)+(x-y)}{8+4} = \frac{2}{3}$, which gives $\frac{2x}{12} = \frac{2}{3}$, so $x = 4$.

(c) We use all three ratios to get $\frac{(3x-y)+(y-3)+3}{15+10+5} = \frac{3}{5}$ or $\frac{3x}{30} = \frac{3}{5}$, so $x = 6$.

7.3 Proportional Segments

If two segments are divided proportionately, (1) the corresponding new segments are in proportion, and (2) the two original segments and either pair of corresponding new segments are in proportion.

Thus if $\overline{AB}$ and $\overline{AC}$ in Fig. 7-1 are divided proportionately by $\overline{DE}$, we may write a proportion such as $\frac{a}{b} = \frac{c}{d}$ using the four segments; or we may write a proportion such as $\frac{a}{AB} = \frac{c}{AC}$ using the two original segments and two of their new segments.

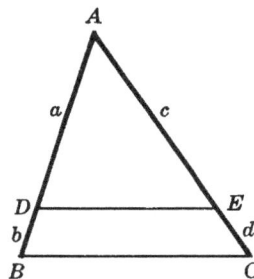

Fig. 7-1

7.3A Obtaining the Eight Arrangements of Any Proportion

A proportion such as $\frac{a}{b} = \frac{c}{d}$ can be arranged in eight ways. To obtain the eight variations, we let each term of the proportion represent one of the new segments of Fig. 7-1. Two of the possible proportions are then obtained from each direction, as follows:

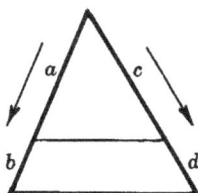

Direction: Down
$\frac{a}{b} = \frac{c}{d}$ or $\frac{c}{d} = \frac{a}{b}$

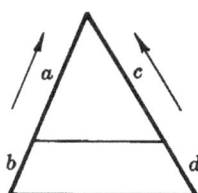

Direction: Up
$\frac{b}{a} = \frac{d}{c}$ or $\frac{d}{c} = \frac{b}{a}$

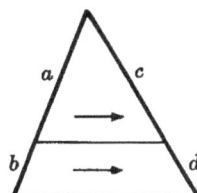

Direction: Right
$\frac{a}{c} = \frac{b}{d}$ or $\frac{b}{d} = \frac{a}{c}$

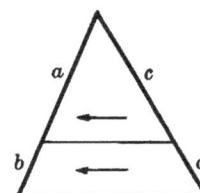

Direction: Left
$\frac{c}{a} = \frac{d}{b}$ or $\frac{d}{b} = \frac{c}{a}$

7.3B Principles of Proportional Segments

PRINCIPLE 1: *If a line is parallel to one side of a triangle, then it divides the other two sides proportionately.*

Thus in $\triangle ABC$ of Fig. 7-2, if $\overline{DE} \parallel \overline{BC}$, then $\frac{a}{b} = \frac{c}{d}$.

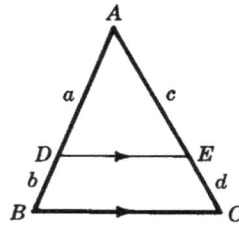
Fig. 7-2

PRINCIPLE 2: *If a line divides two sides of a triangle proportionately, it is parallel to the third side. (Principles 1 and 2 are converses.)*

Thus in $\triangle ABC$ (Fig. 7-2), if $\dfrac{a}{b} = \dfrac{c}{d}$, then $\overline{DE} \parallel \overline{BC}$.

PRINCIPLE 3: *Three or more parallel lines divide any two transversals proportionately.*

Thus if $\overleftrightarrow{AB} \parallel \overleftrightarrow{EF} \parallel \overleftrightarrow{CD}$ in Fig. 7-3, then $\dfrac{a}{b} = \dfrac{c}{d}$.

PRINCIPLE 4: *A bisector of an angle of a triangle divides the opposite side into segments which are proportional to the adjacent sides.*

Thus in $\triangle ABC$ of Fig. 7-4, if $\overline{CD}$ bisects $\angle C$, then $\dfrac{a}{b} = \dfrac{c}{d}$.

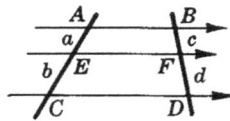
Fig. 7-3

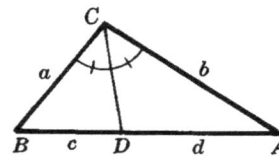
Fig. 7-4

SOLVED PROBLEMS

7.13 Applying principle 1

Find x in each part of Fig. 7-5.

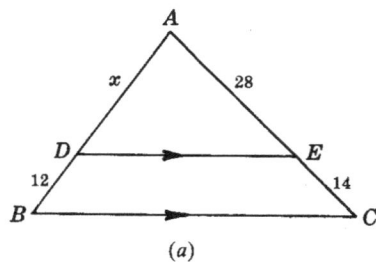
(a)

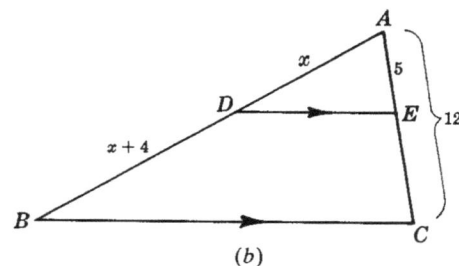
(b)

Fig. 7-5

Solutions

(a) $\overline{DE} \parallel \overline{BC}$; hence $\dfrac{x}{12} = \dfrac{28}{14}$, so that $x = 24$.

(b) We have $EC = 7$ and $\overline{DE} \parallel \overline{BC}$; hence, $\dfrac{x}{x+4} = \dfrac{5}{7}$. Then $7x = 5x + 20$ and $x = 10$.

7.14 Applying principle 3

Find x in each part of Fig. 7-6.

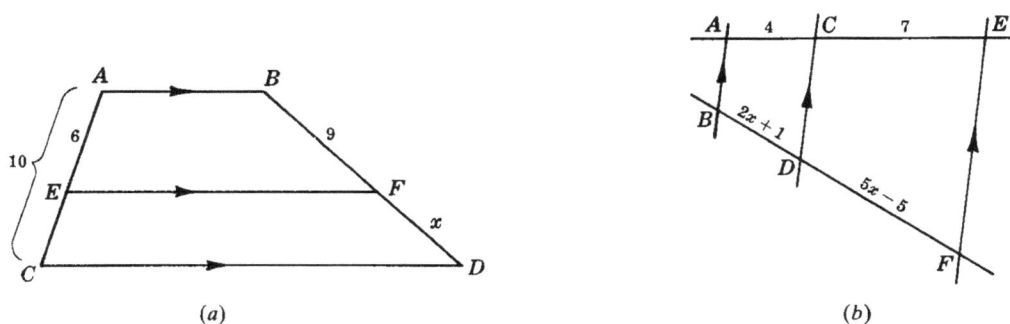

Fig. 7-6

Solutions

(a) We have $EC = 4$ and $\overline{AB} \parallel \overline{EF} \parallel \overline{CD}$; hence, $\dfrac{x}{9} = \dfrac{4}{6}$ and $x = 6$.

(b) $\overleftrightarrow{AB} \parallel \overleftrightarrow{CD} \parallel \overleftrightarrow{EF}$; hence, $\dfrac{5x - 5}{2x + 1} = \dfrac{7}{4}$, from which $20x - 20 = 14x + 7$. Then $6x = 27$ and $x = 4\frac{1}{2}$.

7.15 Applying principle 4

Find x in each part of Fig. 7-7.

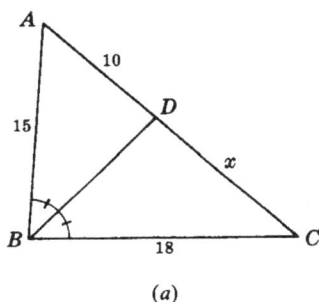

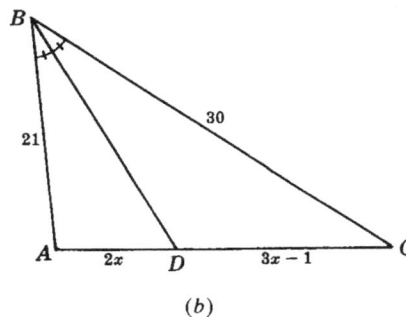

Fig. 7-7

Solutions

(a) $\overline{BD}$ bisects $\angle B$; hence, $\dfrac{x}{10} = \dfrac{18}{15}$ and $x = 12$.

(b) $\overline{BD}$ bisects $\angle B$; hence, $\dfrac{3x - 1}{2x} = \dfrac{30}{21} = \dfrac{10}{7}$. Thus, $21x - 7 = 20x$ and $x = 7$.

7.16 Proving a proportional-segments problem

Given: $\overline{EG} \parallel \overline{BD}, \overline{EF} \parallel \overline{BC}$
To Prove: $\overline{FG} \parallel \overline{CD}$
Plan: Prove that $\overline{FG}$ divides $\overline{AC}$ and $\overline{AD}$ proportionately.

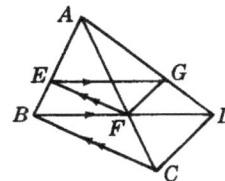

PROOF:

Statements	Reasons
1. $\overline{EG} \parallel \overline{BD}, \overline{EF} \parallel \overline{BC}$	1. Given
2. $\dfrac{AE}{EB} = \dfrac{AG}{GD}, \dfrac{AE}{EB} = \dfrac{AF}{FC}$	2. A line (segment) parallel to one side of a triangle divides the other two sides proportionately.
3. $\dfrac{AF}{FC} = \dfrac{AG}{GD}$	3. Substitution postulate
4. $\overline{FG} \parallel \overline{CD}$	4. If a line divides two sides of a triangle proportionately, it is parallel to the third side.

7.4 Similar Triangles

Similar polygons are polygons whose corresponding angles are congruent and whose corresponding sides are in proportion. Similar polygons have the same shape although not necessarily the same size.

The symbol for "similar" is ~. The notation $\triangle ABC \sim \triangle A'B'C'$ is read "triangle *ABC* is similar to triangle *A*-prime *B*-prime *C*-prime." As in the case of congruent triangles, *corresponding sides of similar triangles are opposite congruent angles.* (Note that corresponding sides and angles are usually designated by the same letter and primes.)

In Fig. 7-8 $\triangle ABC \sim \triangle A'B'C'$ because

$$m\angle A = m\angle A' = 37° \qquad m\angle B = m\angle B' = 53° \qquad m\angle C = m\angle C' = 90°$$

and

$$\frac{a}{ar} = \frac{b}{br} = \frac{c}{cr} \qquad \text{or} \qquad \frac{6}{3} = \frac{8}{4} = \frac{10}{5}$$

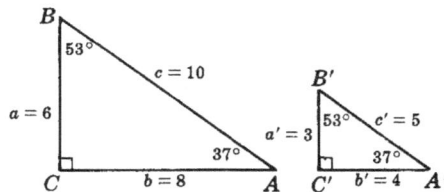

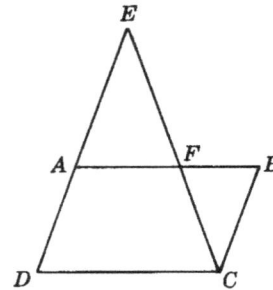

Fig. 7-8   Fig. 7-9

7.4A Selecting Similar Triangles to Prove a Proportion

In Solved Problem 7.25, it is given that *ABCD* in a figure like Fig. 7-9 is a parallelogram, and we must prove that $\frac{AE}{BC} = \frac{AF}{FB}$. To prove this proportion, it is necessary to find similar triangles whose sides are in the proportion. This can be done simply by selecting the triangle whose letters *A, E,* and *F* are in the numerators and the triangle whose letters *B, C,* and *F* are in the denominators. Hence, we would prove $\triangle AEF \sim \triangle BCF$.

Suppose that the proportion to be proved is $\frac{AE}{AF} = \frac{BC}{FB}$. In such a case, interchanging the means leads to $\frac{AE}{BC} = \frac{AF}{FB}$. The needed triangles can then be selected based on the numerators and the denominators.

Suppose that the proportion to be proved is $\frac{AE}{AD} = \frac{AF}{FB}$. Then our method of selecting triangles could not be used until the term *AD* were replaced by *BC*. This is possible, because $\overline{AD}$ and $\overline{BC}$ are opposite sides of the parallelogram *ABCD* and therefore are congruent.

7.4B Principles of Similar Triangles

PRINCIPLE 1: *Corresponding angles of similar triangles are congruent (by the definition).*

PRINCIPLE 2: *Corresponding sides of similar triangles are in proportion (by the definition).*

PRINCIPLE 3: *Two triangles are similar if two angles of one triangle are congruent respectively to two angles of the other.*

Thus in Fig. 7-10, if $\angle A \cong \angle A'$ and $\angle B \cong \angle B'$, then $\triangle ABC \sim \triangle A'B'C'$.

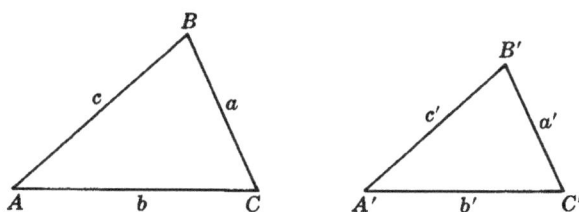

Fig. 7-10

PRINCIPLE 4: *Two triangles are similar if an angle of one triangle is congruent to an angle of the other and the sides including these angles are in proportion.*

Thus in Fig. 7-10, if $\angle C \cong \angle C'$ and $\dfrac{a}{a'} = \dfrac{b}{b'}$, then $\triangle ABC \sim \triangle A'B'C'$.

PRINCIPLE 5: *Two triangles are similar if their corresponding sides are in proportion.*

Thus in Fig. 7-10, if $\dfrac{a}{a'} = \dfrac{b}{b'} = \dfrac{c}{c'}$, then $\triangle ABC \sim A'B'C'$.

PRINCIPLE 6: *Two right triangles are similar if an acute angle of one is congruent to an acute angle of the other (corollary of Principle 3).*

PRINCIPLE 7: *A line parallel to a side of a triangle cuts off a triangle similar to the given triangle.*

Thus in Fig. 7-11, if $\overline{DE} \parallel \overline{BC}$, then $\triangle ADE \sim \triangle ABC$.

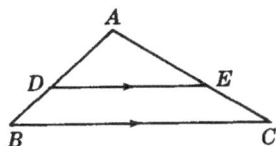

Fig. 7-11

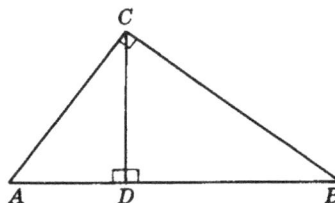

Fig. 7-12

PRINCIPLE 8: *Triangles similar to the same triangle are similar to each other.*

PRINCIPLE 9: *The altitude to the hypotenuse of a right triangle divides it into two triangles which are similar to the given triangle and to each other.*

Thus in Fig. 7-12, $\triangle ABC \sim \triangle ACD \sim \triangle CBD$.

PRINCIPLE 10: *Triangles are similar if their sides are respectively parallel to each other.*

Thus in Fig. 7-13, $\triangle ABC \sim \triangle A'B'C'$.

PRINCIPLE 11: *Triangles are similar if their sides are respectively perpendicular to each other.*

Thus in Fig. 7-14, $\triangle ABC \sim \triangle A'B'C'$.

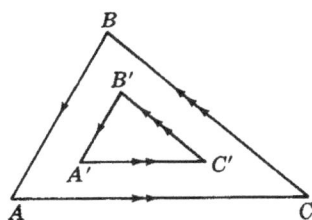

Fig. 7-13

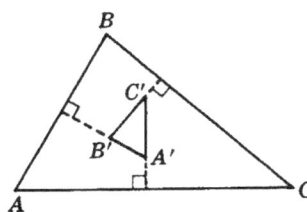

Fig. 7-14

SOLVED PROBLEMS

7.17 Applying principle 2

In similar triangles ABC and $A'B'C'$ (Fig. 7-15), find x and y if $\angle A \cong \angle A'$ and $\angle B \cong \angle B'$.

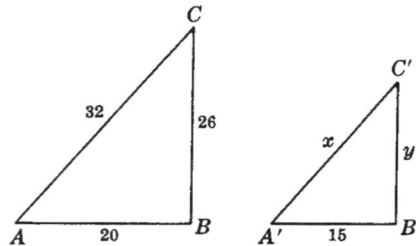

Fig. 7-15

Solution

Since $\angle A \cong \angle A'$ and $\angle B \cong \angle B'$, x and y correspond to 32 and 26, respectively. Hence, $\dfrac{x}{32} = \dfrac{15}{20}$, from which $x = 24$; also $\dfrac{y}{26} = \dfrac{15}{20}$ so $y = 19\frac{1}{2}$.

7.18 Applying principle 3

In each part of Fig. 7-16, two pairs of congruent angles can be used to prove the indicated triangles similar. Determine the congruent angles and state the reason they are congruent.

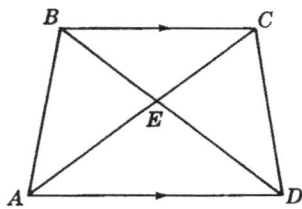

(a) $\triangle BEC \sim \triangle AED$
ABCD is a trapezoid.

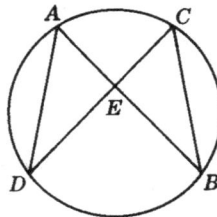

(b) $\triangle AED \sim \triangle CEB$

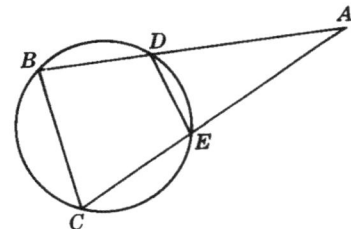

(c) $\triangle ABC \sim \triangle AED$

Fig. 7-16

Solutions

(a) $\angle CBD \cong \angle BDA$ and $\angle BCA \cong \angle CAD$, since alternate interior angles of parallel lines are congruent ($\overline{BC} \parallel \overline{AD}$). Also, $\angle BEC$ and $\angle AED$ are congruent vertical angles.

(b) $\angle A \cong \angle C$ and $\angle B \cong \angle D$, since angles inscribed in the same arc are congruent. Also, $\angle AED$ and $\angle CEB$ are congruent vertical angles.

(c) $\angle ABC \cong \angle AED$, since each is a supplement of $\angle DEC$. $\angle ACB \cong \angle ADE$, since each is a supplement of $\angle BDE$. Also, $\angle A \cong \angle A$.

7.19 Applying principle 6

In each part of Fig. 7-17, determine the angles that can be used to prove the indicated triangles similar.

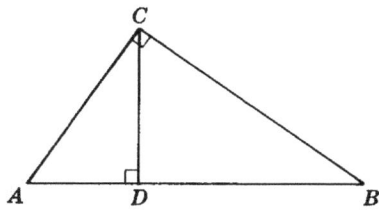

(a) △ACB ~ △ADC

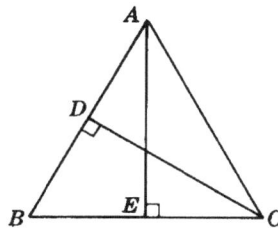

(b) △AEC ~ △CDB
     AB = AC

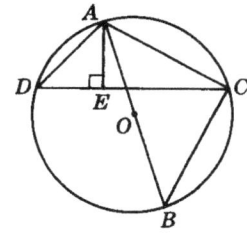

(c) △ADE ~ △ABC
     $\overline{AB}$ is a diameter.

Fig. 7-17

Solutions

(a) ∠ACB and ∠ADC are right angles. ∠A ≅ ∠A.

(b) ∠AEC and ∠BDC are right angles. ∠B ≅ ∠ACE, since angles in a triangle opposite congruent sides are congruent.

(c) ∠ACB is a right angle, since it is inscribed in a semicircle. Hence, ∠AED ≅ ∠ACB. ∠D ≅ ∠B, since angles inscribed in the same arc are congruent.

7.20 Applying principle 4

In each part of Fig. 7-18, determine the pair of congruent angles and the proportion needed to prove the indicated triangles similar.

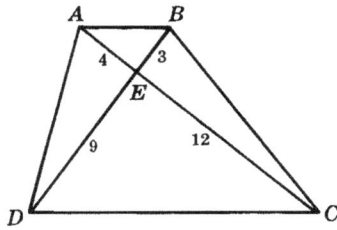

(a) △AEB ~ △DEC

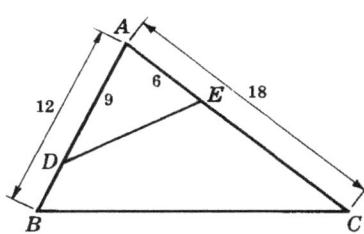

(b) △AED ~ △ABC

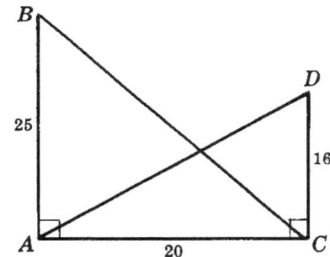

(c) △ABC ~ △CAD

Fig. 7-18

Solutions

(a) ∠AEB ≅ ∠DEC; $\dfrac{3}{9} = \dfrac{4}{12}$

(b) ∠A ≅ ∠A; $\dfrac{6}{12} = \dfrac{9}{18}$

(c) ∠BAC ≅ ∠ACD; $\dfrac{20}{16} = \dfrac{25}{20}$

7.21 Applying principle 5

In each part of Fig. 7-19, determine the proportion needed to prove the indicated triangles similar.

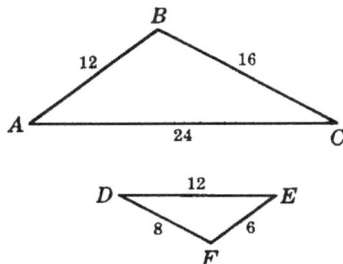

(a) △ABC ~ △EFD

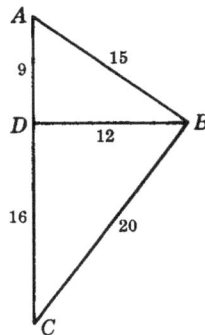

(b) △ABD ~ △BCD

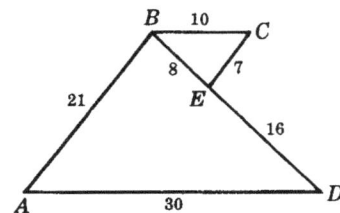

(c) △ABD ~ △CEB

Fig. 7-19

Solutions

(a) $\dfrac{6}{12} = \dfrac{8}{16} = \dfrac{12}{24}$ (c) $\dfrac{7}{21} = \dfrac{8}{24} = \dfrac{10}{30}$

(b) $\dfrac{9}{12} = \dfrac{12}{16} = \dfrac{15}{20}$

7.22   Proportions obtained from similar triangles

Find x in each part of Fig. 7-20.

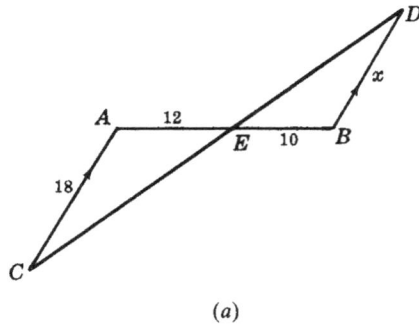

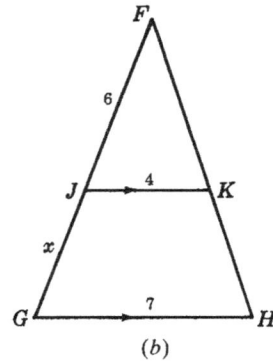

Fig. 7-20

Solutions

(a) Since $\overline{BD} \parallel \overline{AC}$, $\angle A \cong \angle B$ and $\angle C \cong \angle D$; hence, $\triangle AEC \sim \triangle BED$. Then $\dfrac{x}{18} = \dfrac{10}{12}$ and $x = 15$.

(b) Since $\overline{JK} \parallel \overline{GH}$, $\triangle FJK \sim \triangle FGH$ by Principle 7. Hence, $\dfrac{6}{x + 6} = \dfrac{4}{7}$ and $x = 4\frac{1}{2}$.

7.23   Finding heights using ground shadows

A tree casts a 15-ft shadow at a time when a nearby upright pole 6 ft high casts a shadow of 2 ft. Find the height of the tree if both tree and pole make right angles with the ground.

Solution

At the same time in localities near each other, the rays of the sun strike the ground at equal angles; hence, $\angle B \cong \angle B'$ in Fig. 7-21. Since the tree and the pole make right angles with the ground, $\angle C \cong \angle C'$. Hence, $\triangle ABC \sim \triangle A'B'C'$, so $\dfrac{h}{6} = \dfrac{15}{2}$ and $h = 45$ ft.

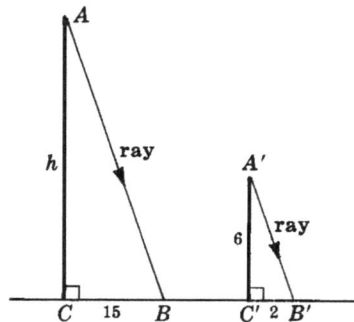

Fig. 7-21

7.24 Proving a similar-triangle problem stated in words

Prove that two isosceles triangles are similar if a base angle of one is congruent to a base angle of the other.

Solution

Given: Isosceles $\triangle ABC$ ($AB = AC$)
Isosceles $\triangle A'B'C'$ ($A'B' = A'C'$)
$\angle B \cong \angle B'$
To Prove: $\triangle ABC \sim \triangle A'B'C'$
Plan: Prove $\angle C \cong \angle C'$ and use Principle 3.

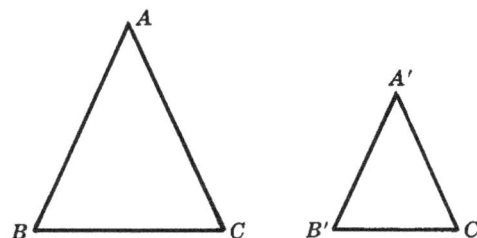

PROOF:

Statements	Reasons
1. $\angle B \cong \angle B'$	1. Given
2. $\angle B \cong \angle C, \angle B' \cong \angle C'$	2. Base angles of an isosceles triangle are congruent.
3. $\angle C \cong \angle C'$	3. Things $\cong$ to $\cong$ things are $\cong$ to each other.
4. $\triangle ABC \sim \triangle A'B'C'$	4. Two triangles are similar if two angles of one triangle are congruent to two angles of the other.

7.25 Proving a proportion problem involving similar triangles

Given: Parallelogram $ABCD$
To Prove: $\dfrac{AE}{BC} = \dfrac{AF}{BF}$
Plan: Prove ($AEF \sim \triangle BCF$)

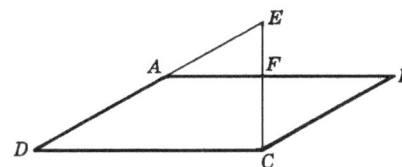

PROOF:

Statements	Reasons
1. $ABCD$ is a parallelogram.	1. Given
2. $\overline{ED} \parallel \overline{BC}$	2. Opposite sides of a parallelogram are parallel.
3. $\angle DEC \cong \angle ECB$	3. Alternate interior angles of parallel lines are congruent.
4. $\angle EFA \cong \angle BFC$	4. Vertical angles are congruent.
5. $\triangle AEF \sim \triangle BCF$	5. Two triangles are similar if two angles of one triangle are congruent to two angles of the other.
6. $\dfrac{AE}{BC} = \dfrac{AF}{BF}$	6. Corresponding sides of similar triangles are in proportion.

7.5 Extending A Basic Proportion Principle

PRINCIPLE 1: *Corresponding sides of similar triangles are in proportion.*

PRINCIPLE 2: *Corresponding segments of similar triangles are in proportion.*

PRINCIPLE 3: *Corresponding segments of similar polygons are in proportion.*

When *segments* replaces *sides*, Principle 1 becomes the more general Principle 2. When *polygons* replaces *triangles*, Principle 2 becomes the even more general Principle 3.

By *segments* we mean straight or curved segments such as altitudes, medians, angle bisectors, radii of inscribed or circumscribed circles, and circumferences of inscribed or circumscribed circles.

The *ratio of similitude* of two similar polygons is the ratio of any pair of corresponding lines.

Corollaries of Principles 2 and 3, such as the following, can be devised for any combination of corresponding lines:

1. Corresponding *altitudes* of similar triangles have the same ratio as any two correponding *medians*.

Thus if $\triangle ABC \sim \triangle A'B'C'$ in Fig. 7-22, then $\dfrac{h}{h'} = \dfrac{m}{m'}$.

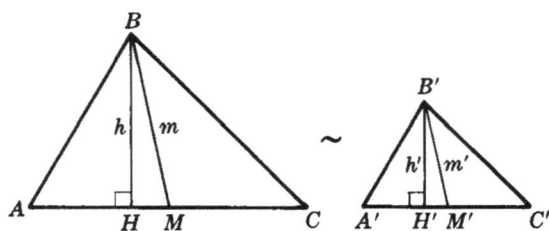

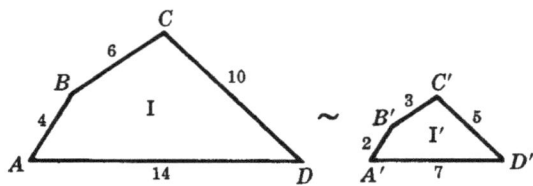

Fig. 7-22   Fig. 7-23

2. *Perimeters* of similar polygons have the same ratio as any two corresponding *sides*.

Thus in Fig. 7-23, if quadrilateral I $\sim$ quadrilateral I', then $\frac{34}{17} = \frac{4}{2} = \frac{6}{3} = \frac{10}{5} = \frac{14}{7}$.

SOLVED PROBLEMS

7.26 Line ratios from similar triangles

(a) In two similar triangles, corresponding sides are in the ratio 3:2. Find the ratio of corresponding medians [see Fig. 7-24(a)].

(b) The sides of a triangle are 4, 6, and 7 [Fig. 7-24(b)]. If the perimeter of a similar triangle is 51, find its longest side.

(c) In $\triangle ABC$ of Fig. 7-24(c), $BC = 25$ and the measure of the altitude to $\overline{BC}$ is 10. A line segment terminating in the sides of the triangle is parallel to $\overline{BC}$ and 3 units from A. Find its length.

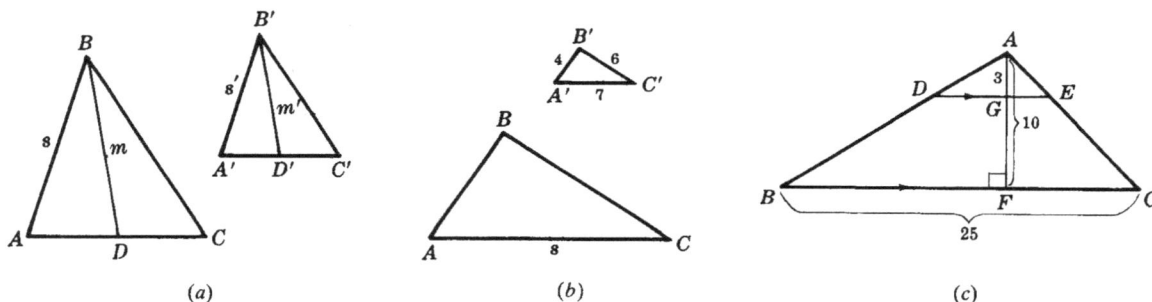

(a)   (b)   (c)

Fig. 7-24

Solutions

(a) If $\triangle ABC \sim \triangle A'B'C'$ and $\frac{s}{s'} = \frac{3}{2}$, then $\frac{m}{m'} = \frac{3}{2}$.

(b) The perimeter of $\triangle A'B'C'$ is $4 + 6 + 7 = 17$. Since $\triangle ABC \sim \triangle A'B'C'$, $\frac{s}{7} = \frac{51}{17}$ and $s = 21$.

(c) Since $\triangle ADE \sim ABC$, $\frac{DE}{25} = \frac{3}{10}$ and $DE = 7\frac{1}{2}$.

7.27 Line ratios from similar polygons

Complete each of the following statements:

(a) If corresponding sides of two similar polygons are in the ratio of 4:3, then the ratio of their perimeters is $\frac{?}{}$.

(b) The perimeters of two similar quadrilaterals are 30 and 24. If a side of the smaller quadrilateral is 8, the corresponding side of the larger is $\frac{?}{}$.

(c) If each side of a pentagon is tripled and the angles remain the same, then each diagonal is $\frac{?}{}$.

Solutions

(a) Since the polygons are similar, $\frac{p}{p'} = \frac{s}{s'} = \frac{4}{3}$.

(b) Since the quadrilaterals are similar, $\frac{s}{s'} = \frac{p}{p'}$. Then $\frac{s}{8} = \frac{30}{24}$ and $s = 10$.

(c) Tripled, since polygons are similar if their corresponding angles are congruent and their corresponding sides are in proportion.

7.6 Proving Equal Products of Lengths of Segments

In a problem, to prove that the product of the lengths of two segments equals the product of the lengths of another pair of segments, it is necessary to set up the proportion which will lead to the two equal products.

SOLVED PROBLEM

7.28 Proving an equal-products problem

Prove that if two secants intersect outside a circle, the product of the lengths of one of the secants and its external segment equals the product of the lengths of the other secant and its external segment.

Solution

Given: Secants $\overline{AB}$ and $\overline{AC}$.
To Prove: $AB \times AD = AC \times AE$
Plan: Prove $\triangle ABE \sim \triangle ACD$ to obtain $\frac{AB}{AC} = \frac{AE}{AD}$.

PROOF:

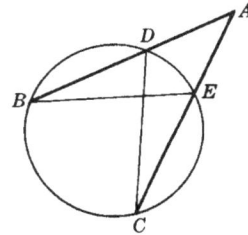

Statements	Reasons
1. Draw $\overline{BE}$ and $\overline{CD}$.	1. A segment may be drawn between any two points.
2. $\angle A \cong \angle A$	2. Reflexive property.
3. $\angle B \cong \angle C$	3. Angles inscribed in the same arc are congruent.
4. $\triangle AEB \sim \triangle ADC$	4. Two triangles are similar if two angles of one triangle are congruent respectively to two angles of the other.
5. $\frac{AB}{AC} = \frac{AE}{AD}$	5. Corresponding sides of similar triangles are in proportion.
6. $AB \times AD = AC \times AE$	6. In a proportion, the product of the means equals the product of the extremes.

7.7 Segments Intersecting Inside and Outside a Circle

PRINCIPLE 1: *If two chords intersect within a circle, the product of the lengths of the segments of one chord equals the product of the lengths of the segments of the other.*

Thus in Fig. 7-25, $AE \times EB = CE \times ED$.

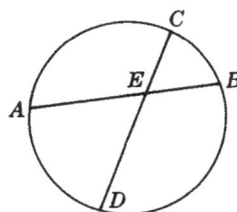

Fig. 7-25

PRINCIPLE 2: *If a tangent and a secant intersect outside a circle, the tangent is the mean proportional between the secant and its external segment.*

Thus in Fig. 7-26, if $\overline{PA}$ is a tangent, then $\dfrac{AB}{AP} = \dfrac{AP}{AC}$.

PRINCIPLE 3: *If two secants intersect outside a circle, the product of the lengths of one of the secants and its external segment equals the product of the lengths of the other secant and its external segment.*

Thus in Fig. 7-27, $AB \times AD = AC \times AE$.

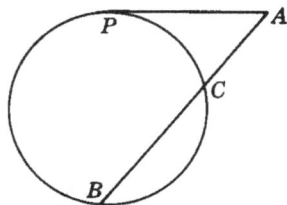
Fig. 7-26

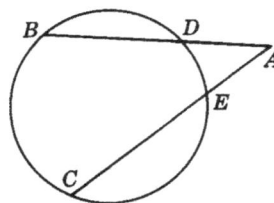
Fig. 7-27

SOLVED PROBLEMS

7.29 Applying principle 1

Find x in each part of Fig. 7-28, if chords $\overline{AB}$ and $\overline{CD}$ intersect in E.

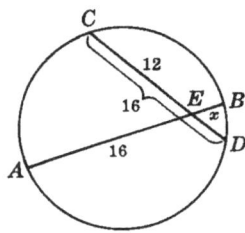
(a)

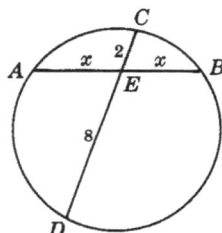
(b)

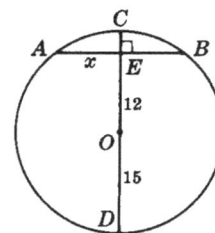
(c) Diameter $CD \perp AB$

Fig. 7-28

Solutions

(a) $ED = 4$. Then $16x = 4(12)$, so that $16x = 48$ or $x = 3$.

(b) $AE = EB = x$. Then $x^2 = 8(2)$, so $x^2 = 16$ and $x = 4$.

(c) $CE = 3$ and $AE = EB = x$. Then $x^2 = 27(3)$ or $x^2 = 81$, and $x = 9$.

7.30 Applying principle 2

Find x in each part of Fig. 7-29 if tangent $\overline{AP}$ and $\overline{AB}$ intersect at A.

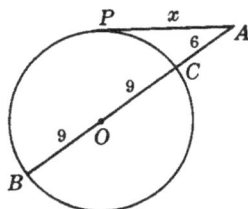
(a)

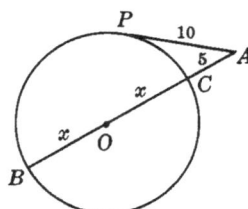
(b)

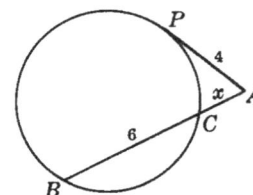
(c)

Fig. 7-29

Solutions

(a) $AB = 9 + 9 + 6 = 24$. Then $x^2 = 24(6)$ or $x^2 = 144$, and $x = 12$.

(b) $AB = 2x + 5$. Then $5(2x + 5) = 100$ and $x = 7\frac{1}{2}$.

(c) $AB = x + 6$. Then $x(x + 6) = 16$ or $x^2 + 6x - 16 = 0$. Factoring gives $(x + 8)(x - 2) = 0$ and $x = 2$.

7.31 Applying principle 3

Find x in each part of Fig. 7-30 if secants $\overline{AB}$ and $\overline{AC}$ intersect in A.

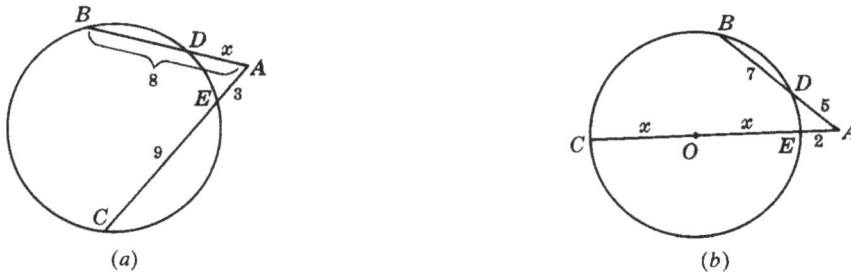

Fig. 7-30

Solutions

(a) $AC = 12$. Then $8x = 12(3)$ and $x = 4\frac{1}{2}$.

(b) $AC = 2x + 2$ and $AB = 12$. Then $2(2x + 2) = 12(5)$ and $x = 14$.

7.8 Mean Proportionals in a Right Triangle

PRINCIPLE 1: *The length of the altitude to the hypotenuse of a right triangle is the mean proportional be-tween the lengths of the segments of the hypotenuse.*

Thus in right $\triangle ABC$ (Fig. 7-31), $\dfrac{BD}{CD} = \dfrac{CD}{DA}$.

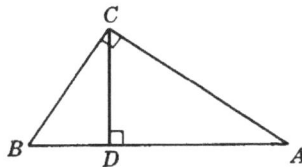

Fig. 7-31

PRINCIPLE 2: *In a right triangle, the length of either leg is the mean proportional between the length of the hypotenuse and the length of the projection of that leg on the hypotenuse.*

Thus in right $\triangle ABC$ (Fig. 7-31), $\dfrac{AB}{BC} = \dfrac{BC}{BD}$ and $\dfrac{AB}{AC} = \dfrac{AC}{AD}$.

A proof of this principle is given in Chapter 16.

SOLVED PROBLEMS

7.32 Finding mean proportionals in a right triangle

In each triangle in Fig. 7-32, find x and y.

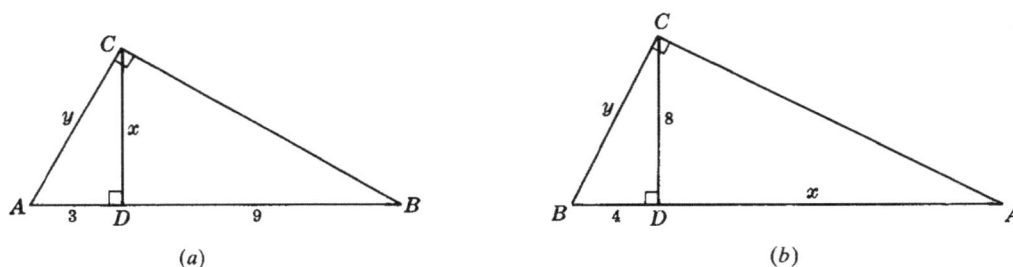

Fig. 7-32

Solutions

(a) By Principle 1, $\frac{3}{x} = \frac{x}{9}$ or $x^2 = 27$, and $x = 3\sqrt{3}$. By Principle 2, $\frac{12}{y} = \frac{y}{3}$, so $y^2 = 36$ and $y = 6$.

(b) By Principle 1, $\frac{x}{8} = \frac{8}{4}$ and $x = 16$. By Principle 2, $\frac{20}{y} = \frac{y}{4}$, so $y^2 = 80$ and $y = 4\sqrt{5}$.

7.9 Pythagorean Theorem

In a right triangle, the square of the length of the hypotenuse equals the sum of the squares of the lengths of the legs.

Thus in Fig. 7-33, $c^2 = a^2 + b^2$.

A proof of the Pythagorean Theorem is given in Chapter 16.

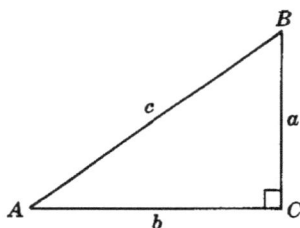

Fig. 7-33

7.9A Tests for Right, Acute, and Obtuse Triangles

If $c^2 = a^2 + b^2$ applies to the three sides of a triangle, then the triangle is a right triangle; but if $c^2 \neq a^2 + b^2$, then the triangle is not a right triangle.

In $\triangle ABC$, if $c^2 < a^2 + b^2$ where c is the longest side of the triangle, then the triangle is an acute triangle. Thus in Fig. 7-34, $9^2 < 6^2 + 8^2$ (that is, $81 < 100$); hence, $\triangle ABC$ is an acute triangle.

In $\triangle ABC$, if $c^2 > a^2 + b^2$ where c is the longest side of the triangle, then the triangle is an obtuse triangle. Thus in Fig. 7-35, $11^2 > 6^2 + 8^2$ (that is, $121 > 100$); hence, $\triangle ABC$ is an obtuse triangle.

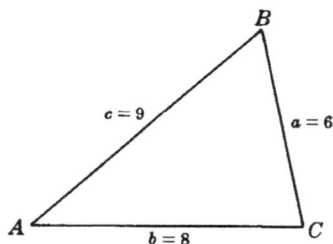

Fig. 7-34

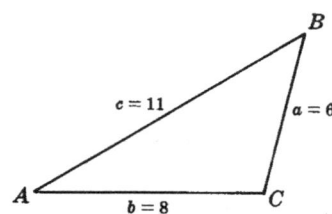

Fig. 7-35

SOLVED PROBLEMS

7.33 Finding the sides of a right triangle

In Fig. 7-36, (a) find the length of hypotenuse c if $a = 12$ and $b = 9$; (b) find a if $b = 6$ and $c = 8$; (c) find b if $a = 4\sqrt{3}$ and $c = 8$.

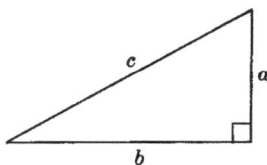

Fig. 7-36

Solutions

(a) $c^2 = a^2 + b^2 = 12^2 + 9^2 = 225$ and $c = 15$.

(b) $a^2 = c^2 - b^2 = 8^2 - 6^2 = 28$ and $a = 2\sqrt{7}$.

(c) $b^2 = c^2 - a^2 = 8^2 - (4\sqrt{3})^2 = 64 - 48 = 16$ and $b = 4$.

7.34 Ratios in a right triangle

In a right triangle, the hypotenuse has length 20 and the ratio of the two arms is 3:4. Find each arm.

Solution

Let the lengths of the two arms be denoted by $3x$ and $4x$. Then $20^2 = (3x)^2 + (4x)^2$. Multiplying out, we get $400 = 9x^2 + 16x^2$ or $400 = 25x^2$, and $x = 4$; hence, the arms have lengths 12 and 16.

7.35 Applying the pythagorean theorem to an isosceles triangle

Find the length of the altitude to the base of an isosceles triangle if the base is 8 and the equal sides are 12.

Solution

The altitude h of an isosceles triangle bisects the base (Fig. 7-37). Then $h^2 = a^2 - (\frac{1}{2}b)^2 = 12^2 - 4^2 = 128$ and $h = 8\sqrt{2}$.

7.36 Applying the pythagorean theorem to a rhombus

In a rhombus, find (a) the length of a side s if the diagonals are 30 and 40; (b) the length of a diagonal d if a side is 26 and the other diagonal is 20.

Solution

The diagonals of a rhombus are perpendicular bisectors of each other; hence, $s^2 = (\frac{1}{2}d)^2 + (\frac{1}{2}d')^2$ in Fig. 7-38.

(a) If $d = 30$ and $d' = 40$, then $s^2 = 15^2 + 20^2 = 625$ or $s = 25$.

(b) If $s = 26$ and $d' = 20$, then $26^2 = (\frac{1}{2}d)^2 + 10^2$ or $576 = (\frac{1}{2}d)^2$. Thus, $\frac{1}{2}d = 24$ or $d = 48$.

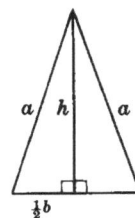

Fig. 7-37

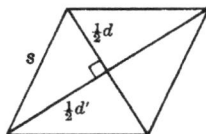

Fig. 7-38

7.37 Applying the pythagorean theorem to a trapezoid

Find x in each part of Fig. 7-39 if $ABCD$ is a trapezoid.

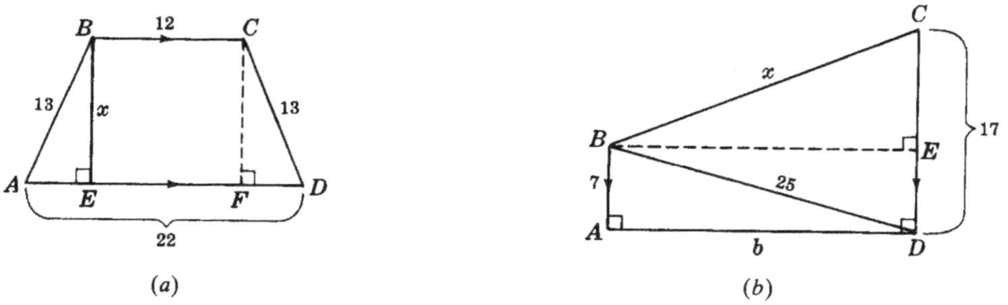

Fig. 7-39

Solutions

The dashed perpendiculars in the diagrams are additional segments needed only for the solutions. Note how rectangles are formed by these added segments.

(a) $EF = BC = 12$ and $AE = \frac{1}{2}(22 - 12) = 5$. Then $x^2 = 13^2 - 5^2 = 144$ or $x = 12$.

(b) $b^2 = 25^2 - 7^2 = 576$ or $b = 24$; also, $BE = b = 24$ and $CE = 17 - 7 = 10$. Then $x^2 = 24^2 + 10^2$ or $x = 26$.

7.38 Applying the pythagorean theorem to a circle

(a) Find the distance d from the center of a circle of radius 17 to a chord whose length is 30 [Fig. 7-40(a)].

(b) Find the length of a common external tangent to two externally tangent circles with radii 4 and 9 [Fig. 7-40(b)].

Fig. 7-40

Solutions

(a) $BC = \frac{1}{2}(30) = 15$. Then $d^2 = 17^2 - 15^2 = 64$ and $d = 8$.

(b) $\overline{OS} \cong \overline{PR}$, $RS = 4$, $OQ = 13$, and $SQ = 9 - 4 = 5$. Then in right $\triangle OSQ$, $(OS)^2 = 13^2 - 5^2 = 144$ so $OS = 12$; hence $PR = 12$.

7.10 Special Right Triangles

7.10A The 30°-60°-90° Triangle

A 30°-60°-90° triangle is one-half an equilateral triangle. Thus, in right $\triangle ABC$ (Fig. 7-41), $a = \frac{1}{2}c$. Consider that $c = 2$; then $a = 1$, and the Pythagorean Theorem gives

$$b^2 = c^2 - a^2 = 2^2 - 1^2 = 3 \quad \text{or} \quad b = \sqrt{3}$$

The ratio of the sides is then $a{:}b{:}c = 1{:}\sqrt{3}{:}2$

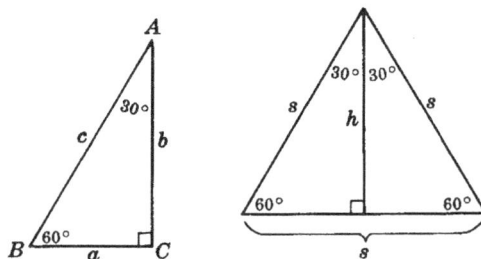

Fig. 7-41

Principles of the 30°-60°-90° Triangle

PRINCIPLE 1: *The length of the leg opposite the 30° angle equals one-half the length of the hypotenuse.*

In Fig. 7-41, $a = \frac{1}{2}c$.

PRINCIPLE 2: *The length of the leg opposite the 60° angle equals one-half the length of the hypotenuse times the square root of 3.*

In Fig. 7-41, $b = \frac{1}{2}c\sqrt{3}$.

PRINCIPLE 3: *The length of the leg opposite the 60° angle equals the length of the leg opposite the 30° angle times the square root of 3.*

In Fig. 7-41, $b = a\sqrt{3}$.

Equilateral-Triangle Principle

PRINCIPLE 4: *The length of the altitude of an equilateral triangle equals one-half the length of a side times the square root of 3. (Principle 4 is a corollary of Principle 2.)*

In Fig. 7-41, $h = \frac{1}{2}s\sqrt{3}$.

7.10B The 45°-45°-90° Triangle

A 45°-45°-90° triangle is one-half a square. In right triangle *ABC* (Fig. 7-42), $c^2 = a^2 + a^2$ or $c = a\sqrt{2}$. Hence, the ratio of the sides is $a{:}a{:}c = 1{:}1{:}\sqrt{2}$.

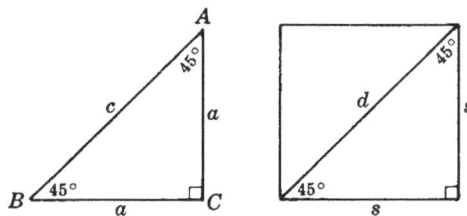

Fig. 7-42

Principles of the 45°-45°-90° Triangle

PRINCIPLE 5: *The length of a leg opposite a 45° angle equals one-half the length of the hypotenuse times the square root of 2.*

In Fig. 7-42, $a = \frac{1}{2}c\sqrt{2}$

PRINCIPLE 6: *The length of the hypotenuse equals the length of a side times the square root of 2.*

In Fig. 7-42, $c = a\sqrt{2}$.

Square Principle

PRINCIPLE 7:   *In a square, the length of a diagonal equals the length of a side times the square root of 2.*

In Fig. 7-42, $d = s\sqrt{2}$

SOLVED PROBLEMS

7.39   Applying principles 1 to 4

   (a) If the length of the hypotenuse of a 30°-60°-90° triangle is 12, find the lengths of its legs [Fig.7-43(a)].

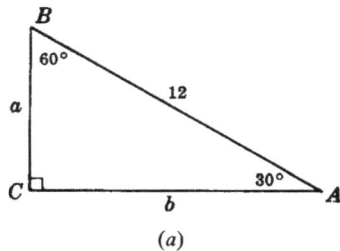

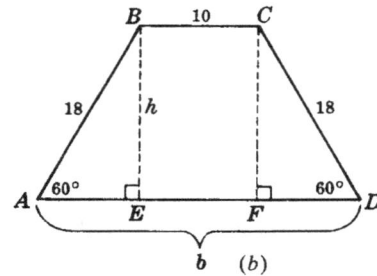

Fig. 7-43

   (b) Each leg of an isosceles trapezoid has length 18. If the base angles are 60° and the upper base is 10, find the lengths of the altitude and the lower base [Fig. 7-43(b)].

Solutions

(a) By Principle 1, $a = \frac{1}{2}(12) = 6$. By Principle 2, $b = \frac{1}{2}(12)\sqrt{3} = 6\sqrt{3}$.

(b) By Principle 2, $h = \frac{1}{2}(18)\sqrt{3} = 9\sqrt{3}$. By Principle 1, $AE = FD = \frac{1}{2}(18) = 9$; hence, $b = 9 + 10 + 9 = 28$.

7.40   Applying principles 5 and 6

   (a) Find the length of the leg of an isosceles right triangle whose hypotenuse has length 28 [Fig. 7-44(a)].

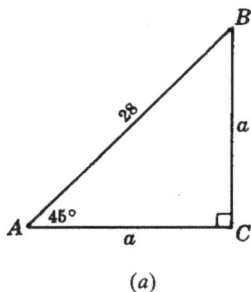

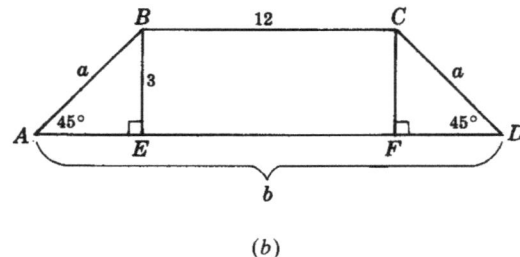

Fig. 7-44

   (b) An isosceles trapezoid has base angles measuring 45°. If the upper base has length 12 and the altitude has length 3, find the lengths of the lower base and each leg [Fig. 7-44(b)].

Solutions

(a) By Principle 5, $a = \frac{1}{2}(28)\sqrt{2} = 14\sqrt{2}$.

(b) By Principle 6, $a = 3\sqrt{2}$, $AE = BE = 3$ and $EF = 12$; hence, $b = 3 + 12 + 3 = 18$.

SUPPLEMENTARY PROBLEMS

7.1. Express each of the following ratios in lowest terms: (7.1)

    (a) 20 cents to 5 cents   (f) 50% to 25%   (k) $\frac{1}{2}$ lb to $\frac{1}{4}$ lb

    (b) 5 dimes to 15 dimes   (g) $15°$ to $75°$ (l) $2\frac{1}{2}$ days to $3\frac{1}{2}$ days

    (c) 30 lb to 25 lb   (h) 33% to 77%   (m) 5 ft to $\frac{1}{4}$ ft

    (d) $20°$ to $14°$ (i) \$2.20 to \$3.30   (n) $\frac{1}{2}$ yd to $1\frac{1}{2}$ yd

    (e) 27 min to 21 min   (j) \$.84 to \$.96   (o) $16\frac{1}{2}$ m to $5\frac{1}{2}$ m

7.2. Express each of the following ratios in lowest terms: (7.2)

    (a) 1 year to 2 months   (e) 2 yd to 2 ft   (i) 100 lb to 1 ton

    (b) 2 weeks to 5 days   (f) $2\frac{1}{3}$ yd to 2 ft   (j) \$2 to 25 cents

    (c) 3 days to 3 weeks   (g) $1\frac{1}{2}$ ft to 9 in   (k) 2 quarters to 3 dimes

    (d) $2\frac{1}{2}$ h to 20 min   (h) 2 g to 8 mg   (l) 1 yd^2 to 2 ft^2

7.3. Express each of the following ratios in lowest terms: (7.3)

    (a) 20 cents to 30 cents to \$1   (f) 2 h to $\frac{1}{2}$ h to 15 min

    (b) \$3 to \$1.50 to 25 cents   (g) 1 ton to 200 lb to 40 lb

    (c) 1 quarter to 1 dime to 1 nickel   (h) 3 lb to 1 lb to 8 oz

    (d) 1 day to 4 days to 1 week   (i) 1 gal to 1 qt to 1 pt

    (e) $\frac{1}{2}$ day to 9 h to 3 h

7.4. Express each of the following ratios in lowest terms: (7.4)

    (a) 60 to 70   (e) 630 to 105   (i) 0.002 to 0.007   (m) $7\frac{1}{2}$ to $2\frac{1}{2}$

    (b) 84 to 7   (f) 1760 to 990   (j) 0.055 to 0.005   (n) $1\frac{1}{2}$ to 10

    (c) 65 to 15   (g) 0.7 to 2.1   (k) 6.4 to 8   (o) $\frac{5}{6}$ to $1\frac{2}{3}$

    (d) 125 to 500   (h) 0.36 to 0.24   (l) 144 to 2.4   (p) $\frac{7}{4}$ to $\frac{1}{8}$

7.5. Express each of the following ratios in lowest terms: (7.4)

    (a) x to $3x$ (d) $2\pi r$ to πD (g) S^3 to $6S^2$ (j) $15y$ to $10y$ to $5y$

    (b) $15c$ to 5   (e) πab to πa^2 (h) $9r^2$ to $6rt$ (k) x^3 to x^2 to x

    (c) $11d$ to 22   (f) $4S$ to S^2 (i) x to $4x$ to $10x$ (l) $12w$ to $10w$ to $8w$ to $2w$

7.6. Use x as the common factor to represent the following numbers and their sum: (7.4)

    (a) Two numbers whose ratio is 5:4   (c) Three numbers whose ratio is 2:5:11

    (b) Two numbers whose ratio is 9:1   (d) Five numbers whose ratio is 1:2:2:3:7

7.7. If two angles in the ratio of 5:4 are represented by $5x$ and $4x$, express each of the following statements as an equation; then find x and the angles: (7.5)

    (a) The angles are adjacent and together form an angle measuring $45°$.

    (b) The angles are complementary.

    (c) The angles are supplementary.

    (d) The angles are two angles of a triangle whose third angle is their difference.

7.8. If three angles in the ratio of $7:6:5$ are represented by $7x, 6x$, and $5x$, express each of the following statements as an equation; then find x and the angles: (7.6)

(a) The first and second are adjacent and together form an angle measuring $91°$.

(b) The first and third are supplementary.

(c) The first and one-half the second are complementary.

(d) The angles are the three angles of a triangle.

7.9. Solve the following proportions for x: (7.7)

(a) $x:6 = 8:3$ (d) $x:2 = 10:x$ (g) $a:b = c:x$

(b) $5:4 = 20:x$ (e) $(x+4):3 = 3:(x-4)$ (h) $x:2y = 18y:x$

(c) $9:x = x:4$ (f) $(2x+8):(x+2) = (2x+5):(x+1)$

7.10. Solve the following proportions for x: (7.7)

(a) $\frac{5}{7} = \frac{15}{x}$ (c) $\frac{3}{x} = \frac{x}{12}$ (e) $\frac{x+2}{5} = \frac{6}{3}$ (g) $\frac{2x}{x+7} = \frac{3}{5}$

(b) $\frac{7}{x} = \frac{3}{2}$ (d) $\frac{x}{5} = \frac{15}{x}$ (f) $\frac{x-1}{3} = \frac{5}{x+1}$ (h) $\frac{a}{x} = \frac{x}{b}$

7.11. Find the fourth proportional for each of the following sets of numbers: (7.8)

(a) $1,3,5$ (c) $2,3,4$ (e) $3,2,5$ (g) $2,8,8$

(b) $8,6,4$ (d) $3,4,2$ (f) $\frac{1}{3},2,5$ (h) $b,2a,3b$

7.12. Find the positive mean proportional between each of the following pairs of numbers: (7.9)

(a) 4 and 9 (c) $\frac{1}{3}$ and 27 (e) 2 and 5 (g) p and q

(b) 12 and 3 (d) $2b$ and $8b$ (f) 3 and 9 (h) a^2 and b

7.13. From each of these equations, form a proportion whose fourth term is x: (a) $cx = bd$; (b) $pq = ax$; (c) $hx = a^2$; (d) $3x = 7$; (e) $x = ab/c$. (7.10)

7.14. In each of the following equations, find the ratio of x to y: (a) $2x = y$; (b) $3y = 4x$; (c) $x = \frac{1}{2}y$; (d) $ax = hy$; (e) $x = by$. (7.10)

7.15. Which of the following is not a proportion? (7.11)

(a) $\frac{4}{3} \stackrel{?}{=} \frac{24}{18}$ (b) $\frac{3}{3} \stackrel{?}{=} \frac{7}{12}$ (c) $\frac{25}{45} \stackrel{?}{=} \frac{10}{18}$ (d) $\frac{.2}{.3} \stackrel{?}{=} \frac{6}{9}$ (e) $\frac{x}{8} \stackrel{?}{=} \frac{3}{4}$ when $x = 6$.

7.16. From each of the following, form a new proportion whose first term is x. Then find x. (7.11)

(a) $\frac{3}{2} = \frac{9}{x}$ (b) $\frac{1}{x} = \frac{5}{4}$ (c) $\frac{a}{x} = \frac{2}{b}$ (d) $\frac{x+5}{5} = \frac{11}{10}$ (e) $\frac{x-20}{20} = \frac{1}{4}$

7.17. Find x in each of these pairs of proportions: (7.11)

(a) $a:b = c:x$ and $a:b = c:d$ (c) $2:3x = 4:5y$ and $2:15 = 4:5y$

(b) $5:7 = x:42$ and $5:7 = 35:42$ (d) $7:5x-2 = 14:3y$ and $7:18 = 14:3y$

7.18. Find x in each of the following proportions: (7.12)

(a) $\frac{x-7}{8} = \frac{7}{4}$ (b) $\frac{x+y}{6} = \frac{x-y}{3} = \frac{1}{3}$ (c) $\frac{2x-y}{8} = \frac{y-1}{10} = \frac{1}{2}$

7.19. Find x in each part of Fig. 7-45. (7.13)

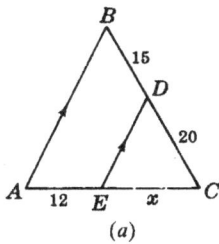 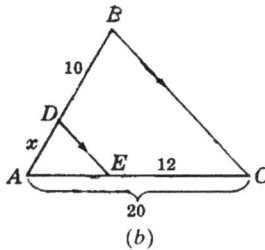 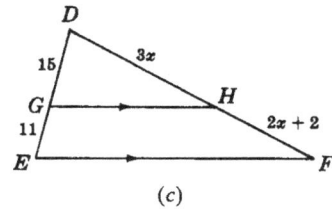

          (a)   (b)   (c)

Fig. 7-45

7.20. In which parts of Fig. 7-46 is a line parallel to one side of the triangle? (7.13)

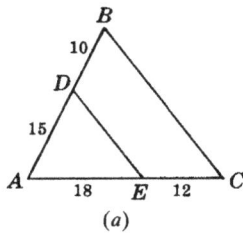 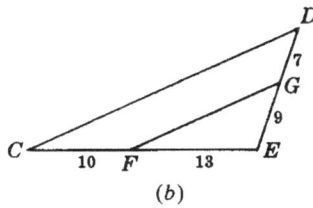 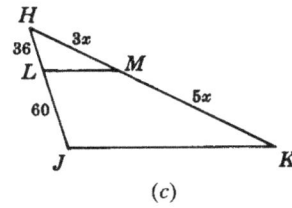

          (a)   (b)   (c)

Fig. 7-46

7.21. Find x in each part of Fig. 7-47. (7.14)

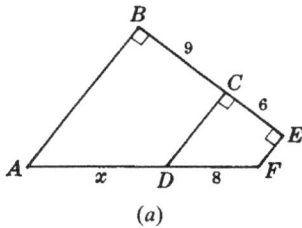 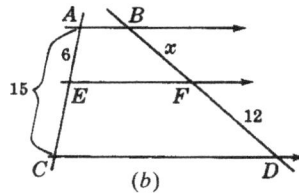 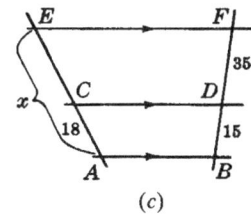

          (a)   (b)   (c)

Fig. 7-47

7.22. Find x in each part of Fig. 7-48. (7.15)

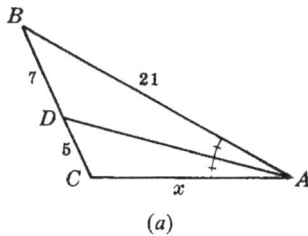 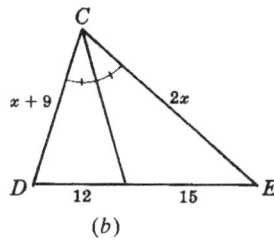 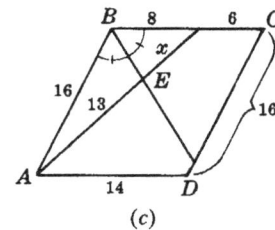

          (a)   (b)   (c)

Fig. 7-48

7.23. Prove that three or more parallel lines divide any two transversals proportionately. (7.16)

7.24. In similar triangles ABC and $A'B'C'$ of Fig. 7-49, $\angle B$ and $\angle B'$ are corresponding angles. Find $m\angle B$ if
(a) $m\angle A' = 120°$ and $m\angle C' = 25°$; (b) $m\angle A' + m\angle C' = 127°$. (7.17)

7.25. In similar triangles *ABC* and *A'B'C'* of Fig. 7-50, $\angle A \cong \angle A'$ and $\angle B \cong \angle B'$. (a) Find *a* if *c* = 24; (b) find *b* if *a* = 20; (c) find *c* if *b* = 63.   (7.17)

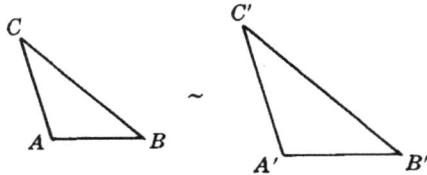

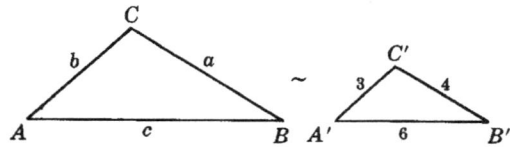

Fig. 7-49   Fig. 7-50

7.26. In each part of Fig. 7-51, show that the indicated triangles are similar.   (7.18)

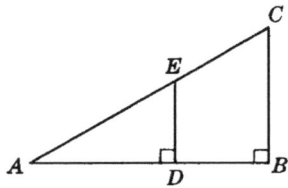

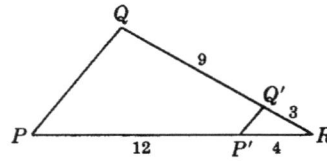

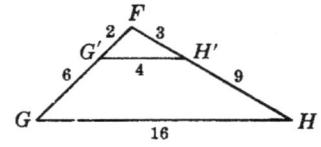

(a)   △*ADE* ~ △*ABC*   (b)   △*RQP* ~ △*RQ'P'*   (c)   △*FG'H'* ~ △*FGH*

Fig. 7-51

7.27. In each part of Fig. 7-52, two pairs of congruent angles can be used to prove the indicated triangles similar. Find the congruent angles.   (7.18)

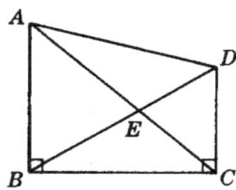

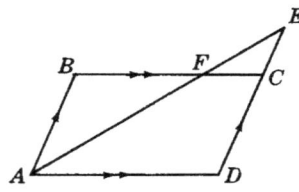

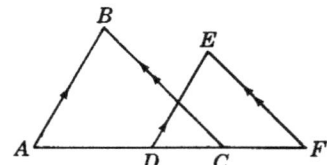

(a)   △*AEB* ~ △*DEC*   (b)   △*BAF* ~ △*AED*   (c)   △*ABC* ~ △*DEF*

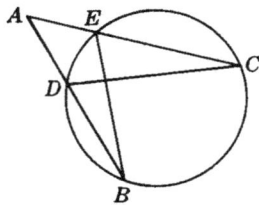

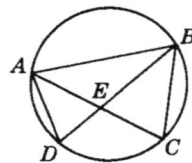

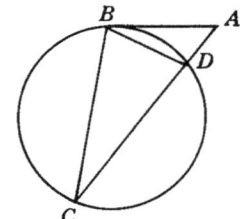

(d)   △*AEB* ~ △*ADC*   (e)   △*ADE* ~ △*ACB*   (f)   △*ABC* ~ △*ABD*
                                    BC ≅ *CD* $\overline{AB}$ is a tangent

Fig. 7-52

7.28. In each part of Fig. 7-53, determine the angles that can be used to prove the indicated triangles similar.   (7.19)

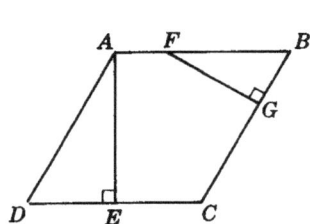

(*a*)   △*AED* ∼ △*FGB*
        ABCD is a parallelogram.

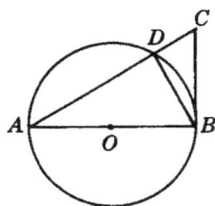

(*b*)   △*ADB* ∼ △*ABC*
        $\overline{AB}$ is a diameter.
        $\overline{BC}$ is a tangent.

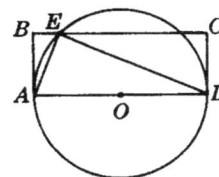

(*c*)   △*ABE* ∼ △*AED*
        AD is a diameter.
        ABCD is a rectangle.

Fig. 7-53

7.29. In each part of Fig. 7-54, determine the pair of congruent angles and the proportion needed to prove the indicated triangles similar.   (7.20)

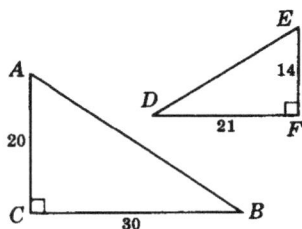

(*a*)   △*ABC* ∼ △*EDF*

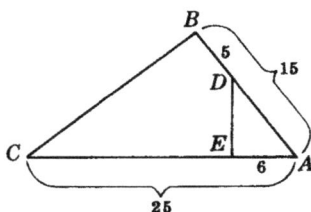

(*b*)   △*ADE* ∼ △*ACB*

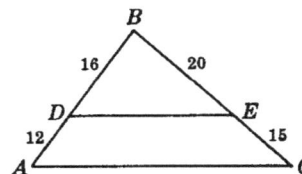

(*c*)   △*BDE* ∼ △*BAC*

Fig. 7-54

7.30. In each part of Fig. 7-55, state the proportion needed to prove the indicated triangles similar.   (7.21)

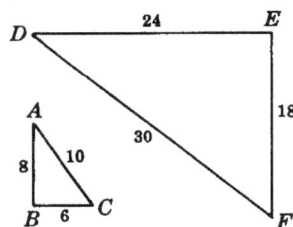

(*a*)   △*ABC* ∼ △*DEF*

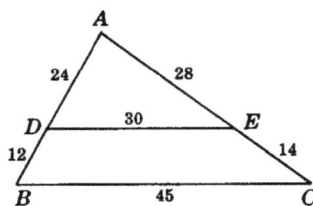

(*b*)   △*ADE* ∼ △*ABC*

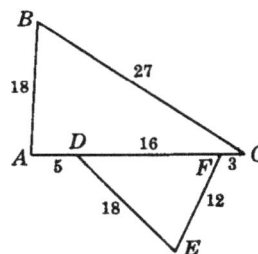

(*c*)   △*DEF* ∼ △*CBA*

Fig. 7-55

7.31. In each part of Fig. 7-56, prove the indicated proportion.   (7.25)

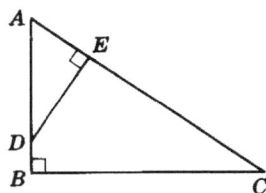

(*a*)   *AD*:*AC* = *DE*:*BC*

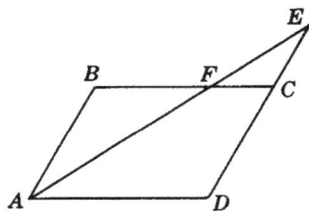

(*b*)   *AB*:*EC* = *BF*:*FC*
        ABCD is a parallelogram.

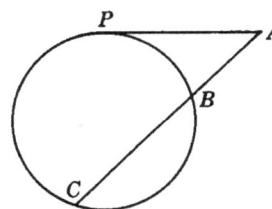

(*c*)   *AC*:*AP* = *AP*:*AB*
        $\overrightarrow{AP}$ is a tangent.

Fig. 7-56

7.32. In △*ABC* (Fig. 7-57), $\overline{DE} \parallel \overline{BC}$. (7.22)

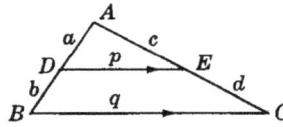

Fig. 7-57

(a) Let $a = 4, AB = 8, p = 10$. Find q. (d) Let $b = 9, p = 20, q = 35$. Find a.

(b) Let $c = 5, AC = 15, q = 24$. Find p. (e) Let $a = 10, p = 24, q = 84$. Find AB.

(c) Let $a = 7, p = 11, q = 22$. Find b. (f) Let $c = 3, p = 4, q = 7$. Find d.

7.33. Find x in each part of Fig. 7-58. (7.22)

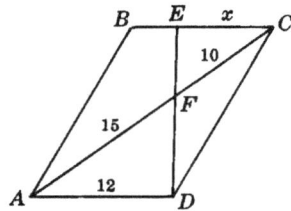

(a)   *ABCD* is a parallelogram.   (b)   *ABCD* is a trapezoid.   (c)

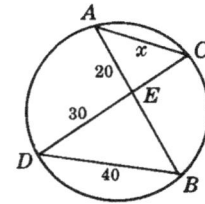

Fig. 7-58

7.34. A 7-ft upright pole near a vertical tree casts a 6-ft shadow. At that same time, find (a) the height of the tree if its shadow is 36 ft long; (b) the length of the shadow of the tree if its height is 77 ft. (7.23)

7.35. Prove each of the following: (7.23)

(a) In △*ABC*, if $\overline{AD}$ and $\overline{CE}$ are altitudes, then $AD:CE = AB:BC$.

(b) In circle O, diameter $\overline{AB}$ and tangent $\overline{BC}$ are sides of △*ABC*. If $\overline{AC}$ intersects the circle in D, then $AD:AB = AB:AC$.

(c) The diagonals of a trapezoid divide each other into proportional segments.

(d) In right △*ABC*, $\overline{CD}$ is the altitude to the hypotenuse $\overline{AB}$, then $AC:CD = AB:BC$.

7.36. Prove each of the following: (7.24)

(a) A line parallel to one side of a triangle cuts off a triangle similar to the given triangle.

(b) Isosceles right triangles are similar to each other.

(c) Equilateral triangles are similar to each other.

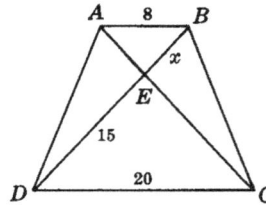

(d) The bases of a trapezoid form similar triangles with the segments of the diagonals.

7.37. Complete each of the following statements: (7.26)

(a) In similar triangles, if corresponding sides are in the ratio $8:5$, then corresponding altitudes are in the ratio _?_.

(b) In similar triangles, if corresponding angle bisectors are in the ratio $3:5$, then their perimeters are in the ratio _?_.

(c) If the sides of a triangle are halved, then the perimeter is _?_ , the angle bisectors are _?_ , the medians are _?_, and the radii of the circumscribed circle are _?_.

7.38. (a) Corresponding sides of two similar triangles have lengths 18 and 12. If an altitude of the smaller has length 10, find the length of the corresponding altitude of the larger. (7.26)

(b) Corresponding medians of two similar triangles have lengths 25 and 15. Find the perimeter of the larger if the perimeter of the smaller is 36.

(c) The sides of a triangle have lengths 5, 7, and 8. If the perimeter of a similar triangle is 100, find its sides.

(d) The bases of a trapezoid have lengths 5 and 20, and the altitude has length 12. Find the length of the altitude of the triangle formed by the shorter base and the nonparallel sides extended to meet.

(e) The bases of a trapezoid have lengths 11 and 22. Its altitude has length 9. Find the distance from the point of intersection of the diagonals to each of the bases.

7.39. Complete each of the following statements: (7.27)

(a) If corresponding sides of two similar polygons are in the ratio $3:7$, then the ratio of their corresponding altitudes is __?__.

(b) If the perimeters of two similar hexagons are in the ratio of 56 to 16, then the ratio of their corresponding diagonals is __?__.

(c) If each side of an octagon is quadrupled and the angles remain the same, then its perimeter is __?__.

(d) The base of a rectangle is twice that of a similar rectangle. If the radius of the circumscribed circle of the first rectangle is 14, then the radius of the circumscribed circle of the second is __?__.

7.40. Prove each of the following: (7.28)

(a) Corresponding angle bisectors of two triangles have the same ratio as a pair of corresponding sides.

(b) Corresponding medians of similar triangles have the same ratio as a pair of corresponding sides.

7.41. Provide the proofs requested in Fig. 7-59. (7.28)

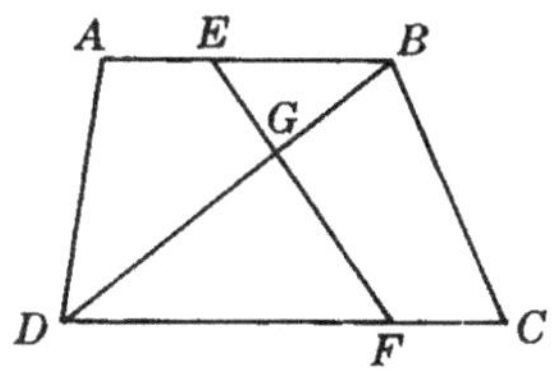

(a) **Given:** Trapezoid $ABCD$
   To Prove:
   $GB \times DF = GD \times EB$

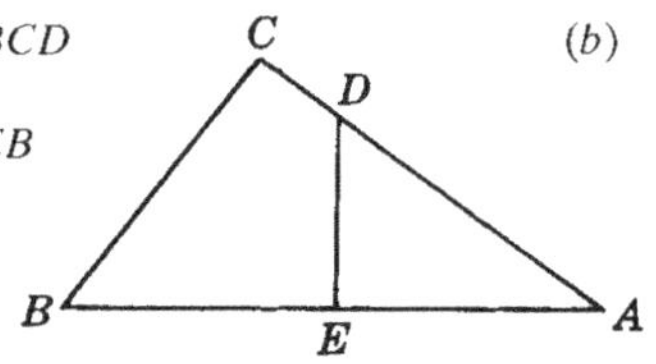

(b) **Given:** $\overline{BC} \perp \overline{AC}$
   $\overline{DE} \perp \overline{AB}$
   To Prove:
   $DE \times AC = BC \times AE$

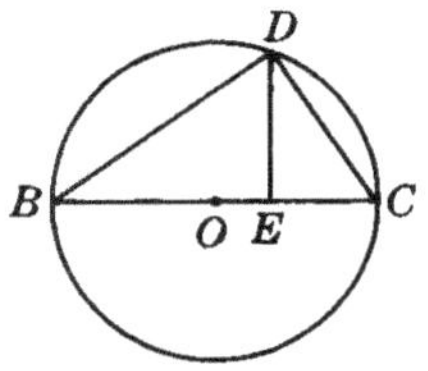

(c) **Given:** Diameter $\overline{BC}$
   $\overline{DE} \perp \overline{BC}$
   To Prove:
   $(BD^2) = BE \times BC$

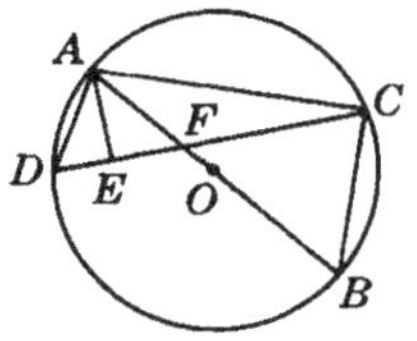

(d) **Given:** Circle O
   Diameter $\overline{AB}$
   $\overline{AE} \perp \overline{CD}$
   To Prove:
   $AD \times BC = AB \times DE$

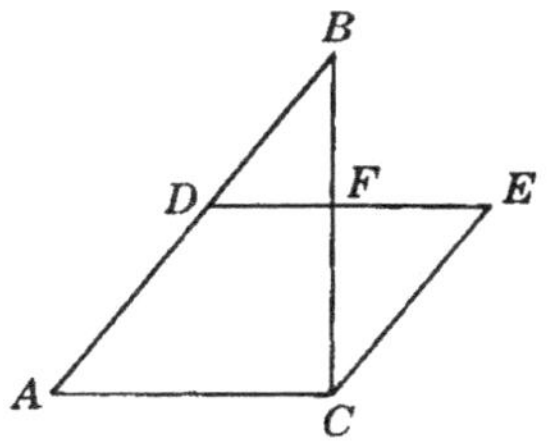

(e) **Given:** $\overline{AB} \parallel \overline{CE}$
   $\overline{AC} \perp \overline{BC}$
   $\overline{DE} \perp \overline{BC}$
   To Prove:
   $AB \times CF = BC \times EC$

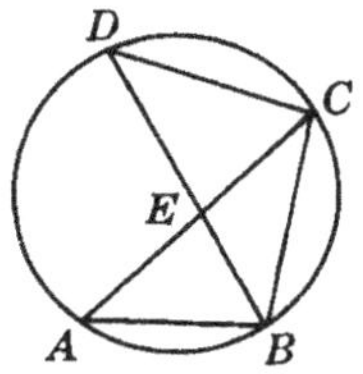

(f) **Given:** $BC \cong CD$
   To Prove:
   $(BC^2) = AC \times EC$

Fig. 7-59

7.42. Prove each of the following: (7.28)

(a) If two chords intersect in a circle, the product of the lengths of the segments of one chord equals the product of the lengths of the segments of the other.

(b) In a right triangle, the product of the lengths of the hypotenuse and the altitude upon it equals the product of the lengths of the legs.

(c) If in inscribed $\triangle ABC$ the bisector of $\angle A$ intersects $\overline{BC}$ in D and the circle in E, then $BD \times AC = AD \times EC$.

7.43. In Fig. 7-60   In Fig. 7-61, diameter $\overline{CD} \perp$ chord $\overline{AB}$ (7.29)

(a) Let $AE = 10, EB = 6, CE = 12$. Find ED.   (e) Let $OD = 10, OE = 8$. Find AB.

(b) Let $AB = 15, EB = 8, ED = 4$. Find CE.   (f) Let $AB = 24, OE = 5$. Find OD.

(c) Let $AE = 6, ED = 4, CD = 13$. Find EB.   (g) Let $OD = 25, EC = 18$. Find AB.

(d) Let $ED = 5, EB = 2(AE), CD = 15$. Find AE.   (h) Let $AB = 8, OD = 5$. Find EC.

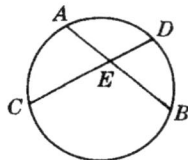
Fig. 7-60

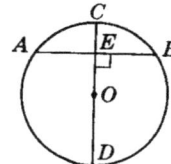
Fig. 7-61

7.44. A point is 12 in from the center of a circle whose radius is 15 in. Find the lengths of the longest and shortest chords that can be drawn through this point. (*Hint:* The longest chord is a diameter, and the shortest chord is perpendicular to this diameter.) (7.29)

7.45. In Fig. 7-62, $\overline{AB}$ is a tangent   In Fig. 7-63, $\overline{CD}$ is a diameter; $\overline{AB}$ is a tangent (7.29)

(a) Let $AC = 16, AD = 4$. Find AB.   (f) Let $AD = 6, OD = 9$. Find AB.

(b) Let $CD = 5, AD = 4$. Find AB.   (g) Let $AD = 2, AB = 8$. Find CD.

(c) Let $AB = 6, AD = 3$. Find AC.   (h) Let $AD = 5, AB = 10$. Find OD.

(d) Let $AC = 20, AB = 10$. Find AD.   (i) Let $AB = 12, AC = 18$. Find OD.

(e) Let $AB = 12, AD = 9$. Find CD.   (j) Let $OD = 5, AB = 12$. Find AD.

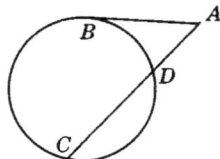
Fig. 7-62

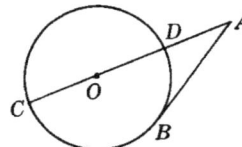
Fig. 7-63

7.46. In Fig. 7-64   In Fig. 7-65, $\overline{CE}$ is a diameter (7.31)

(a) Let $AB = 14, AD = 4, AE = 7$. Find AC.   (e) Let $OC = 3, AE = 6, AD = 8$. Find AB.

(b) Let $AC = 8, AE = 6, AD = 3$. Find BD.   (f) Let $BD = 7, AD = 5, AE = 2$. Find OC.

(c) Let $BD = 5, AD = 7, AE = 4$. Find AC.   (g) Let $OC = 11, AB = 15, AD = 5$. Find AE.

(d) Let $AD = DB, EC = 14, AE = 4$. Find AD. (h) Let $OC = 5, AE = 6, BD = 4$. Find AD.

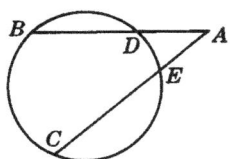

Fig. 7-64

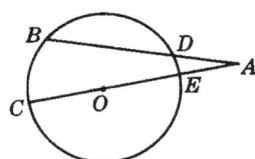

Fig. 7-65

7.47. $\overline{CD}$ is the altitude to hypotenuse $\overline{AB}$ in Fig. 7-66.   (7.32)

(a) If $p = 2$ and $q = 6$, find a and h.   (c) If $p = 16$ and $h = 8$, find q and b.

(b) If $p = 4$ and $a = 6$, find c and h.   (d) If $b = 12$ and $q = 6$, find p and h.

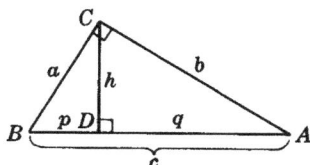

Fig. 7-66

7.48. In a right triangle whose arms have lengths a and b, find the length of the hypotenuse c when   (7.33)

(a) $a = 15, b = 20$ (c) $a = 5, b = 4$ (e) $a = 7, b = 7$

(b) $a = 15, b = 36$ (d) $a = 5, b = 5\sqrt{3}$

7.49. In the right triangle in Fig. 7-67, find the length of each missing arm when   (7.33)

(a) $a = 12, c = 20$ (c) $b = 15, c = 17$ (e) $a = 5\sqrt{2}, c = 10$

(b) $b = 6, c = 8$ (d) $a = 2, c = 4$ (f) $a = \sqrt{5}, c = 2\sqrt{2}$

Fig. 7-67

7.50. Find the lengths of the arms of a right triangle whose hypotenuse has length c if these arms have a ratio of (a) $3:4$ and $c = 15$; (b) $5:12$ and $c = 26$; (c) $8:15$ and $c = 170$; (d) $1:2$ and $c = 10$.   (7.34)

7.51. In a rectangle, find the length of the diagonal if its sides have lengths (a) 9 and 40; (b) 5 and 10.   (7.33)

7.52. In a rectangle, find the length of one side if the diagonal has length 15 and the other side has length (a) 9; (b) 5; (c) 10.   (7.33)

7.53. Of triangles having sides with lengths as follows, which are right triangles?

(a) $33, 55, 44$ (c) $4, 7\frac{1}{2}, 8\frac{1}{2}$ (e) 5 in, 1 ft, 1 ft 1 in   (g) 11 mi, 60 mi, 61 mi

(b) $120, 130, 50$ (d) $25, 7, 24$ (f) 1 yd, 1 yd 1 ft, 1 yd 2 ft   (h) 5 cm, 5 cm, 7 cm

7.54. Is a triangle a right triangle if its sides have the ratio of (a) $3:4:5$; (b) $2:3:4$?

7.55. Find the length of the altitude of an isosceles triangle if each of its two congruent sides has length 10 and its base has length (a) 12; (b) 16; (c) 18; (d) 10.   (7.35)

7.56. In a rhombus, find the length of a side if the diagonals have lengths (a) 18 and 24; (b) 4 and 8; (c) 6 and $6\sqrt{3}$.
(7.36)

7.57. In a rhombus, find the length of a diagonal if a side and the other diagonal have lengths, respectively, (a) 10 and 12; (b) 17 and 16; (c) 4 and 4; (d) 10 and $10\sqrt{3}$. (7.36)

7.58. In isosceles trapezoid *ABCD* in Fig. 7-68 (7.37)

    (a) Find a if $b = 32, b' = 20$, and $h = 8$.　　(c) Find b if $a = 15, b' = 10$, and $h = 12$.

    (b) Find h if $b = 24, b' = 14$, and $a = 13$.　　(d) Find b' if $a = 6, b = 21$, and $h = 3\sqrt{3}$.

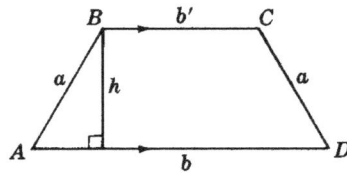
Fig. 7-68

7.59. In a trapezoid *ABCD* in Fig. 7-69 (7.37)

    (a) Find d if $a = 11, b = 3$, and $c = 15$.　　(c) Find d if $a = 5, p = 13$, and $c = 14$.

    (b) Find a if $d = 20, b = 12$, and $c = 36$.　　(d) Find p if $a = 20, c = 28$, and $d = 17$.

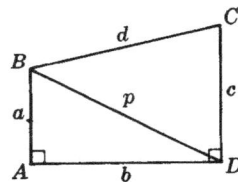
Fig. 7-69

7.60. The radius of a circle is 15. Find (a) the distance from its center to a chord whose length is 18; (b) the length of a chord whose distance from its center is 9. (7.38)

7.61. In a circle, a chord whose length is 16 is at a distance of 6 from the center. Find the length of a chord whose distance from the center is 8. (7.38)

7.62. Two externally tangent circles have radii of 25 and 9. Find the length of a common external tangent. (7.38)

7.63. In a 30°-60°-90° triangle, find the lengths of (a) the legs if the hypotenuse has length 20; (b) the other leg and hypotenuse if the leg opposite 30° has length 7; (c) the other leg and hypotenuse if the leg opposite 60° has length $5\sqrt{3}$. (7.39)

7.64. In an equilateral triangle, find the length of the altitude if the side has length (a) 22; (b) $2a$. Find the side if the altitude has length (c) $24\sqrt{3}$; (d) 24. (7.39)

7.65. In a rhombus which has an angle measuring 60°, find the lengths of (a) the diagonals if a side has length 25; (b) the side and larger diagonal if the smaller diagonal has length 35. (7.39)

7.66. In an isosceles trapezoid which has base angles measuring 60°, find the lengths of (a) the lower base and altitude if the upper base has length 12 and the legs have length 16; (b) the upper base and altitude if the lower base has length 45 and the legs have length 28. (7.39)

7.67. In an isosceles right triangle, find the length of each leg if the hypotenuse has length (a) 34; (b) $2a$. Find the length of the hypotenuse if each leg has length (c) 34; (d) $15\sqrt{2}$. (7.40)

7.68. In a square, find the length of (a) the side if the diagonal has length 40; (b) the diagonal if the side has length 40. (7.40)

7.69. In an isosceles trapezoid which has base angles of measure 45°, find the lengths of (a) the lower base and each leg if the altitude has length 13 and the upper base has length 19; (b) the upper base and each leg if the altitude has length 27 and the lower base has length 65; (c) each leg and the lower base if the upper base has length 25 and the altitude has length 15. (7.40)

7.70. A parallelogram has an angle measuring 45°. Find the distances between its pairs of opposite sides if its sides have lengths 10 and 12. (7.40)

CHAPTER 8

Trigonometry

8.1 Trigonometric Ratios

Trigonometry means "measurement of triangles." Consider its parts: *tri* means "three," *gon* means "angle," and *metry* means "measure." Thus, in trigonometry we study the measurement of triangles.

The following ratios relate the sides and acute angles of a *right triangle*:

1. *Tangent ratio*: The tangent (abbreviated "tan") of an acute angle equals the length of the leg opposite the angle divided by the length of the leg adjacent to the angle.
2. *Sine ratio*: The sine (abbreviated "sin") of an acute angle equals the length of the leg opposite the angle divided by the length of the hypotenuse.
3. *Cosine ratio*: The cosine (abbreviated "cos") of an acute angle equals the length of the leg adjacent to the angle divided by the length of the hypotenuse.

Thus in right triangle *ABC* of Fig. 8-1,

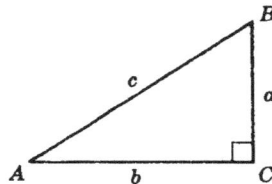

Fig. 8-1

$$\tan A = \frac{\text{length of leg opposite } A}{\text{length of leg adjacent to } A} = \frac{a}{b} \qquad \tan B = \frac{\text{length of leg opposite } B}{\text{length of leg adjacent to } B} = \frac{b}{a}$$

$$\sin A = \frac{\text{length of leg opposite } A}{\text{length of hypotenuse}} = \frac{a}{c} \qquad \sin B = \frac{\text{length of leg opposite } B}{\text{length of hypotenuse}} = \frac{b}{c}$$

$$\cos A = \frac{\text{length of leg adjacent to } A}{\text{length of hypotenuse}} = \frac{b}{c} \qquad \cos B = \frac{\text{length of leg adjacent to } B}{\text{length of hypotenuse}} = \frac{a}{c}$$

If *A* and *B* are the acute angles of a right triangle, then

$$\sin A = \cos B \qquad \cos A = \sin B \qquad \tan A = \frac{1}{\tan B} \qquad \tan B = \frac{1}{\tan A}$$

A scientific calculator can compute the sine, cosine, and tangent of an angle with the "SIN," "COS," and "TAN" buttons, respectively. Make sure the calculator is set to degrees (DEG). For those without a calculator, a table of sines, cosines, and tangents is in the back of this book.

SOLVED PROBLEMS

8.1 Using the table of sines, cosines, and tangents

The following values were taken from a table of sines, cosines, and tangents. State, in equation form, what the values on the first three lines mean. Then use the table at the back of this book to complete the last line.

	Angle	Sine	Cosine	Tangent
(a)	1°	0.0175	0.9998	0.0175
(b)	30°	0.5000	0.8660	0.5774
(c)	60°	0.8660	0.5000	1.7321
(d)	?	?	0.3420	?

Solutions

(a) $\sin 1° = 0.0175$; $\cos 1° = 0.9998$; $\tan 1° = 0.0175$

(b) $\sin 30° = 0.5000$; $\cos 30° = 0.8660$; $\tan 30° = 0.5774$

(c) $\sin 60° = 0.8660$; $\cos 60° = 0.5000$; $\tan 60° = 1.7321$

(d) In the table of trigonometric functions, the cosine value 0.3420 is on the 70° line; hence, the angle measures 70°. Then, from the table, $\sin 70° = 0.9397$ and $\tan 70° = 2.7475$.

8.2 Finding angle measures to the nearest degree

Find the measure of x to the nearest degree if (a) $\sin x = 0.9235$; (b) $\cos x = \frac{21}{25}$ or 0.8400; (c) $\tan x = \sqrt{5}/10$ or 0.2236. Use the table of trigonometric functions.

Solutions

Differences

(a) $\sin 68° = 0.9272$
    $\sin x = 0.9235$ $\to 0.0037$
    $\sin 67° = 0.9205$ $\to 0.0030$

Since $\sin x$ is nearer to $\sin 67°$, $m\angle x = 67°$ to the nearest degree.

(b) $\cos 32° = 0.8480$
    $\cos x = 0.8400$ $\to 0.0080$
    $\cos 33° = 0.8387$ $\to 0.0013$

Since $\cos x$ is nearer to $\cos 33°$, $m\angle x = 33°$ to the nearest degree.

(c) $\tan 13° = 0.2309$
    $\tan x = 0.2236$ $\to 0.0073$
    $\tan 12° = 0.2126$ $\to 0.0110$

Since $\tan x$ is nearer to $\tan 13°$, $m\angle x = 13°$ to the nearest degree.

With a calculator, the above answers can be found with the inverse sine (sin⁻¹), inverse cosine (cos⁻¹), and inverse tangent (tan⁻¹). These usually require first pressing the "2nd" or "INV" button and then "SIN," "COS," or "TAN."

8.3 Finding trigonometric ratios

For each right triangle in Fig. 8-2, find the trigonometric ratios of each acute angle.

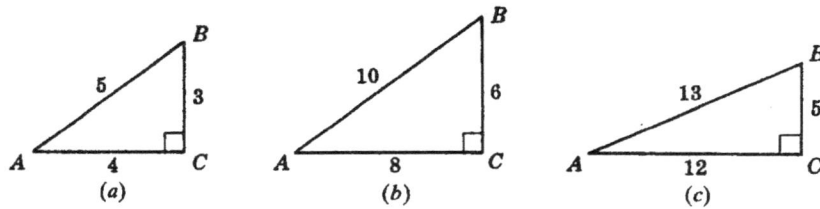

Fig. 8-2

Solutions

Formulas	(a) $a = 3, b = 4, c = 5$	(b) $a = 6, b = 8, c = 10$	(c) $a = 5, b = 12, c = 13$
$\tan A = \dfrac{a}{b}$	$\tan A = \dfrac{3}{4}$	$\tan A = \dfrac{6}{8} = \dfrac{3}{4}$	$\tan A = \dfrac{5}{12}$
$\tan B = \dfrac{b}{a}$	$\tan B = \dfrac{4}{3}$	$\tan B = \dfrac{8}{6} = \dfrac{4}{3}$	$\tan B = \dfrac{12}{5}$
$\sin A = \dfrac{a}{c}$	$\sin A = \dfrac{3}{5}$	$\sin A = \dfrac{6}{10} = \dfrac{3}{5}$	$\sin A = \dfrac{5}{13}$
$\sin B = \dfrac{b}{c}$	$\sin B = \dfrac{4}{5}$	$\sin B = \dfrac{8}{10} = \dfrac{4}{5}$	$\sin B = \dfrac{12}{13}$
$\cos A = \dfrac{b}{c}$	$\cos A = \dfrac{4}{5}$	$\cos A = \dfrac{8}{10} = \dfrac{4}{5}$	$\cos A = \dfrac{12}{13}$
$\cos B = \dfrac{a}{c}$	$\cos B = \dfrac{3}{5}$	$\cos B = \dfrac{6}{10} = \dfrac{3}{5}$	$\cos B = \dfrac{5}{13}$

8.4 Finding measures of angles by trigonometric ratios

Find the measure of angle A, to the nearest degree, in each part of Fig. 8-3.

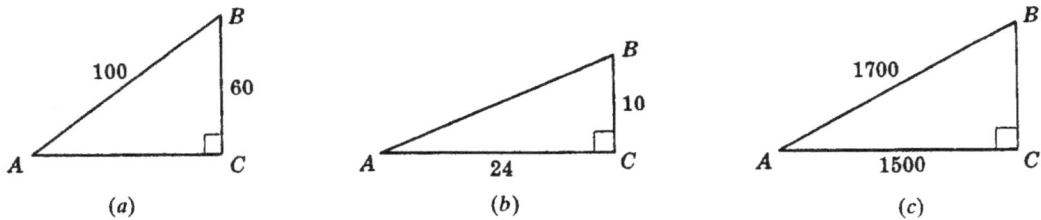

Fig. 8-3

Solutions

(a) $\sin A = \dfrac{60}{100} = 0.6000$. Since $\sin 37° = 0.6018$ is the nearest-degree sine value, $m\angle A = 37°$.

(b) $\tan A = \dfrac{10}{24} = 0.4167$. Since $\tan 23° = 0.4245$ is the nearest-degree tangent value, $m\angle A = 23°$.

(c) $\cos A = \dfrac{1500}{1700} = 0.8824$. Since $\cos 28° = 0.8829$ is the nearest-degree cosine value, $m\angle A = 28°$.

8.5 Trigonometric ratios of 30° and 60°

Show that

(a) $\tan 30° = 0.577$ (c) $\cos 30° = 0.866$ (e) $\sin 60° = 0.866$

(b) $\sin 30° = 0.500$ (d) $\tan 60° = 1.732$ (f) $\cos 60° = 0.500$

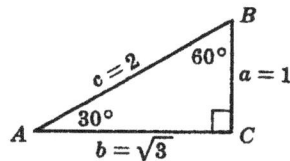

Fig. 8-4

Solutions

The trigonometric ratios for 30° and 60° may be obtained by using a 30°-60°-90° triangle (Fig. 8-4); in such a triangle, the ratio of the sides is $a{:}b{:}c = 1{:}\sqrt{3}{:}2$. Thus:

(a) $\tan 30° = \dfrac{1}{\sqrt{3}} = \dfrac{1}{\sqrt{3}} \cdot \dfrac{\sqrt{3}}{\sqrt{3}} = \dfrac{\sqrt{3}}{3} = 0.577$

(d) $\tan 60° = \dfrac{\sqrt{3}}{1} = 1.732$

(b) $\sin 30° = \dfrac{1}{2} = 0.500$

(e) $\sin 60° = \dfrac{\sqrt{3}}{2} = 0.866$

(c) $\cos 30° = \dfrac{\sqrt{3}}{2} = 0.866$

(f) $\cos 60° = \dfrac{1}{2} = 0.500$

8.6 Finding lengths of sides by trigonometric ratios

In each triangle of Fig. 8-5, solve for x and y to the nearest integer.

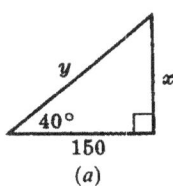
(a)

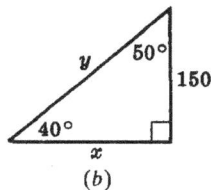
(b)

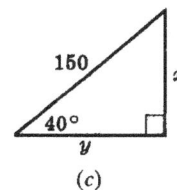
(c)

Fig. 8-5

Solutions

(a) Since $\tan 40° = \dfrac{x}{150}, x = 150 \tan 40° = 150(0.8391) = 126.$

Since $\cos 40° = \dfrac{150}{y}, y = \dfrac{150}{\cos 40°} = \dfrac{150}{0.766} = 196.$

(b) Since $\tan 50° = \dfrac{x}{150}, x = 150 \tan 50° = 150(1.1918) = 179.$

Since $\sin 40° = \dfrac{150}{y}, y = \dfrac{150}{\sin 40°} = \dfrac{150}{0.6428} = 233.$

(c) Since $\sin 40° = \dfrac{x}{150}, x = 150 \sin 40° = 150 (0.6428) = 96.$

Since $\cos 40° = \dfrac{y}{150}, y = 150 \cos 40° = 150(0.776) = 115.$

8.7 Solving trigonometry problems

(a) An aviator flew 70 mi east from A to C. From C, he flew 100 mi north to B. Find the measure of the angle of the turn (to the nearest degree) that must be made at B to return to A.

(b) A road is to be constructed so that it will rise 105 ft for each 1000 ft of horizontal distance. Find the measure of the angle of rise to the nearest degree, and the length of road to the nearest foot for each 1000 ft of horizontal distance.

Solutions

(a) The required angle is $\angle EBA$ in Fig. 8-6(a). In rt. $\triangle ABC$, $\tan B = \dfrac{70}{100} = 0.7000$; hence, $m\angle B = 35°$ and $m\angle EBA = 180° - 35° = 145°$.

(b) We need to find $m\angle A$ and x in Fig. 8-6(b). Since $\tan A = \dfrac{105}{1000} = 0.1050$, $m\angle A = 6°$. Then $\cos 6° = \dfrac{1000}{x}$, so $x = \dfrac{1000}{\cos 6°} = \dfrac{1000}{0.9945} = 1006\,\text{ft}$.

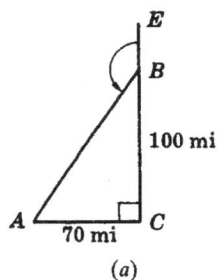

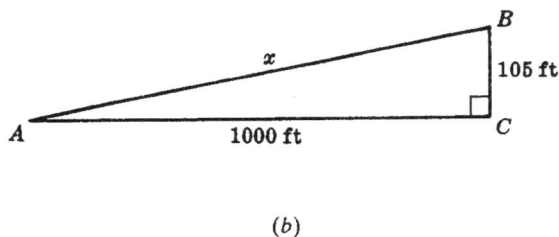

Fig. 8-6

8.2   Angles of Elevation and Depression

Here are some definitions that are involved in observed-angle problems:

The *line of sight* is the line from the eye of the observer to the object sighted.

A *horizontal line* is a line that is parallel to the surface of water.

An *angle of elevation (or depression)* is an angle formed by a horizontal line and a line of sight above (or below) the horizontal line and in the same vertical plane.

Thus in Fig. 8-7, the observer is sighting an airplane above the horizontal, and the angle formed by the horizontal and the line of sight is an *angle of elevation*. In sighting the car, the angle her line of sight makes with the horizontal is an *angle of depression*.

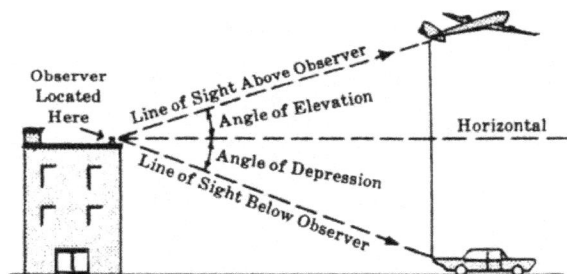

Fig. 8-7

SOLVED PROBLEMS

8.8   Using an angle of elevation

(a) Sighting to the top of a building, Henry found the angle of elevation to measure 21°. The ground is level. The transit is 5 ft above the ground and 200 ft from the building. Find the height of the building to the nearest foot.

(b) If the angle of elevation of the sun at a certain time measures 42°, find to the nearest foot the height of a tree whose shadow is 25 ft long.

Solutions

(a) If x is the height of the part of the building above the transit [Fig. 8-8(a)], then $\tan 21° = \dfrac{x}{200}$ and
$x = 200 \tan 21° = 200(0.3839) = 77$ ft.
Thus, the height of the building is $h = x + 5 = 77 + 5 = 82$ ft.

(b) If h is the height of the tree [Fig. 8-8(b)], then we have $\tan 42° = \dfrac{h}{25}$ and $h = 25 \tan 42° = 25(0.9004) = 23$ ft.

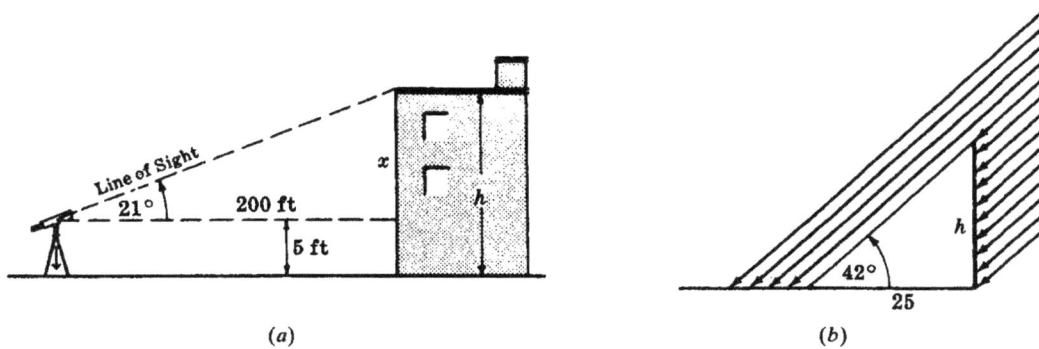

(a)   (b)

Fig. 8-8

8.9 Using both an angle of elevation and an angle of depression

Standing at the top of a lighthouse 200 ft high, a lighthouse keeper sighted both an airplane and a ship directly beneath the plane. The angle of elevation of the plane measured 25°; the angle of depression of the ship measured 32°. Find (a) the distance d of the boat from the foot of the lighthouse, to the nearest 10 ft; (b) the height of the plane above the water, to the nearest 10 ft.

Solutions

(a) See Fig. 8-9. In $\triangle$III, $\tan 58° = \dfrac{d}{200}$ and $d = 200 \tan 58° = 200(1.6003) = 320$ ft.

(b) In $\triangle$I, $\tan 25° = \dfrac{x}{320}$ and $x = 320(0.4663) = 150$ ft. Since the height of the tower is 200 ft, the height of the airplane is $200 + 150 = 350$ ft.

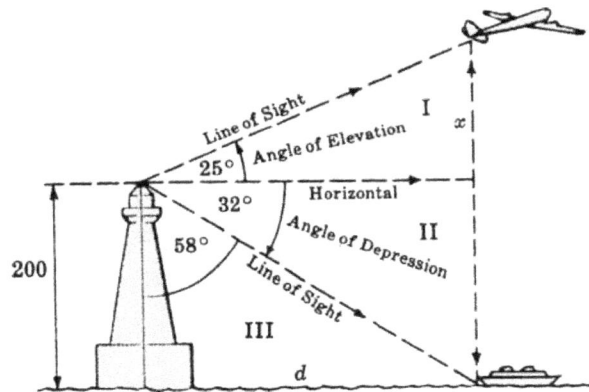

Fig. 8-9

8.10 Using two angles of depression

An observer on the top of a hill 250 ft above the level of a lake sighted two boats directly in line. Find, to the nearest foot, the distance between the boats if the angles of depression noted by the observer measured 11° and 16°.

Solutions

In $\triangle AB'C$ of Fig. 8-10, $m\angle B'AC = 90° - 11° = 79$. Then $CB' = 250 \tan 79°$.

In $\triangle ABC$, $m\angle BAC = 90° - 16° = 74°$. Then $CB = 250 \tan 74°$.

Hence, $BB' = CB' - CB = 250(\tan 79° - \tan 74°) = 250(5.1446 - 3.4874) = 414$ ft.

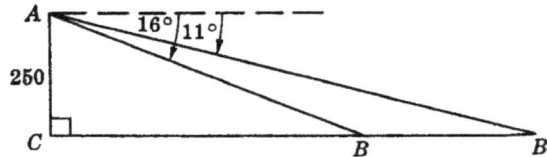

Fig. 8-10

SUPPLEMENTARY PROBLEMS

8.1. Using the table of trigonometric functions at the end of the book, find   (8.1)

(a) sin 25°, sin 48°, sin 59°, and sin 89°

(b) cos 15°, cos 52°, cos 74°, and cos 88°

(c) tan 4°, tan 34°, tan 55°, and tan 87°

(d) which trigonometric ratios increase as the measure of the angle increases from 0° to 90°

(e) which trigonometric ratio decreases as the measure of the angle increases from 0° to 90°

(f) which trigonometric ratio has values greater than 1

8.2. Using the table at the end of the book, find the angle for which   (8.1)

(a) $\sin x = 0.3420$ (c) $\sin B = 0.9455$ (e) $\cos y = 0.7071$ (g) $\tan W = 0.3443$

(b) $\sin A = 0.4848$ (d) $\cos A' = 0.9336$ (f) $\cos Q = 0.3584$ (h) $\tan B' = 2.3559$

8.3. Using the table at the end of the book, find the measure of x to the nearest degree if   (8.2)

(a) $\sin x = 0.4400$ (e) $\cos x = 0.7650$ (i) $\tan x = 5.5745$ (m) $\cos x = \dfrac{3}{8}$

(b) $\sin x = 0.7280$ (f) $\cos x = 0.2675$ (j) $\sin x = \dfrac{11}{50}$ (n) $\cos x = \dfrac{\sqrt{3}}{2}$

(c) $\sin x = 0.9365$ (g) $\tan x = 0.1245$ (k) $\sin x = \dfrac{\sqrt{2}}{2}$ (o) $\tan x = \dfrac{2}{7}$

(d) $\cos x = 0.9900$ (h) $\tan x = 0.5200$ (l) $\cos x = \dfrac{13}{25}$ (p) $\tan x = \dfrac{\sqrt{3}}{10}$

8.4. In each right triangle of Fig. 8-11, find $\sin A$, $\cos A$, and $\tan A$.   (8.3)

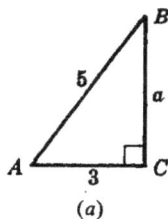

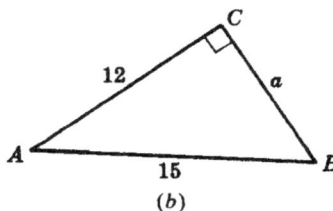

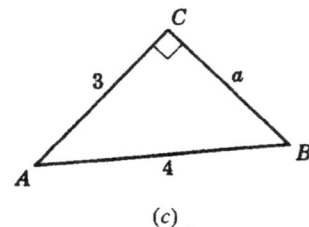

Fig. 8-11

8.5. Find $m\angle A$ to the nearest degree in each part of Fig. 8-12.   (8.4)

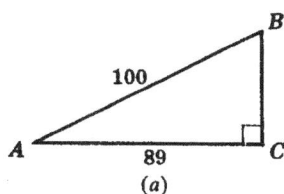
(a)

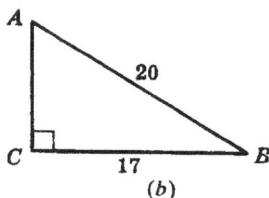
(b)

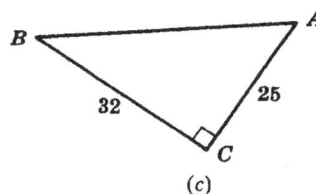
(c)

Fig. 8-12

8.6. Find $m\angle B$ to the nearest degree if (a) $b = 67$ and $c = 100$; (b) $a = 14$ and $c = 50$; (c) $a = 22$ and $b = 55$; (d) $a = 3$ and $b = \sqrt{3}$ in Fig. 8-13.   (8.4)

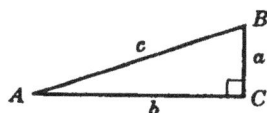
Fig. 8-13

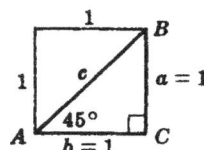
Fig. 8-14

8.7. Making use of a square with a side of 1 (Fig. 8-14), show that (a) the diagonal $c = \sqrt{2}$; (b) $\tan 45° = 1$; (c) $\sin 45° = \cos 45° = 0.707$.   (8.5)

8.8. To the nearest degree, find the measure of each acute angle of any right triangle whose sides are in the ratio of (a) 5:12:13; (b) 8:15:17; (c) 7:24:25; (d) 11:60:61.   (8.4)

8.9. In each triangle of Fig. 8-15, solve for x and y to the nearest integer.   (8.5)

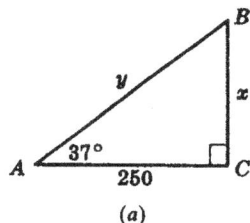
(a)

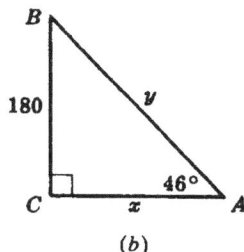
(b)

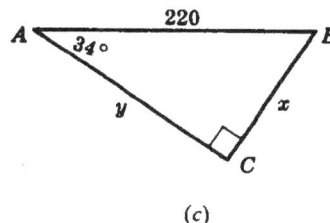
(c)

Fig. 8-15

8.10. A ladder leans against the side of a building and makes an angle measuring 70° with the ground. The foot of the ladder is 30 ft from the building. Find, to the nearest foot, (a) how high up on the building the ladder reaches; (b) the length of the ladder.   (8.6)

8.11. To find the distance across a swamp, a surveyor took measurements as shown in Fig. 8-16. $\overline{AC}$ is at right angles to $\overline{BC}$. If $m\angle A = 24°$ and $AC = 350$ ft, find the distance BC across the swamp.   (8.6)

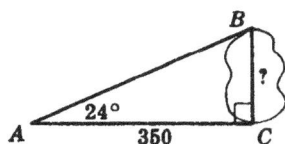
Fig. 8-16

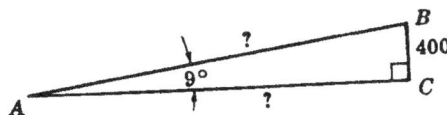
Fig. 8-17

8.12.   A plane rises from take-off and flies at a fixed angle measuring 9° with the horizontal ground (Fig. 8-17). When it has gained 400 ft in altitude, find, to the nearest 10 ft, (a) the horizontal distance flown; (b) the distance the plane has actually flown.   (8.6)

8.13.   The base angle of an isosceles triangle measures 28°, and each leg has length 45 in (Fig. 8-18). Find, to the nearest inch, (a) the length of the altitude drawn to the base; (b) the length of the base.   (8.6)

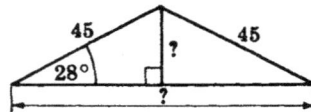

Fig. 8-18

8.14.   In a triangle, an angle measuring 50° is included between sides of length 12 and 18. Find the length of the altitude to the side of length 12, to the nearest integer.   (8.6)

8.15.   Find the lengths of the sides of a rectangle to the nearest inch if a diagonal of length 24 in makes an angle measuring 42° with a side.   (8.6)

8.16.   A rhombus has an angle measuring 76° and a long diagonal of length 40 ft. Find the length of the short diagonal to the nearest foot.   (8.6)

8.17.   Find the length of the altitude to the base of an isosceles triangle to the nearest yard if its base has length 40 yd and its vertex angle measures 106°.   (8.6)

8.18.   A road is inclined uniformly at an angle measuring 6° with the horizontal (Fig. 8-19). After a car is driven 10,000 ft along this road, find, to the nearest 10 ft, the (a) increase in the altitude of the car and driver; (b) horizontal distance that has been driven.   (8.7)

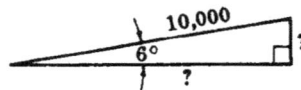

Fig. 8-19

8.19.   An airplane travels 15,000 ft through the air at a uniform angle of climb, thereby gaining 1900 ft in altitude. Find its angle of climb.   (8.7)

8.20.   Sighting to the top of a monument. William found the angle of elevation to measure 16° (Fig. 8-20). The ground is level, and the transit is 5 ft above the ground. If the monument is 86 ft high, find, to the nearest foot, the distance from William to the foot of the monument.   (8.8)

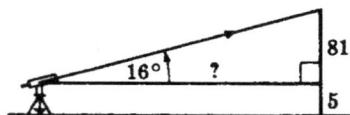

Fig. 8-20

8.21.   Find to the nearest degree the measure of the angle of elevation of the sun when a tree 60 ft high casts a shadow of (a) 10 ft; (b) 60 ft.

8.22. At a certain time of day, the angle of elevation of the sun measures 34°. Find, to the nearest foot, the length of the shadow cast by (a) a 15-ft vertical pole; (b) a building 70 ft high. (8.8)

8.23. A light at C is projected vertically to a cloud at B. An observer at A, 1000 ft horizontally from C, notes the angle of elevation of B. Find the height of the cloud, to the nearest foot, if $m\angle A = 37°$. (8.8)

8.24. A lighthouse built at sea level is 180 ft high (Fig. 8-21). From its top, the angle of depression of a buoy measures 24°. Find, to the nearest foot, the distance from the buoy to the foot of the lighthouse. (8.8)

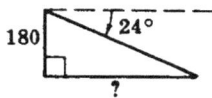

Fig. 8-21

8.25. An observer, on top of a hill 300 ft above the level of a lake, sighted two ships directly in line. Find, to the nearest foot, the distance between the boats if the angles of depression noted by the observer measured (a) 20° and 15°; (b) 35° and 24°; (c) 9° and 6°. (8.10)

8.26. In Fig. 8-22, $m\angle A = 43°$, $m\angle BDC = 54°$, $m\angle C = 90°$, and $DC = 170$ ft. (a) Find the length of $\overline{BC}$. (b) Using the result of (a), find the length of $\overline{AB}$. (8.10)

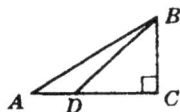

Fig. 8-22   Fig. 8-23

8.27. In Fig. 8-23, $m\angle B = 90°$, $m\angle ACB = 58°$, $m\angle D = 23°$, and $BC = 60$ ft. (a) Find the length of $\overline{AB}$. (b) Using the result of part (a), find the length of $\overline{CD}$.

8.28. Tangents $\overline{PA}$ and $\overline{PB}$ are drawn to a circle from external point P. $m\angle APB = 40°$, and $PA = 25$. Find to the nearest tenth the radius of the circle.

CHAPTER 9

Areas

9.1 Area of a Rectangle and of a Square

A *square unit* is the surface enclosed by a square whose side is 1 unit (Fig. 9-1).

The *area of a closed plane figure*, such as a polygon, is the number of square units contained in its surface. Since a rectangle 5 units long and 4 units wide can be divided into 20 unit squares, its area is 20 square units (Fig. 9-2).

The *area of a rectangle equals the product of the length of its base and the length of its altitude* (Fig. 9-3). Thus if $b = 8$ in and $h = 3$ in, then $A = 24$ in^2.

The *area of a square equals the square of the length of a side* (Fig. 9-4). Thus if $s = 6$, then $A = s^2 = 36$.

It follows that the area of a square also equals one-half the square of the length of a diagonal. Since $A = s^2$ and $s = d/\sqrt{2}$, $A = \frac{1}{2}d^2$.

Note that we sometimes use the letter A for both a vertex of a figure and its area. You should have no trouble determining which is meant.

The reader should feel free to use a calculator for the work in this chapter.

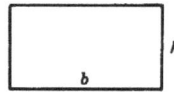

Fig. 9-1   Fig. 9-2   Rectangle: $A = bh$ Fig. 9-3   Square: $A = s^2$ $A = \frac{1}{2}d^2$ Fig. 9-4

SOLVED PROBLEMS

9.1 Area of a rectangle

(a) Find the area of a rectangle if the base has length 15 and the perimeter is 50.

(b) Find the area of a rectangle if the altitude has length 10 and the diagonal has length 26.

(c) Find the lengths of the base and altitude of a rectangle if its area is 70 and its perimeter is 34.

Solutions

See Fig. 9-5.

(a) Here $p = 50$ and $b = 15$. Since $p = 2b + 2h$, we have $50 = 2(15) + 2h$ so $h = 10$.
Hence, $A = bh = 15(10) = 150$.

(b) Here $d = 26$ and $h = 10$. In right $\triangle ACD$, $d^2 = b^2 + h^2$, so $26^2 = b^2 + 10^2$ or $b = 24$.
Hence, $A = bh = 24(10) = 240$.

164

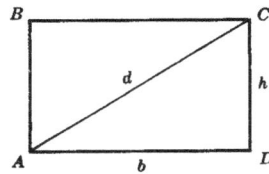

Fig. 9-5

(c) Here $A = 70$ and $p = 34$. Since $p = 2b + 2h$, we have $34 = 2(b + h)$ or $h = 17 - b$.
Since $A = bh$, we have $70 = b(17 - b)$, so $b^2 - 17b + 70 = 0$ and $b = 7$ or 10. Then since $h = 17 - b$,
we obtain $h = 10$ or 7.
Ans. 10 and 7, or 7 and 10.

9.2 Area of a Square

(a) Find the area of a square whose perimeter is 30.

(b) Find the area of a square if the radius of the circumscribed circle is 10.

(c) Find the side and the perimeter of a square whose area is 20.

(d) Find the number of square inches in a square foot.

Solutions

(a) Since $p = 4s = 30$ in Fig. 9-6(a), $s = 7\frac{1}{2}$. Then $A = s^2 = (7\frac{1}{2})^2 = 56\frac{1}{4}$.

(b) Since $r = 10$ in Fig. 9-6(b), $d = 2r = 20$. Then $A = \frac{1}{2}d^2 = \frac{1}{2}(20)^2 = 200$.

(c) In Fig. 9-6(a), $A = s^2 = 20$; hence, $s = 2\sqrt{5}$. Then perimeter $= 4s = 8\sqrt{5}$.

(d) $A = s^2$. Since 1ft $= 12$ in, 1 ft^2 = 1ft $\times$ 1ft $= 12$ in $\times$ 12 in $= 144$ in^2.

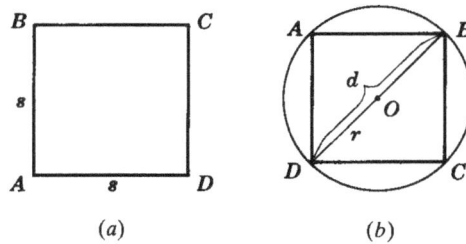

(a)   (b)

Fig. 9-6

9.2 Area of a Parallelogram

The area of a parallelogram equals the product of the length of a side and the length of the altitude to that side. (A proof of this theorem is given in Chapter 16.) Thus in $\square ABCD$ (Fig. 9-7), if $b = 10$ and $h = 2.7$,
then $A = 10(2.7) = 27$.

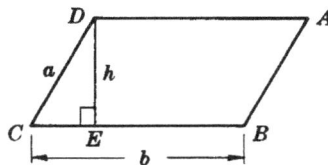

Parallelogram: $A = bh$

Fig. 9-7

SOLVED PROBLEMS

9.3   Area of a parallelogram

(a) Find the area of a parallelogram if the area is represented by x^2-4, the length of a side by $x + 4$, and the length of the altitude to that side by $x-3$.

(b) In a parallelogram, find the length of the altitude if the area is 54 and the ratio of the altitude to the base is 2:3.

Solutions

See Fig. 9-7.

(a) $A = x^2-4$, $b = x + 4$, $h = x-3$. Since $A = bh$, $x^2-4 = (x + 4)(x-3)$ or $x^2-4 = x^2 + x-12$ and $x = 8$. Hence, $A = x^2-4 = 64-4 = 60$.

(b) Let $h = 2x$, $b = 3x$. Then $A = bh$ or $54 = (3x)(2x) = 6x^2$, so $9 = x^2$ and $x = 3$. Hence, $h = 2x = 2(3) = 6$.

9.3   Area of a Triangle

The area of a triangle equals one-half the product of the length of a side and the length of the altitude to that side. (A proof of this theorem is given in Chapter 16.)

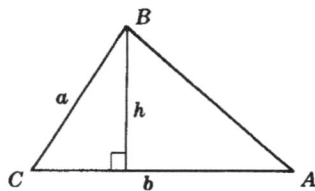

Triangle: $A = \frac{1}{2}bh$

Fig. 9-8

SOLVED PROBLEMS

9.4   Area of a triangle

Find the area of the triangle in Fig. 9-9.

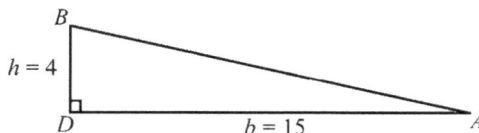

Fig. 9-9

Solution

Here, $b = 15$ and $h = 4$. Thus, $A = \frac{1}{2}bh = \frac{1}{2}(15)(4) = 30$.

9.5   Formulas for the area of an equilateral triangle

Derive the formula for the area of an equilateral triangle (a) whose side has length s; (b) whose altitude has length h.

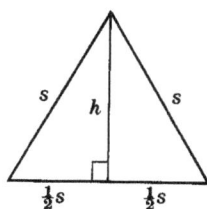

Equilateral Triangle:
$A = \frac{1}{4}s^2\sqrt{3}$
$A = \frac{1}{3}h^2\sqrt{3}$

Fig. 9-10

Solutions

See Fig. 9-10.

(a) Here $A = \frac{1}{2}bh$, where $b = s$ and $h^2 = s^2 - (\frac{1}{2}s)^2$ or $h = \frac{1}{2}s\sqrt{3}$.
   Then $A = \frac{1}{2}bh = \frac{1}{2}s(\frac{1}{2}s\sqrt{3}) = \frac{1}{4}s^2\sqrt{3}$.

(b) Here $A = \frac{1}{2}bh$, where $b = s$ and $h = \frac{1}{2}s\sqrt{3}$ or $s = \frac{2h}{\sqrt{3}}$.
   Then $A = \frac{1}{2}bh = \frac{1}{2}sh = \frac{1}{2}(\frac{2h}{\sqrt{3}}) \cdot h$

9.6 Area of an equilateral triangle

In Fig. 9-11, find the area of (a) an equilateral triangle whose perimeter is 24; (b) a rhombus in which the shorter diagonal has length 12 and an angle measures 60°; (c) a regular hexagon with a side of length 6.

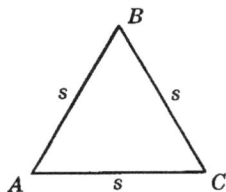 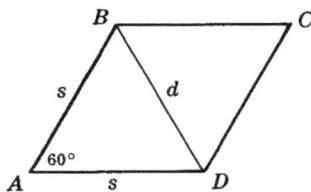 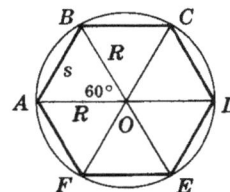

(a)   (b)   (c)

Fig. 9-11

Solutions

(a) Since $p = 3s = 24$, $s = 8$. Then $A = \frac{1}{4}s^2\sqrt{3} = \frac{1}{4}(64)\sqrt{3} = 16\sqrt{3}$.

(b) Since $m\angle A = 60°$, $\triangle ADB$ is equilateral and $s = d = 12$. The area of the rhombus is twice the area of $\triangle ABD$. Hence $A = 2(\frac{1}{4}s^2\sqrt{3}) = 2(\frac{1}{4})(144)\sqrt{3} = 72\sqrt{3}$.

(c) A side s of the inscribed hexagon subtends a central angle of measure $\frac{1}{6}(360°) = 60°$. Then, since $OA = OB$ = radius R of the circumscribed circle, $m\angle OAB = m\angle OBA = 60°$. Thus $\triangle AOB$ is equilateral.
   Area of hexagon = 6(area of $\triangle AOB$) = $6(\frac{1}{4}s^2\sqrt{3}) = 6(\frac{1}{4})(36\sqrt{3}) = 54\sqrt{3}$.

9.4 Area of a Trapezoid

The area of a trapezoid equals one-half the product of the length of its altitude and the sum of the lengths of its bases. (A proof of this theorem is given in Chapter 16.) Thus if $h = 20$, $b = 27$, and $b' = 23$ in Fig. 9-12, then $A = \frac{1}{2}(20)(27 + 23) = 500$.

The area of a trapezoid equals the product of the lengths of its altitude and median. Since $A = \frac{1}{2}h(b + b')$ and $m = \frac{1}{2}(b + b'), A = hm$.

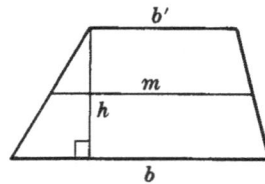

Trapezoid: $A = \frac{1}{2}h(b + b')$
$A = hm$

Fig. 9-12

SOLVED PROBLEMS

9.7 Area of a trapezoid

(a) Find the area of a trapezoid if the bases have lengths 7.3 and 2.7, and the altitude has length 3.8.

(b) Find the area of an isosceles trapezoid if the bases have lengths 22 and 10, and the legs have length 10.

(c) Find the bases of an isosceles trapezoid if the area is $52\sqrt{3}$, the altitude has length $4\sqrt{3}$, and each leg has length 8.

Solutions

See Fig. 9-13.

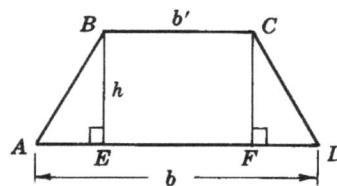

Fig. 9-13

(a) Here b = 7.3, $b' = 2.7$, h = 3.8. Then $A = \frac{1}{2}h (b + b') = \frac{1}{2} (3.8 + 2.7) = 19$.

(b) Here $b = 22, b' = 10, AB = 10$. Also $EF = b' = 10$ and $AE = \frac{1}{2}(22-10) = 6$.

In $\triangle BEA, h^2 = 10^2 - 6^2 = 64$ so $h = 8$. Then $A = \frac{1}{2}h(b + b') = \frac{1}{2}(8)(22 + 10) = 128$.

(c) $AE = \sqrt{(AB)^2 - h^2} = \sqrt{64 - 48} = 4$. Also $FD = AE = 4$, and $b' = b-(AE + FD) = b-8$. Then $A = \frac{1}{2}h(b + b') = \frac{1}{2}h(2b-8)$ or $52\sqrt{3} = \frac{1}{2}(4\sqrt{3})(2b - 8)$, from which $26 = 2b-8$ or $b = 17$. Then $b' = b-8 = 17-8 = 9$.

9.5   Area of a Rhombus

The area of a rhombus equals one-half the product of the lengths of its diagonals.

Since each diagonal is the perpendicular bisector of the other, the area of triangle I in Fig. 9-14 is $\frac{1}{2}(\frac{1}{2}d)(\frac{1}{2}d') = \frac{1}{8}dd'$. Thus the rhombus, which consists of four triangles congruent to $\triangle$I, has an area of $4(\frac{1}{8}dd')$ or $\frac{1}{2}dd'$.

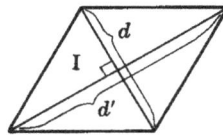

Rhombus: $A = \frac{1}{2}dd'$

Fig. 9-14

SOLVED PROBLEMS

9.8   Area of a rhombus

(a) Find the area of a rhombus if one diagonal has length 30 and a side has length 17.

(b) Find the length of a diagonal of a rhombus if the other diagonal has length 8 and the area of the rhombus is 52.

Solutions

See Fig. 9-15.

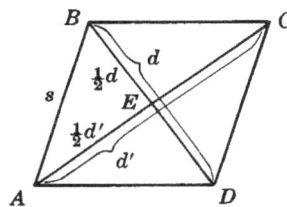

Fig. 9-15

(a) In right $\triangle AEB$, $s^2 = (\frac{1}{2}d)^2 + (\frac{1}{2}d')^2$ or $17^2 = (\frac{1}{2}d)^2 + 15^2$. Then $\frac{1}{2}d = 8$ and $d = 16$. Now $A = \frac{1}{2}dd' = \frac{1}{2}(16)(30) = 240$.

(b) We have $d' = 8$ and $A = 52$. Then $A = \frac{1}{2}dd'$ or $52 = \frac{1}{2}(d)(8)$ and $d = 13$.

9.6   Polygons of the Same Size or Shape

Figure 9-16 shows what we mean when we say that two polygons are of equal area, or are similar, or are congruent.

Equal Polygons

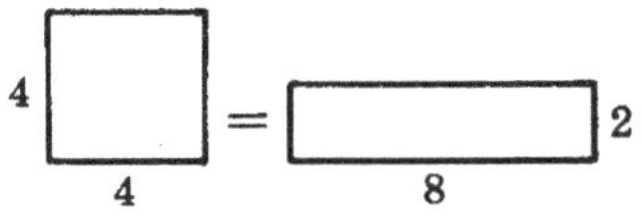

Polygons of the same size have the same area.

Similar Polygons

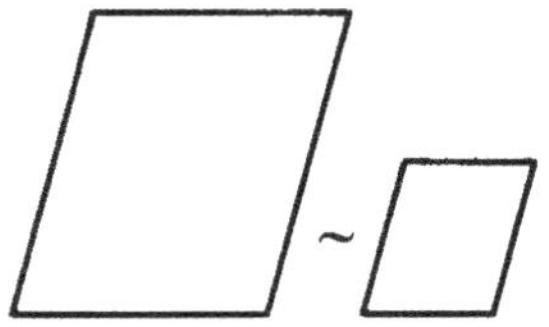

Similar polygons have the same shape.

Congruent Polygons

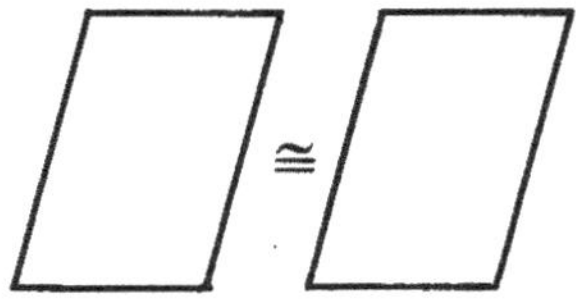

Congruent polygons have the same size and the same shape.

Fig. 9-16

PRINCIPLE 1: *Parallelograms have equal areas if they have congruent bases and congruent altitudes.*

Thus, the two parallelograms shown in Fig. 9-17 are equal.

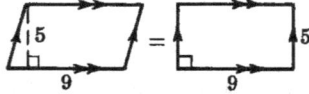

Fig. 9-17

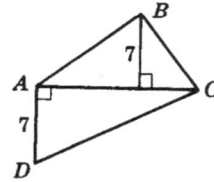

Fig. 9-18

PRINCIPLE 2: *Triangles have equal areas if they have congruent bases and congruent altitudes.*

Thus in Fig. 9-18, the area of $\triangle CAB$ equals the area of $\triangle CAD$.

PRINCIPLE 3: *A median divides a triangle into two triangles with equal areas.*

Thus in Fig. 9-19, where $\overline{BM}$ is a median, the area of $\triangle AMB$ equals the area of $\triangle BMC$ since they have congruent bases ($\overline{AM} \cong \overline{MC}$) and common altitude $\overline{BD}$.

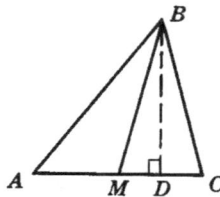

Fig. 9-19

PRINCIPLE 4: *Triangles are equal in area if they have a common base and their vertices lie on a line parallel to the base.*

Thus in Fig. 9-20, the area of $\triangle ABC$ is equal to the area of $\triangle ADC$.

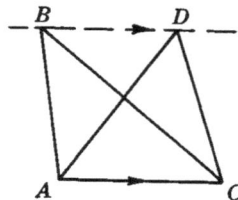

Fig. 9-20

SOLVED PROBLEMS

9.9 Proving an equal-areas problem

Given: Trapezoid $ABCD$ ($\overline{BC} \parallel \overline{AD}$)

Diagonals $\overline{AC}$ and $\overline{BD}$

To Prove: Area($\triangle AEB$) = area($\triangle DEC$)

Plan: Use Principle 4 to obtain area($\triangle ABD$) = area($\triangle ACD$).
Then use the Subtraction Postulate.

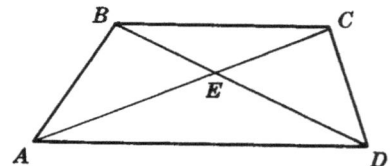

PROOF:

Statements	Reasons
1. $\overline{BC} \parallel \overline{AD}$	1. Given
2. Area($\triangle ABD$) = area($\triangle ACD$)	2. Triangles have equal area if they have a common base and their vertices lie on a line parallel to the base.
3. Area($\triangle AED$) = area($\triangle AED$)	3. Identity Postulate
4. Area($\triangle AEB$) = area($\triangle DEC$)	4. Subtraction Postulate

9.10 Proving an equal-areas problem stated in words

Prove that if M is the midpoint of diagonal $\overline{AC}$ in quadrilateral $ABCD$, and $\overline{BM}$ and $\overline{DM}$ are drawn, then the area of quadrilateral $ABMD$ equals the area of quadrilateral $CBMD$.

Solution

Given: Quadrilateral $ABCD$
          M is midpoint of diagonal AC.
To Prove: Area of quadrilateral $ABMD$
          equals area of quadrilateral $CBMD$.
Plan: Use Principle 3 to obtain two pairs of
        triangles which are equal in area.
        Then use the Addition Postulate.

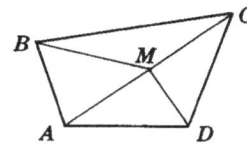

PROOF:

Statements	Reasons
1. M is the midpoint of $\overline{AC}$.	1. Given
2. $\overline{BM}$ is a median of $\triangle ACB$. $\overline{DM}$ is a median of $\triangle ACD$.	2. A line from a vertex of a triangle to the midpoint of the opposite side is a median.
3. Area ($\triangle AMB$) = area($\triangle BMC$), Area($\triangle AMD$) = area($\triangle DMC$).	3. A median divides a triangle into two triangles of equal area.
4. Area of quadrilateral $ABMD$ equals area of quadrilateral $CBMD$.	4. If equals are added to equals, the results are equal.

9.7 Comparing Areas of Similar Polygons

The areas of similar polygons are to each other as the squares of any two corresponding segments.
    Thus if $\triangle ABC \sim \triangle A'B'C'$ and the area of $\triangle ABC$ is 25 times the area of $\triangle A'B'C'$, then the ratio of the lengths of any two corresponding sides, medians, altitudes, radii of inscribed or circumscribed circles, and such is 5:1.

SOLVED PROBLEMS

9.11 Ratio of areas and segments of similar triangles

Find the ratio of the areas of two similar triangles (a) if the ratio of the lengths of two corresponding sides is 3:5; (b) if their perimeters are 12 and 7. Find the ratio of the lengths of a pair of (c) corresponding sides if the ratio of the areas is 4:9; (d) corresponding medians if the areas are 250 and 10.

Solutions

(a) $\frac{A}{A'} = \left(\frac{s}{s'}\right)^2 = \left(\frac{3}{5}\right)^2 = \frac{9}{25}$

(c) $\left(\frac{s}{s'}\right)^2 = \frac{A}{A'} = \frac{4}{9}$ or $\frac{s}{s'} = \frac{2}{3}$

(b) $\frac{A}{A'} = \left(\frac{p}{p'}\right)^2 = \left(\frac{12}{7}\right)^2 = \frac{144}{49}$

(d) $\left(\frac{m}{m'}\right)^2 = \frac{A}{A'} = \frac{250}{10}$ or $\frac{m}{m'} = 5$

9.12 Proportions derived from similar polygons

(a) The areas of two similar polygons are 80 and 5. If a side of the smaller polygon has length 2, find the length of the corresponding side of the larger polygon.

(b) The corresponding diagonals of two similar polygons have lengths 4 and 5. If the area of the larger polygon is 75, find the area of the smaller polygon.

Solutions

(a) $\left(\frac{s}{s'}\right)^2 = \;\; = \frac{A}{A'}$, so $\left(\frac{s}{2}\right)^2 = \frac{80}{5} = 16$. Then $\frac{s}{2} = 4$ and $s = 8$.

(b) $\frac{A}{A'} = \left(\frac{d}{d'}\right)^2$, so $\frac{A}{75} = \left(\frac{4}{5}\right)^2$. Then $A = 75\left(\frac{16}{25}\right) = 48$.

SUPPLEMENTARY PROBLEMS

9.1. Find the area of a rectangle $\hspace{12cm}$ (9.1)

(a) If the base has length 11 in and the altitude has length 9 in

(b) If the base has length 2 ft and the altitude has length 1 ft 6 in

(c) If the base has length 25 and the perimeter is 90

(d) If the base has length 15 and the diagonal has length 17

(e) If the diagonal has length 12 and the angle between the diagonal and the base measures 60°

(f) If the diagonal has length 20 and the angle between the diagonal and the base measures 30°

(g) If the diagonal has length 25 and the lengths of the sides are in the ratio of 3:4

(h) If the perimeter is 50 and the lengths of the sides are in the ratio of 2:3

9.2. Find the area of a rectangle inscribed in a circle $\hspace{8cm}$ (9.1)

(a) If the radius of the circle is 5 and the base has length 6

(b) If the radius of the circle is 15 and the altitude has length 24

(c) If the radius and the altitude both have length 5

(d) If the diameter has length 26 and the base and altitude are in the ratio of 5:12

9.3. Find the base and altitude of a rectangle $\hspace{9cm}$ (9.1)

(a) If its area is 28 and the base has a length of 3 more than the altitude

(b) If its area is 72 and the base is twice the altitude

(c) If its area is 54 and the ratio of the base to the altitude is 3:2

(d) If its area is 12 and the perimeter is 16

(e) If its area is 70 and the base and altitude are represented by $2x$ and $x + 2$

(f) If its area is 160 and the base and altitude are represented by $3x-4$ and x

9.4. Find the area of (a) a square yard in square inches; (b) a square meter in square decimeters (1 m = 10 dm).   (9.2)

9.5. Find the area of a square if (a) a side has length 15; (b) a side has length $3\frac{1}{2}$; (c) a side has length 1.8; (d) a side has length $8a$; (e) the perimeter is 44; (f) the perimeter is 10; (g) the perimeter is $12b$; (h) the diagonal has length 8; (i) the diagonal has length 9; (j) the diagonal has length $8\sqrt{2}$.   (9.2)

9.6. Find the area of a square if (a) the radius of the circumscribed circle is 8; (b) the diameter of the circumscribed circle is 12; (c) the diameter of the circumscribed circle is $10\sqrt{2}$; (d) the radius of the inscribed circle is $3\frac{1}{2}$; (e) the diameter of the inscribed circle is 20.   (9.2)

9.7. If a floor is 20 m long and 80 m wide, how many tiles are needed to cover it if (a) each tile is 1 m²; (b) each tile is a square 2 m on a side; (c) each tile is a square 4 m on a side.   (9.2)

9.8. If the area of a square is 81, find the length of (a) its side; (b) its perimeter; (c) its diagonal; (d) the radius of the inscribed circle; (e) the radius of the circumscribed circle.   (9.2)

9.9. (a) Find the length of the side of a square whose area is $6\frac{1}{4}$.   (9.2)

(b) Find the perimeter of a square whose area is 169.

(c) Find the length of the diagonal of a square whose area is 50.

(d) Find the length of the diagonal of a square whose area is 25.

(e) Find the radius of the inscribed circle of a square whose area is 144.

(f) Find the radius of the circumscribed circle of a square whose area is 32.

9.10. Find the area of a parallelogram if the base and altitude have lengths, respectively, of (a) 3 ft and $5\frac{1}{3}$ ft; (b) 4 ft and 1 ft 6 in; (c) 20 and 3.5; (d) 1.8 m and 0.9 m.   (9.3)

9.11. Find the area of a parallelogram if the base and altitude have lengths, respectively, of (a) $3x$ and x; (b) $x + 3$ and x; (c) $x-5$ and $x + 5$; (d) $4x + 1$ and $3x + 2$.   (9.3)

9.12. Find the area of a parallelogram if

(a) The area is represented by x^2, the base by $x + 3$, and the altitude by $x-2$

(b) The area is represented by x^2-10, the base by x, and the altitude by $x-2$

(c) The area is represented by $2x^2-34$, the base by $x + 3$, and the altitude by $x-3$

9.13. In a parallelogram, find   (9.3)

(a) The base if the area is 40 and the altitude has length 15

(b) The length of the altitude if the area is 22 and the base has length 1.1

(c) The length of the base if the area is 27 and the base is three times the altitude

(d) The length of the altitude if the area is 21 and the base has length four more than the altitude

(e) The base if the area is 90 and the ratio of the base to the altitude is 5:2

(f) The length of the altitude to a side of length 20 if the altitude to a side of length 15 is 16

(g) The length of the base if the area is 48, the base is represented by $x + 3$, and the altitude is represented by $x + 1$

(h) The length of the base if the area is represented by $x^2 + 17$, the base by $2x-3$, and the altitude by $x + 1$

9.14. Find the area of a triangle if the lengths of the base and altitude are, respectively, (a) 6 in and $3\frac{2}{3}$ in; (b) 1 yd and 2 ft; (c) 8 and $x-7$; (d) $5x$ and $4x$; (e) $4x$ and $x + 9$; (f) $x + 4$ and $x-4$; (g) $2x-6$ and $x + 3$.   (9.4)

9.15. Find the area of   (9.4)

(a) A triangle if two sides have lengths 13 and 15 and the altitude to the third side has length 12

(b) A triangle whose sides have lengths 10, 10, and 16

(c) A triangle whose sides have lengths 5, 12, and 13

(d) An isosceles triangle whose base has length 30 and whose legs each have length 17

(e) An isosceles triangle whose base has length 20 and whose vertex angle measures 68°

(f) An isosceles triangle whose base has length 30 and whose base angle measures 62°

(g) A triangle inscribed in a circle of radius 4 if one side is a diameter and another side makes an angle measuring 30° with the diameter

(h) A triangle cut off by a line parallel to the base of a triangle if the base and altitude of the larger triangle have lengths 10 and 5, respectively, and the line parallel to the base is 6

9.16. Find the altitude of a triangle if   (9.4)

(a) Its base has length 10 and the triangle is equal in area to a parallelogram whose base and altitude have lengths 15 and 8.

(b) Its base has length 8 and the triangle is equal in area to a square whose diagonal has length 4.

(c) Its base has length 12 and the triangle is equal in area to another triangle whose sides have lengths 6, 8, and 10.

9.17. In a triangle, find the length of   (9.4)

(a) A side if the area is 40 and the altitude to that side has length 10

(b) An altitude if the area is 25 and the side to which the altitude is drawn has length 5

(c) A side if the area is 24 and the side has length 2 more than its altitude

(d) A side if the area is 108 and the side and its altitude are in the ratio 3:2

(e) The altitude to a side of length 20, if the sides of the triangle have lengths 12, 16, and 20

(f) The altitude to a side of length 12 if another side and its altitude have lengths 10 and 15

(g) A side represented by $4x$ if the altitude to that side is represented by $x + 7$ and the area is 60

(h) A side if the area is represented by $x^2 - 55$, the side by $2x - 2$, and its altitude by $x - 5$

9.18. Find the area of an equilateral triangle if (a) a side has length 10; (b) the perimeter is 36; (c) an altitude has length 6; (d) an altitude has length $5\sqrt{3}$; (e) a side has length 2b; (f) the perimeter is 12x; (g) an altitude has length 3r. (9.6)

9.19. Find the area of a rhombus having an angle of 60° if (a) a side has length 2; (b) the shorter diagonal has length 7; (c) the longer diagonal has length 12; (d) the longer diagonal has length $6\sqrt{3}$. (9.6)

9.20. Find the area of a regular hexagon if (a) a side is 4; (b) the radius of the circumscribed circle is 6; (c) the diameter of the circumscribed circle is 20. (9.6)

9.21. Find the side of an equilateral triangle whose area equals (9.6)

(a) The sum of the areas of two equilateral triangles whose sides have lengths 9 and 12

(b) The difference of the areas of two equilateral triangles whose sides have lengths 17 and 15

(c) The area of a trapezoid whose bases have lengths 6 and 2 and whose altitude has length $9\sqrt{3}$

(d) Twice the area of a right triangle having a hypotenuse of length 5 and an acute angle of measure 30°

9.22. Find the area of trapezoid *ABCD* in Fig. 9-21, if: (9.7)

(a) $b = 25, b' = 15,$ and $h = 7$

(b) $m = 10$ and $h = 6.9$

(c) $AB = 30, m\angle A = 30°, b = 24,$ and $b' = 6$

(d) $AB = 12, m\angle A = 45°, b = 13,$ and $b' = 7$

(e) $AB = 10, m\angle A = 70°,$ and $b + b' = 20$

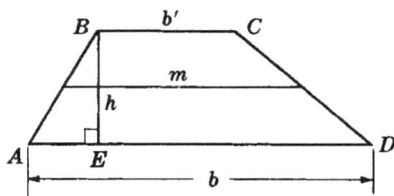

Fig. 9-21

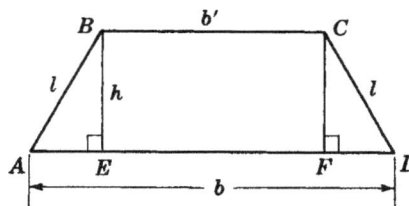

Fig. 9-22

9.23. Find the area of isosceles trapezoid *ABCD* in Fig. 9-22, if (9.7)

(a) $b' = 17, l = 10,$ and $h = 6$ (d) $b = 20, l = 8,$ and $m\angle A = 60°$

(b) $b = 22, b' = 12,$ and $l = 13$ (e) $b = 40, b' = 20,$ and $m\angle A = 28°$

(c) $b = 16, b' = 10,$ and $m\angle A = 45°$

9.24. (a) Find the length of the altitude of a trapezoid if the bases have lengths 13 and 7 and the area is 40. (9.7)

(b) Find the length of the altitude of a trapezoid if the sum of the lengths of the bases is twice the length of the altitude and the area is 49.

(c) Find the sum of the lengths of the bases and the median of a trapezoid if the area is 63 and the altitude has length 7.

1884

   (d) Find the lengths of the bases of a trapezoid if the upper base has length 3 less than the lower base, the altitude has length 4, and the area is 30.

   (e) Find the lengths of the bases of a trapezoid if the lower base has length twice that of the upper base, the altitude has length 6, and the area is 45.

9.25. In an isosceles trapezoid   (9.7)

   (a) Find the lengths of the bases if each leg has length 5, the altitude has length 3, and the area is 39.

   (b) Find the lengths of the bases if the altitude has length 5, each base angle measures 45°, and the area is 90.

   (c) Find the lengths of the bases if the area is $42\sqrt{3}$, the altitude has length $3\sqrt{3}$, and each base angle measures 60°.

   (d) Find the length of each leg if the bases have lengths 24 and 32 and the area is 84.

   (e) Find the length of each leg if the area is 300, the median has length 25, and the lower base has length 30.

9.26. Find the area of a rhombus if   (9.8)

   (a) The diagonals have lengths 8 and 9.

   (b) The diagonals have lengths 11 and 7.

   (c) The diagonals have lengths 4 and $6\sqrt{3}$.

   (d) The diagonals have lengths $3x$ and $8x$.

   (e) One diagonal has length 10 and a side has length 13.

   (f) The perimeter is 40 and a diagonal has length 12.

   (g) The side has length 6 and an angle measures 30°.

   (h) The perimeter is 28 and an angle measures 45°.

   (i) The perimeter is 32 and the length of the short diagonal equals a side in length.

   (j) A side has length 14 and an angle measures 120°.

9.27. Find the area of a rhombus to the nearest integer if (a) the side has length 30 and an angle measures 55°; (b) the perimeter is 20 and an angle measures 33°; (c) the side has length 10 and an angle measures 130°.   (9.8)

9.28. In a rhombus, find the length of   (9.8)

   (a) A diagonal if the other diagonal has length 7 and the area is 35   (9.8)

   (b) The diagonals if their ratio is 4:3 and the area is 54

   (c) The diagonals if the longer is twice the shorter and the area is 100

   (d) The side if the area is 24 and one diagonal has length 6

   (e) The side if the area is 6 and one diagonal has length 4 more than the other

9.29. A rhombus is equal to a trapezoid whose lower base has length 26 and whose other three sides have length 10. Find the length of the altitude of the rhombus if its perimeter is 36.   (9.8)

9.30. Provide the proofs requested in Fig. 9-23. (9.9)

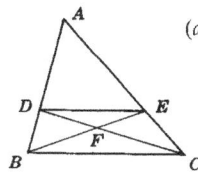

(a) **Given:** $\triangle ABC$
$DB = \frac{1}{3}AB$
$EC = \frac{1}{3}AC$
To Prove: Area($\triangle DFB$)
$=$ area($\triangle FEC$)

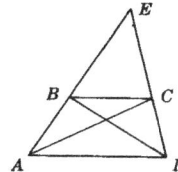

(b) **Given:** Trapezoid $ABCD$
$\overline{AB}$ and $\overline{CD}$ extended
meet at E.
To Prove: Area($\triangle ECA$)
$=$ area($\triangle EBD$)

Fig. 9-23

9.31. Provide the proofs requested in Fig. 9-24. (9.9)

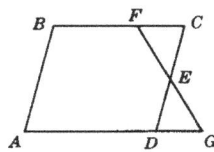

(a) **Given:** $\square ABCD$
E is midpoint of $\overline{CD}$.
To Prove:
Area($\square ABCD$) $=$
area(trapezoid $BFGA$)

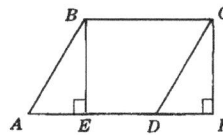

(b) **Given:** $\square ABCD$
$\overline{BE}$ and $\overline{CF} \perp \overline{AF}$.
To Prove:
$BCFE$ is a rectangle and
equal in area to $\square ABCD$.

Fig. 9-24

9.32. Prove each of the following: (9.10)

(a) A median divides a triangle into two triangles having equal areas.

(b) Triangles are equal in area if they have a common base and their vertices lie in a line parallel to the base.

(c) In a triangle, if lines are drawn from a vertex to the trisection points of the opposite sides, the area of the triangle is trisected.

(d) In trapezoid $ABCD$, base $\overline{AD}$ is twice base $\overline{BC}$. If M is the midpoint of $\overline{AD}$, then $ABCM$ and $BCDM$ are parallelograms which are equal in area.

9.33. (a) In $\triangle ABC$, E is a point on $\overline{BM}$, the median to $\overline{AC}$. Prove that area($\triangle BEA$) = area($\triangle BEC$).

(b) In $\triangle ABC$, Q is a point on $\overline{BC}$, M is the midpoint of $\overline{AB}$, and P is the midpoint of $\overline{AC}$. Prove that area($\triangle BQM$) + area($\triangle PQC$) = area(quadrilateral $APQM$).

(c) In quadrilateral $ABCD$, diagonal $\overline{AC}$ bisects diagonal $\overline{BD}$. Prove that area($\triangle ABC$) = area ($\triangle ACD$).

(d) Prove that the diagonals of a parallelogram divide the parallelogram into four triangles which are equal in area. (9.10)

9.34. Find the ratio of the areas of two similar triangles if the ratio of two corresponding sides is (a) 1:7; (b) 7:2; (c) $1:\sqrt{3}$; (d) $a:5a$; (e) $9:x$; (f) $3:\sqrt{x}$; (g) $s:s\sqrt{2}$. (9.11)

9.35. Find the ratio of the areas of two similar triangles (9.11)

(a) If the ratio of the lengths of two corresponding medians is 7:10

(b) If the length of an altitude of the first is two-thirds of a corresponding altitude of the second

(c) If two corresponding angle bisectors have lengths 10 and 12

(d) If the length of each side of the first is one-third the length of each corresponding side of the second

(e) If the radii of their circumscribed circles are $7\frac{1}{2}$ and 5

(f) If their perimeters are 30 and $30\sqrt{2}$

9.36. Find the ratio of any two corresponding sides of two similar triangles if the ratio of their areas is (a) 100:1; (b) 1:49; (c) 400:81; (d) 25:121; (e) $4:y^2$; (f) $9x^2:1$; (g) 3:4; (h) 1:2; (i) $x^2:5$; (j) $x:16$.   (9.11)

9.37. In two similar triangles, find the ratio of the lengths of   (9.11)

(a) Corresponding sides if the areas are 72 and 50

(b) Corresponding medians if the ratio of the areas is 9:49

(c) Corresponding altitudes if the areas are 18 and 6

(d) The perimeters if the areas are 50 and 40

(e) Radii of the inscribed circles if the ratio of the areas is 1:3

9.38. The areas of two similar triangles are in the ratio of 25:16. Find   (9.11)

(a) The length of a side of the larger if the corresponding side of the smaller has length 80

(b) The length of a median of the larger if the corresponding median of the smaller has length 10

(c) The length of an angle bisector of the smaller if the corresponding angle bisector of the larger has length 15

(d) The perimeter of the smaller if the perimeter of the larger is 125

(e) The circumference of the inscribed circle of the larger if the circumference of the inscribed circle of the smaller is 84

(f) The diameter of the circumscribed circle of the smaller if the diameter of the circumscribed circle of the larger is 22.5

(g) The length of an altitude of the larger if the corresponding altitude of the smaller has length $16\sqrt{3}$

9.39. (a) The areas of two similar triangles are 36 and 25. If a median of the smaller triangle has length 10, find the length of the corresponding median of the larger.   (9.12)

(b) Corresponding altitudes of two similar triangles have lengths 3 and 4. If the area of the larger triangle is 112, find the area of the smaller.

(c) Two similar polygons have perimeters of 32 and 24. If the area of the smaller is 27, find the area of the larger.

(d) The areas of two similar pentagons are 88 and 22. If a diagonal of the larger has length 5, find the length of the corresponding diagonal of the smaller.

(e) In two similar polygons, the ratio of the lengths of two corresponding sides is $\sqrt{3}:1$. If the area of the smaller is 15, find the area of the larger.

CHAPTER 10

Regular Polygons and the Circle

10.1 Regular Polygons

A *regular polygon* is an equilateral and equiangular polygon.

The *center of a regular polygon* is the common center of its inscribed and circumscribed circles.

A *radius of a regular polygon* is a segment joining its center to any vertex. A radius of a regular polygon is also a radius of the circumscribed circle. (Here, as for circles, we may use the word *radius* to mean the number that is "the length of the radius.")

A *central angle of a regular polygon* is an angle included between two radii drawn to successive vertices.

An *apothem of a regular polygon* is a segment from its center perpendicular to one of its sides. An apothem is also a radius of the inscribed circle.

Thus for the regular pentagon shown in Fig. 10-1, $AB = BC = CD = DE = EA$ and $m\angle A = m\angle B = m\angle C = m\angle D = m\angle E$. Also, its center is O, $\overline{OA}$ and $\overline{OB}$ are its radii; $\angle AOB$ is a central angle; and $\overline{OG}$ and $\overline{OF}$ are apothems.

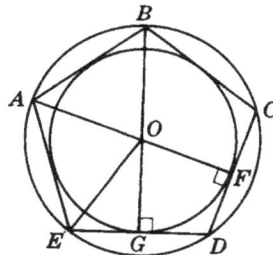

Fig. 10-1

10.1A Regular-Polygon Principles

PRINCIPLE 1: *If a regular polygon of n sides has a side of length s, the perimeter is p = ns.*

PRINCIPLE 2: *A circle may be circumscribed about any regular polygon.*

PRINCIPLE 3: *A circle may be inscribed in any regular polygon.*

PRINCIPLE 4: *The center of the circumscribed circle of a regular polygon is also the center of its inscribed circle.*

PRINCIPLE 5: *An equilateral polygon inscribed in a circle is a regular polygon.*

PRINCIPLE 6: *Radii of a regular polygon are congruent.*

PRINCIPLE 7:   *A radius of a regular polygon bisects the angle to which it is drawn.*

Thus in Fig. 10-1, $\overline{OB}$ bisects $\angle ABC$.

PRINCIPLE 8:   *Apothems of a regular polygon are congruent.*

PRINCIPLE 9:   *An apothem of a regular polygon bisects the side to which it is drawn.*

Thus in Fig. 10-1, $\overline{OF}$ bisects $\overline{CD}$, and $\overline{OG}$ bisects $\overline{ED}$.

PRINCIPLE 10:   *For a regular polygon of n sides:*

1. *Each central angle c measures $\dfrac{360°}{n}$.*

2. *Each interior angle i measures $\dfrac{(n-2)180°}{n}$.*

3. *Each exterior angle e measures $\dfrac{360°}{n}$.*

Thus for the regular pentagon *ABCDE* of Fig. 10-2,

$$m\angle AOB = m\angle ABS = \frac{360°}{n} = \frac{360°}{5} = 72° \qquad m\angle ABC = \frac{(n-2)180°}{n} = \frac{(5-2)180°}{5} = 108°$$

and
$$m\angle ABC + m\angle ABS = 180°$$

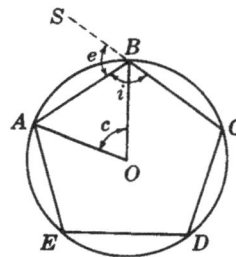

Fig. 10-2

SOLVED PROBLEMS

10.1 Finding measures of lines and angles in a regular polygon

(a) Find the length of a side s of a regular pentagon if the perimeter p is 35.

(b) Find the length of the apothem a of a regular pentagon if the radius of the inscribed circle is 21.

(c) In a regular polygon of five sides, find the measures of the central angle c, the exterior angle e, and the interior angle i.

(d) If an interior angle of a regular polygon measures 108°, find the measures of the exterior angle and the central angle and the number of sides.

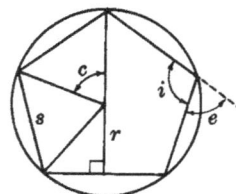

Fig. 10-3

Solutions

(a) $p = 35$. Since $p = 5s$, we have $35 = 5s$ and $s = 7$.

(b) Since an apothem r is a radius of the inscribed circle, it has length 21.

(c) $n = 5$. Then $m\angle c = \dfrac{360°}{n} = \dfrac{360°}{5} = 72°$; $m\angle e = \dfrac{360°}{n} = 72°$; $m\angle i = 180° - m\angle e = 108°$.

(d) $m\angle i = 108°$. Then $m\angle c = 180° - m\angle i = 72°$. Since $m\angle c = \dfrac{360°}{n}$, $n = 5$. (See Fig. 10-3.)

10.2   Proving a regular-polygon problem stated in words

Prove that a vertex angle of a regular pentagon is trisected by diagonals drawn from that vertex.

Solution

Given: Regular pentagon *ABCDE*

Diagonals $\overline{AC}$ and $\overline{AD}$

To Prove: $\overline{AC}$ and $\overline{AD}$ trisect $\angle A$.

Plan: Circumscribe a circle and show that angles *BAC*, *CAD*, and *DAE* are congruent.

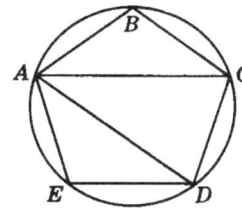

PROOF:

Statements	Reasons
1. *ABCDE* is a regular pentagon.	1. Given
2. Circumscribe a circle about *ABCDE*.	2. A circle may be circumscribed about any regular polygon
3. $BC = CD = DE$	3. A regular polygon is equilateral.
4. $\overset{\frown}{BC} \cong \overset{\frown}{CD} \cong \overset{\frown}{DE}$	4. In a circle, equal chords have equal arcs.
5. $\angle BAC \cong \angle CAD \cong \angle DAE$	5. In a circle, inscribed angles having congruent arcs are congruent.
6. $\angle A$ is trisected.	6. To divide into three congruent parts is to trisect.

10.2   Relationships of Segments in Regular Polygons of 3, 4, and 6 Sides

In the regular hexagon, square, and equilateral triangle, special right triangles are formed when the apothem r and a radius R terminating in the same side are drawn. In the case of the square we obtain a 45°-45°-90° triangle, while in the other two cases we obtain a 30°-60°-90° triangle. The formulas in Fig. 10-4 relate the lengths of the sides and radii of these regular polygons.

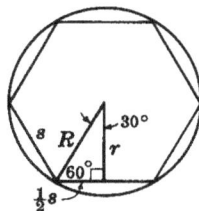

Regular Hexagon
$s = R$
$r = \frac{1}{2}R\sqrt{3}$

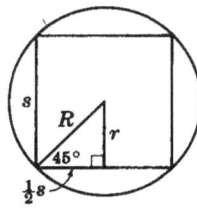

Square
$s = R\sqrt{2}$
$r = \frac{1}{2}s = \frac{1}{2}R\sqrt{2}$

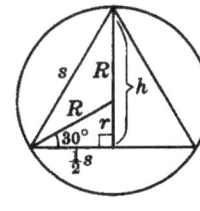

Equilateral Triangle
$s = R\sqrt{3},\ h = r + R$
$r = \frac{1}{3}h,\ R = \frac{2}{3}h,\ r = \frac{1}{2}R$

Fig. 10-4

SOLVED PROBLEMS

10.3   Applying line relationships in a regular hexagon

In a regular hexagon, (a) find the lengths of the side and apothem if the radius is 12; (b) find the radius and length of the apothem if the side has length 8.

Solutions

(a) Since $R = 12$, $s = R = 12$ and $r = \frac{1}{2}R\sqrt{3} = 6\sqrt{3}$.

(b) Since $s = 8$, $R = s = 8$ and $r = \frac{1}{2}R\sqrt{3} = 4\sqrt{3}$.

10.4   Applying line relationships in a square

In a square, (a) find the lengths of the side and apothem if the radius is 16; (b) find the radius and the length of the apothem if a side has length 10.

Solutions

(a) Since $R = 16$, $s = R\sqrt{2} = 16\sqrt{2}$ and $r = \frac{1}{2}s = 8\sqrt{2}$.

(b) Since $s = 10$, $r = \frac{1}{2}s = 5$ and $R = \dfrac{s}{\sqrt{2}} = \frac{1}{2}s\sqrt{2} = 5\sqrt{2}$.

10.5   Applying line relationships in an equilateral triangle

In an equilateral triangle, (a) find the lengths of the radius, apothem, and side if the altitude has length 6; (b) find the lengths of the side, apothem, and altitude if the radius is 9.

Solutions

(a) Since $h = 6$, we have $r = \frac{1}{3}h = 2$; $R = \frac{2}{3}h = 4$; and $s = R\sqrt{3} = 4\sqrt{3}$.

(b) Since $R = 9$, $s = R\sqrt{3} = 9\sqrt{3}$; $r = \frac{1}{2}R = 4\frac{1}{2}$; and $h = \frac{3}{2}R = 13\frac{1}{2}$.

10.3   Area of a Regular Polygon

The area of a regular polygon equals one-half the product of its perimeter and the length of its apothem.

As shown in Fig. 10-5, by drawing radii we can divide a regular polygon of n sides and perimeter $p = ns$ into n triangles, each of area $\frac{1}{2}rs$. Hence, the area of the regular polygon is $n(\frac{1}{2}rs) = \frac{1}{2}nsr = \frac{1}{2}pr$.

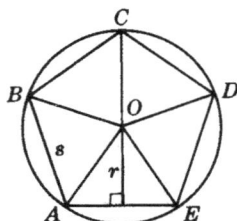

Regular Polygon
$A = \frac{1}{2}nsr = \frac{1}{2}pr$

Fig. 10-5

SOLVED PROBLEMS

10.6 Finding the area of a regular polygon

(a) Find the area of a regular hexagon if the length of the apothem is $5\sqrt{3}$.

(b) Find the area of a regular pentagon to the nearest integer if the length of the apothem is 20.

Solutions

(a) In a regular hexagon, $r = \frac{1}{2}s\sqrt{3}$. Since $r = 5\sqrt{3}$, $s = 10$ and $p = 6(10) = 60$.

Then $A = \frac{1}{2}pr = \frac{1}{2}(60)(5\sqrt{3}) = 150\sqrt{3}$.

(b) In Fig. 10-6, $m\angle AOE = 360°/5 = 72°$ and $m\angle AOF = \frac{1}{2}m\angle AOE = 36°$. Then $\tan 36° = \frac{1}{2}s/20 = s/40$ or $s = 40\tan 36°$.

Now $A = \frac{1}{2}pr = \frac{1}{2}nsr = \frac{1}{2}(5)(40\tan 36°)(20) = 1453$.

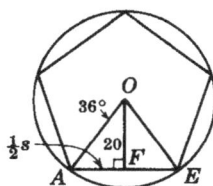

Fig. 10-6

10.4 Ratios of Segments and Areas of Regular Polygons

PRINCIPLE 1: *Regular polygons having the same number of sides are similar.*

PRINCIPLE 2: *Corresponding segments of regular polygons having the same number of sides are in proportion. "Segments" here includes sides, perimeters, radii or circumferences of circumscribed or inscribed circles, and such.*

PRINCIPLE 3: *Areas of regular polygons having the same number of sides are to each other as the squares of the lengths of any two corresponding segments.*

SOLVED PROBLEMS

10.7 Ratios of lines and areas of regular polygons

(a) In two regular polygons having the same number of sides, find the ratio of the lengths of the apothems if the perimeters are in the ratio 5:3.

(b) In two regular polygons having the same number of sides, find the length of a side of the smaller if the lengths of the apothems are 20 and 50 and a side of the larger has length 32.5.

(c) In two regular polygons having the same number of sides, find the ratio of the areas if the lengths of the sides are in the ratio 1:5.

(d) In two regular polygons having the same number of sides, find the area of the smaller if the sides have lengths 4 and 12 and the area of the larger is 10,260.

Solutions

(a) By Principle 2, $r:r'=p:p'=5:3$.

(b) By Principle 2, $s:s'=r:r'$; thus, $s:32.5=20:50$ and $s=13$.

(c) By Principle 3, $\dfrac{A}{A'}=\left(\dfrac{s}{s'}\right)^2=\left(\dfrac{1}{5}\right)^2=\dfrac{1}{25}$.

(d) By Principle 3, $\dfrac{A}{A'}=\left(\dfrac{s}{s'}\right)^2$. Then $\dfrac{A}{10{,}260}=\left(\dfrac{4}{12}\right)^2$ and $A=1140$.

10.5　Circumference and Area of a Circle

π (pi) is the ratio of the circumference C of any circle to its diameter d; that is, $\pi=C/d$. Hence,

$$C=\pi d \quad \text{or} \quad C=2\pi r$$

Approximate values for π are 3.1416 or 3.14 or $\frac{22}{7}$. Unless you are told otherwise, we shall use 3.14 for π in solving problems.

A circle may be regarded as a regular polygon having an infinite number of sides. If a square is inscribed in a circle, and the number of sides is continually doubled (to form an octagon, a 16-gon, and so on), the perimeters of the resulting polygons will get closer and closer to the circumference of the circle (Fig. 10-7).

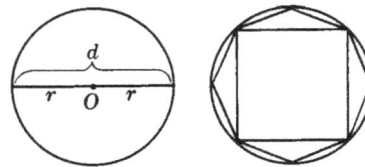

Circle:　$C=2\pi r$
$A=\pi r^2$

Fig. 10-7

Thus to find the area of a circle, the formula $A=\frac{1}{2}pr$ can be used with C substituted for p; doing so, we get

$$A=\tfrac{1}{2}Cr=\tfrac{1}{2}(2\pi r)(r)=\pi r^2$$

All circles are similar figures, since they have the same shape. Because they are similar figures, (1) corresponding segments of circles are in proportion and (2) the areas of two circles are to each other as the squares of their radii or circumferences.

SOLVED PROBLEMS

10.8　Finding the circumference and area of a circle

In a circle, (a) find the circumference and area if the radius is 6; (b) find the radius and area if the circumference is 18π; (c) find the radius and circumference if the area is 144π. (Answer both in terms of π and to the nearest integer.)

Solutions

(a) $r = 6$. Then $C = 2\pi r = 12\pi = 38$ and $A = \pi r^2 = 36\pi = 36(3.14) = 113$.

(b) $C = 18\pi$. Since $C = 2\pi r$, we have $18\pi = 2\pi r$ and $r = 9$. Then $A = \pi r^2 = 81\pi = 254$.

(c) $A = 144\pi$. Since $A = \pi r^2$, we have $144\pi = \pi r^2$ and $r = 12$. Then $C = 2\pi r = 24\pi = 75$.

10.9 Circumference and area of circumscribed and inscribed circles

Find the circumference and area of the circumscribed circle and inscribed circle (a) of a regular hexagon whose side has length 8; (b) of an equilateral triangle whose altitude has length $9\sqrt{3}$. (See Fig. 10-8.)

Solutions

(a) Here $R = s = 8$. Then $C = 2\pi R = 16\pi$ and $A = \pi R^2 = 64\pi$.
Also $r = \frac{1}{2}R\sqrt{3} = 4\sqrt{3}$. Then $C = 2\pi r = 8\pi\sqrt{3}$ and $A = \pi r^2 = 48\pi$.

(b) Here $R = \frac{2}{3}h = 6\sqrt{3}$. Then $C = 2\pi R = 12\pi\sqrt{3}$ and $A = \pi R^2 = 108\pi$.
Also $r = \frac{1}{3}h = 3\sqrt{3}$. Then $C = 2\pi r = 6\pi\sqrt{3}$ and $A = \pi r^2 = 27\pi$.

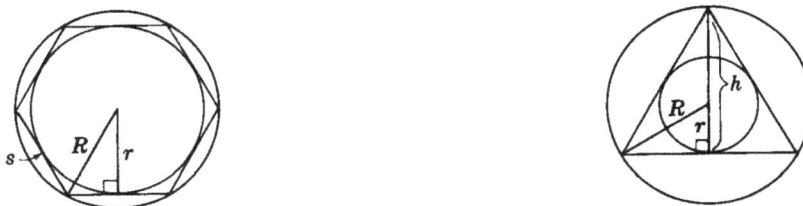

Fig. 10-8

10.10 Ratios of segments and areas in circles

(a) If the circumferences of two circles are in the ratio 2:3, find the ratio of the diameters and the ratio of the areas.

(b) If the areas of two circles are in the ratio 1:25, find the ratio of the diameters and the ratio of the circumferences.

Solutions

(a) $\dfrac{d}{d'} = \dfrac{C}{C'} = \dfrac{2}{3}$ and $\dfrac{A}{A'} = \left(\dfrac{C}{C'}\right)^2 = \left(\dfrac{2}{3}\right)^2 = \dfrac{4}{9}$.

(b) Since $\dfrac{A}{A'} = \left(\dfrac{d}{d'}\right)^2$, $\dfrac{1}{25} = \left(\dfrac{d}{d'}\right)^2$ and $\dfrac{d}{d'} = \dfrac{1}{5}$. Also, $\dfrac{C}{C'} = \dfrac{d}{d'} = \dfrac{1}{5}$.

10.6 Length of an Arc; Area of a Sector and a Segment

A *sector* of a circle is a part of a circle bounded by two radii and their intercepted arc. Thus in Fig. 10-9, the shaded section of circle O is sector OAB.

A *segment of a circle* is a part of a circle bounded by a chord and its arc. A *minor segment* of a circle is the smaller of the two segments thus formed. Thus in Fig. 10-10, the shaded section of circle Q is minor segment ACB.

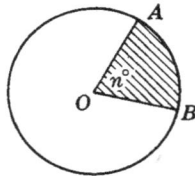

Fig. 10-9

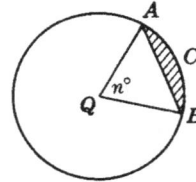

Fig. 10-10

PRINCIPLE 1:   *In a circle of radius r, the length l of an arc of measure $n°$ equals $\dfrac{n}{360}$ of the circumference of the circle, or $l = \dfrac{n}{360}2\pi r = \dfrac{\pi n r}{180}$.*

PRINCIPLE 2:   *In a circle of radius r, the area K of a sector of measure $n°$ equals $\dfrac{n}{360}$ of the area of the circle, or $K = \dfrac{n}{360}\pi r^2$.*

PRINCIPLE 3: $\dfrac{\textit{Area of a sector of } n°}{\textit{Area of the circle}} = \dfrac{\textit{length of an arc of measure } n°}{\textit{circumference of the circle}} = \dfrac{n}{360}$

PRINCIPLE 4:   *The area of a minor segment of a circle equals the area of its sector less the area of the triangle formed by its radii and chord.*

PRINCIPLE 5:   *If a regular polygon is inscribed in a circle, each segment cut off by the polygon has area equal to the difference between the area of the circle and the area of the polygon divided by the number of sides.*

SOLVED PROBLEMS

10.11   Length of an arc

    (a) Find the length of a $36°$ arc in a circle whose circumference is 45π.

    (b) Find the radius of a circle if a $40°$ arc has a length of 4π.

Solutions

    (a) Here $n° = 36°$ and $C = 2\pi r = 45\pi$. Then $l = \dfrac{n}{360}2\pi r = \dfrac{36}{360}45\pi = \dfrac{9}{2}\pi$.

    (b) Here $l = 4\pi$ and $n° = 40°$. Then $l = \dfrac{n}{360}2\pi r$ yields $4\pi = \dfrac{40}{360}2\pi r$, and $r = 18$.

10.12   Area of a sector

    (a) Find the area K of a $300°$ sector of a circle whose radius is 12.

    (b) Find the measure of the central angle of a sector whose area is 6π if the area of the circle is 9π.

    (c) Find the radius of a circle if an arc of length 2π has a sector of area 10π.

Solutions

(a) $n° = 300°$ and $r = 12$. Then $K = \frac{n}{360}\pi r^2 = \frac{300}{360}144\pi = 120\pi$.

(b) $\frac{\text{Area of sector}}{\text{Area of circle}} = \frac{n}{360}$, so $\frac{6\pi}{9\pi} = \frac{n}{360}$, and $n = 240$. Thus, the central angle measures $240°$.

(c) $\frac{\text{Length of arc}}{\text{Circumference}} = \frac{\text{area of sector}}{\text{area of circle}}$, so $\frac{2\pi}{2\pi r} = \frac{10\pi}{\pi r^2}$ and $r = 10$.

10.13   Area of a segment of a circle

(a) Find the area of a segment if its central angle measures $60°$ and the radius of the circle is 12.

(b) Find the area of a segment if its central angle measures $90°$ and the radius of the circle is 8.

(c) Find each segment formed by an inscribed equilateral triangle if the radius of the circle is 8.

Solutions

See Fig. 10-11.

(a) $n° = 60°$ and $r = 12$. Then area of sector $OAB = \frac{n}{360}\pi r^2 = \frac{60}{360}144\pi = 24\pi$.

Also, area of equilateral $\triangle OAB = \frac{1}{4}s^2\sqrt{3} = \frac{1}{4}(144)\sqrt{3} = 36\sqrt{3}$.

Hence, area of segment $ACB = 24\pi - 36\sqrt{3}$.

(b) $n° = 90°$ and $r = 8$. Then area of sector $OAB = \frac{n}{360}\pi r^2 = \frac{90}{360}64\pi = 16\pi$.

Also, area of rt. $\triangle OAB = \frac{1}{2}bh = \frac{1}{2}(8)(8) = 32$.

Hence, area of segment $ACB = 16\pi - 32$.

(c) $R = 8$. Since $s = R\sqrt{3} = 8\sqrt{3}$, the area of $\triangle ABC$ is $\frac{1}{4}s^2\sqrt{3} = 48\sqrt{3}$.

Also, area of circle $O = \pi R^2 = 64\pi$.

Hence, area of segment $BDC = \frac{1}{3}(64\pi - 48\sqrt{3})$.

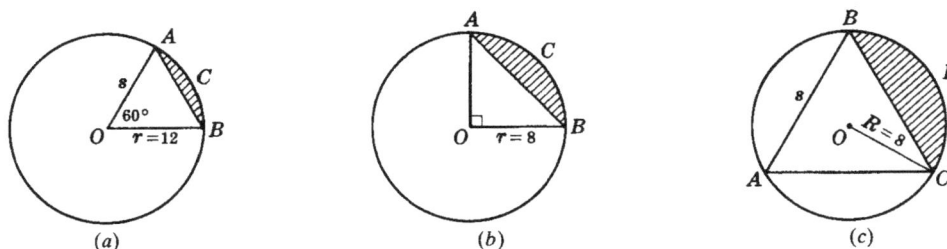

(a)   (b)   (c)

Fig. 10-11

10.14   Area of a segment formed by an inscribed regular polygon

Find the area of each segment formed by an inscribed regular polygon of 12 sides (dodecagon) if the radius of the circle is 12. (See Fig. 10-12.)

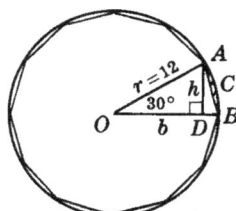

Fig. 10-12

Solution

Area of sector $OAB = \dfrac{n}{360}\pi r^2 = \dfrac{30}{360}144\pi = 12\pi.$

To find the area $\triangle OAB$, we draw altitude $\overline{AD}$ to base $\overline{OB}$. Since $m\angle AOB = 30°, h = AD = \frac{1}{2}r = 6.$

Then the area of $\triangle OAB$ is $\frac{1}{2}bh = \frac{1}{2}(12)(6) = 36.$

Hence, the area of segment ACB is $12\pi - 36.$

10.7 Areas of Combination Figures

The areas of combination figures like that in Fig. 10-13 may be found by determining individual areas and then adding or substracting as required. Thus, the shaded area in the figure equals the sum of the aeas of the square and the semicircle: $A = 8^2 + \frac{1}{2}(16\pi) = 64 + 8\pi.$

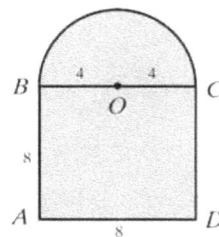

Fig. 10-13

SOLVED PROBLEMS

10.15 Finding areas of combination figures

Find the shaded area in each part of Fig. 10-14. In (a), circles A, B, and C are tangent externally and each has radius 3. In (b), each arc is part of a circle of radius 9.

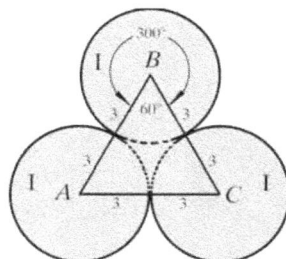

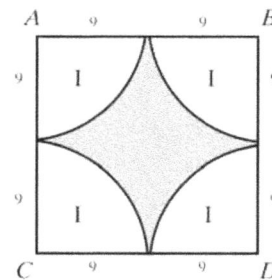

(a)   (b)

Fig. 10-14

Solutions

(a) Area of $\triangle ABC = \frac{1}{4}s^2\sqrt{3} = \frac{1}{4}(6^2)\sqrt{3} = 9\sqrt{3}.$ Area of sector I $= \dfrac{n°}{360°}(\pi r^2) = \dfrac{300}{360}(9\pi) = \dfrac{15}{2}\pi.$

Shaded area $= 9\sqrt{3} + 3\left(\frac{15}{2}\pi\right) = 9\sqrt{3} + \frac{45}{2}\pi.$

(b) Area of square $= 18^2 = 324.$ Area of sector I $= \dfrac{n°}{360°}(\pi r^2) = \dfrac{90}{360}(81\pi) = \dfrac{81}{4}\pi.1$

Shaded area $= 324 - 4\left(\frac{81}{4}\pi\right) = 324 - 81\pi.$

SUPPLEMENTARY PROBLEMS

10.1. In a regular polygon, find (10.1)

    (a) The perimeter if the length of a side is 8 and the number of sides is 25

    (b) The perimeter if the length of a side is 2.45 and the number of sides is 10

    (c) The perimeter if the length of a side is $4\frac{2}{3}$ and the number of sides is 24

    (d) The number of sides if the perimeter is 325 and the length of a side is 25

    (e) The number of sides if the perimeter is $27\sqrt{3}$ and the length of a side is $3\sqrt{3}$

    (f) The length of a side if the number of sides is 30 and the perimeter is 100

    (g) The length of a side if the perimeter is 67.5 and the number of sides is 15

10.2. In a regular polygon, find (10.1)

    (a) The length of the apothem if the diameter of an inscribed circle is 25

    (b) The length of the apothem if the radius of the inscribed circle is 23.47

    (c) The radius of the inscribed circle if the length of the apothem is $7\sqrt{3}$

    (d) The radius of the regular polygon if the diameter of the circumscribed circle is 37

    (e) The radius of the circumscribed circle if the radius of the regular polygon is $3\sqrt{2}$

10.3. In a regular polygon of 15 sides, find the measure of (a) the central angle; (b) the exterior angle; (c) the interior angle. (10.1)

10.4. If an exterior angle of a regular polygon measures 40°, find (a) the measure of the central angle; (b) the number of sides; (c) the measure of the interior angle. (10.1)

10.5. If an interior angle of a regular polygon measures 165°, find (a) the measure of the exterior angle; (b) the measure of the central angle; (c) the number of sides. (10.1)

10.6. If a central angle of a regular polygon measures 5°, find (a) the measure of the exterior angle; (b) the number of sides; (c) the measure of the interior angle. (10.1)

10.7. Name the regular polygon whose (10.1)

    (a) Central angle measures 45°   (d) Exterior angle measures 36°

    (b) Central angle measures 60°   (e) Interior angle is congruent to its central angle

    (c) Exterior angle measures 120°   (f) Interior angle measures 150°

10.8. Prove each of the following: (10.2)

    (a) The diagonals of a regular pentagon are congruent.

    (b) A diagonal of a regular pentagon forms an isosceles trapezoid with three of its sides.

    (c) If two diagonals of a regular pentagon intersect, the longer segment of each diagonal is congruent to a side of the regular pentagon.

10.9. In a regular hexagon, find (10.3)

    (a) The length of a side if its radius is 9

    (b) The perimeter if its radius is 5

(c) The length of the apothem if its radius is 12

(d) Its radius if the length of a side is 6

(e) The length of the apothem if the length of a side is 26

(f) Its radius if the length of the apothem is $3\sqrt{3}$

(g) The length of a side if the length of the apothem is 30

(h) The perimeter if the length of the apothem is $5\sqrt{3}$

10.10. In a square, find   (10.4)

(a) The length of a side if the radius is 18

(b) The length of the apothem if the radius is 14

(c) The perimeter if the radius is $5\sqrt{2}$

(d) The radius if the length of a side is 16

(e) The length of a side if the length of the apothem is 1.7

(f) The perimeter if the length of the apothem is $3\frac{1}{2}$

(g) The radius if the perimeter is 40

(h) The length of the apothem if the perimeter is $16\sqrt{2}$

10.11. In an equilateral triangle, find   (10.5)

(a) The length of a side if its radius is 30

(b) The length of the apothem if its radius is 28

(c) The length of an altitude if its radius is 18

(d) The perimeter if its radius is $2\sqrt{3}$

(e) Its radius if the length of a side is 24

(f) The length of the apothem if the length of a side is 24

(g) The length of its altitude if the length of a side is 96

(h) Its radius if the length of the apothem is 21

(i) The length of a side if the length of the apothem is $\sqrt{3}$

(j) The length of the altitude if the length of the apothem is $3\frac{1}{3}$

(k) The length of the altitude if the perimeter is 15

(l) The length of the apothem if the perimeter is 54

10.12. (a) Find the area of a regular pentagon to the nearest integer if the length of the apothem is 15.   (10.6)

(b) Find the area of a regular decagon to the nearest integer if the length of a side is 20.

10.13. Find the area of a regular hexagon, in radical form, if (a) the length of a side is 6; (b) its radius is 8; (c) the length of the apothem is $10\sqrt{3}$.   (10.6)

10.14. Find the area of a square if (a) the length of the apothem is 12; (b) its radius is $9\sqrt{2}$; (c) its perimeter is 40.   (10.6)

10.15. Find the area of an equilateral triangle, in radical form, if   (10.6)

    (a) The length of the apothem is $2\sqrt{3}$.   (d) The length of the altitude is $12\sqrt{3}$.

    (b) Its radius is 6.   (e) The perimeter is $6\sqrt{3}$.

    (c) The length of the altitude is 4.   (f) The length of the apothem is 4.

10.16. If the area of a regular hexagon is $150\sqrt{3}$, find (a) the length of a side; (b) its radius; (c) the length of the apothem.   (10.6)

10.17. If the area of an equilateral triangle is $81\sqrt{3}$, find (a) the length of a side; (b) the length of the altitude; (c) its radius; (d) the length of the apothem.   (10.6)

10.18. Find the ratio of the perimeters of two regular polygons having the same number of sides if   (10.7)

    (a) The ratio of the sides is 1:8.

    (b) The ratio of their radii is 4:9.

    (c) Their radii are 18 and 20.

    (d) Their apothems have lengths 16 and 22.

    (e) The length of the larger side is triple that of the smaller.

    (f) The length of the smaller apothem is two-fifths that of the larger.

    (g) The lengths of the apothems are $20\sqrt{2}$ and 15.

    (h) The circumference of the larger circumscribed circle is $2\frac{1}{2}$ times that of the smaller.

10.19. Find the ratio of the perimeters of two equilateral triangles if (a) the sides have lengths 20 and 8; (b) their radii are 12 and 60; (c) their apothems have lengths $2\sqrt{3}$ and $6\sqrt{3}$; (d) the circumferences of their inscribed circles are 120 and 160; (e) their altitudes have lengths $5x$ and x.   (10.7)

10.20. Find the ratio of the lengths of the sides of two regular polygons having the same number of sides if the ratio of their areas is (a) 25:1; (b) 16:49; (c) x^2:4; (d) 2:1; (e) 3:y^2; (f) x:18.   (10.7)

10.21. Find the ratio of the areas of two regular hexagons if (a) their sides have lengths 14 and 28; (b) their apothems have lengths 3 and 15; (c) their radii are $6\sqrt{3}$ and $\sqrt{3}$; (b) their perimeters are 75 and 250; (e) the circumferences of the circumscribed circles are 28 and 20.   (10.7)

10.22. Find the circumference of a circle in terms of π if (a) the radius is 6; (b) the diameter is 14; (c) the area is 25π; (d) the area is 3π.   (10.8)

10.23. Find the area of a circle in terms of π if (a) the radius is 3; (b) the diameter is 10; (c) the circumference is 16π; (d) the circumference is π; (e) the circumference is $6\pi\sqrt{2}$.   (10.8)

10.24. In a circle, (a) find the circumference and area if the radius is 5; (b) find the radius and area if the circumference is 16π; (c) find the radius and circumference if the area is 16π.   (10.8)

10.25. In a regular hexagon, find the circumference of the circumscribed circle if (a) the length of the apothem is $3\sqrt{3}$; (b) the perimeter is 12; (c) the length of a side is $3\frac{1}{2}$. Also find the circumference of its inscribed circle if (d) the length of the apothem is 13; (e) the length of a side is 8; (f) the perimeter is $6\sqrt{3}$.   (10.9)

10.26. For a square, find the area in terms of π of the ____ (10.9)

    (a) Circumscribed circle if the length of the apothem is 7

    (b) Circumscribed circle if the perimeter is 24

    (c) Circumscribed circle if the length of a side is 8

    (d) Inscribed circle if the length of the apothem is 5

    (e) Inscribed circle if the length of a side is $12\sqrt{2}$

    (f) Inscribed circle if the perimeter is 80

10.27. Find the circumference and area of the (1) circumscribed circle and (2) inscribed circle of ____ (10.9)

    (a) A regular hexagon if the length of a side is 4

    (b) A regular hexagon if the length of the apothem is $4\sqrt{3}$

    (c) An equilateral triangle if the length of the altitude is 9

    (d) An equilateral triangle if the length of the apothem is 4

    (e) A square if the length of a side is 20

    (f) A square if the length of the apothem is 3

10.28. Find the radius of a pipe having the same capacity as two pipes whose radii are (a) 6 ft and 8 ft; (b) 8 ft and 15 ft; (c) 3 ft and 6 ft. (*Hint*: Find the areas of their circular cross-sections.) (10.10)

10.29. In a circle, find the length of a $90°$ arc if ____ (10.11)

    (a) The radius is 4.

    (b) The diameter is 40.

    (c) The circumference is 32.

    (d) The circumference is 44π.

    (e) An inscribed hexagon has a side of length 12.

    (f) An inscribed equilateral triangle has an altitude of length 30.

10.30. Find the length of ____ (10.11)

    (a) A $90°$ arc if the radius of the circle is 6

    (b) A $180°$ arc if the circumference is 25

    (c) A $30°$ arc if the circumference is 60π

    (d) A $40°$ arc if the diameter is 18

    (e) An arc intercepted by the side of a regular hexagon inscribed in a circle of radius 3

    (f) An arc intercepted by a chord of length 12 in a circle of radius 12.

10.31. In a circle, find the area of a $60°$ sector if ____ (10.12)

    (a) The radius is 6.   (e) The area of the circle is 27.

    (b) The diameter is 2.   (f) The area of a $240°$ sector is 52.

    (c) The circumference is 10π.   (g) An inscribed hexagon has a side of length 12.

    (d) The area of the circle is 150π.   (h) An inscribed hexagon has an area of $24\sqrt{3}$.

10.32. Find the area of a   (10.12)

    (a) $60°$ sector if the radius of the circle is 6   (c) $15°$ sector if the area of the circle is 72π

    (b) $240°$ sector if the area of the circle is 30   (d) $90°$ sector if its arc length is 4π

10.33. Find the measure of a central angle of an arc whose length is   (10.12)

    (a) 3 m if the circumference is 9 m   (d) 6π if the circumference is 12π

    (b) 2 ft if the circumference is 1 yd   (e) Three-eighths of the circumference

    (c) 25 if the circumference is 250   (f) Equal to the radius

10.34. Find the measure of a central angle of a sector whose area is   (10.12)

    (a) 10 if the area of the circle is 50   (d) 5π if the area of the circle is 12π

    (b) 15 cm² if the area of the circle is 20 cm²   (e) Eight-ninths of the area of the circle

    (c) 1 ft² if the area of the circle is 1 yd²

10.35. Find the measure of a central angle of   (10.11 and 10.12)

    (a) An arc whose length is 5π if the area of its sector is 25π

    (b) An arc whose length is 12π if the area of its sector is 48π

    (c) A sector whose area is 2π if the length of its arc is π

    (d) A sector whose area is 10π if the length of its arc is 2π

10.36. Find the radius of a circle if a   (10.11 and 10.12)

    (a) $120°$ arc has a length of 8π (d) $30°$ sector has an area of 3π

    (b) $40°$ arc has a length of 2π (e) $36°$ sector has an area of $2\frac{1}{2}\pi$

    (c) $270°$ arc has a length of 15π (f) $120°$ sector has an area of 6π

10.37. Find the radius of a circle if a sector of area   (10.12)

    (a) 12π has an arc of length 6π

    (b) 10π has an arc of length 2π

    (c) 25 cm² has an arc of length 5 cm

    (d) 162 has an arc of length 36

10.38. Find the area of a segment if its central angle is $60°$ and the radius of the circle is (a) 6; (b) 12; (c) 3; (d) r; (e) $2r$.   (10.13)

10.39. Find the area of a segment of a circle if   (10.13)

    (a) The radius of the circle is 4 and the central angle measures $90°$.

    (b) The radius of the circle is 30 and the central angle measures $60°$.

    (c) The radius of the circle and the chord of the segment each have length 12.

    (d) The central angle is $90°$ and the length of the arc is 4π.

    (e) Its chord of length 20 units is 10 units from the center of the circle.

10.40. Find the area of a segment of a circle if the radius of the circle is 8 and the central angle measures (a) $120°$; (b) $135°$; (c) $150°$.   (10.13)

10.41. If the radius of a circle is 4, find the area of each segment formed by an inscribed (a) equilateral triangle; (b) regular hexagon; (c) square.   (10.13 and 10.14)

10.42. Find the area of each segment of a circle if the segments are formed by an inscribed   (10.13)

     (a) Equilateral triangle and the radius of the circle is 6

     (b) Regular hexagon and the radius of the circle is 3

     (c) Square and the radius of the circle is 6

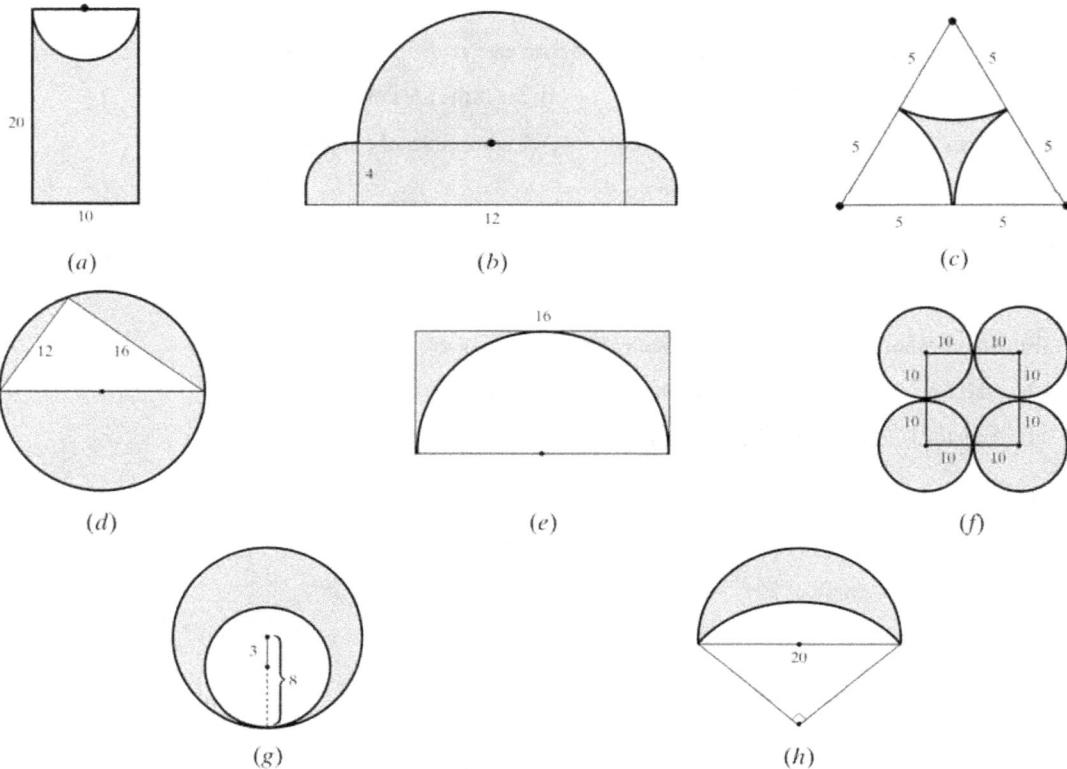

(a)   (b)   (c)

(d)   (e)   (f)

(g)   (h)

Fig. 10-15

10.43. Find the shaded area in each part of Fig. 10-15. Each heavy dot represents the center of an arc or a circle.   (10.15)

10.44. Find the shaded area in each part of Fig. 10-16. Each dot represents the center of an arc or circle.   (10.15)

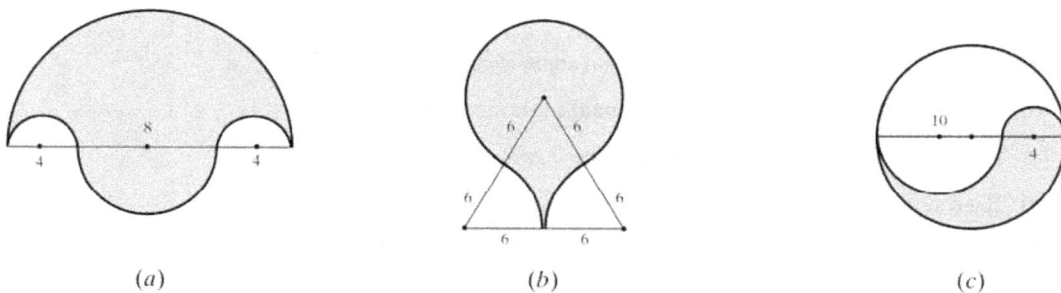

(a)   (b)   (c)

Fig. 10-16

CHAPTER 11

Locus

11.1 Determining a Locus

Locus, in Latin, means *location*. The plural is *loci*. A *locus of points* is the set of points, and only those points, that satisfy given conditions.

Thus, the locus of points that are 1 in from a given point P is the set of points 1 in from P. These points lie on a circle with its center at P and a radius of 1 in, and hence this circle is the required locus (Fig. 11-1). Note that we show loci as long-short dashed figures.

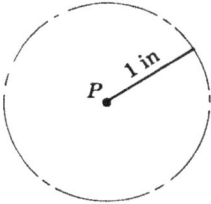

Fig. 11-1

To determine a locus, (1) state what is given and the condition to be satisfied; (2) find several points satisfying the condition which indicate the shape of the locus; then (3) connect the points and describe the locus fully.

All geometric constructions require the use of straightedges and compasses. Hence if a locus is to be *constructed*, such drawing instruments can be used.

11.1A Fundamental Locus Theorems

PRINCIPLE 1: *The locus of points equidistant from two given points is the perpendicular bisector of the line segment joining the two points* (Fig. 11-2).

Fig. 11-2

PRINCIPLE 2: *The locus of points equidistant from two given parallel lines is a line parallel to the two lines and midway between them* (Fig. 11-3).

PRINCIPLE 3: *The locus of points equidistant from the sides of a given angle is the bisector of the angle* (Fig. 11-4).

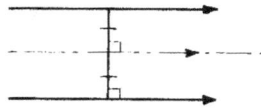

Fig. 11-3

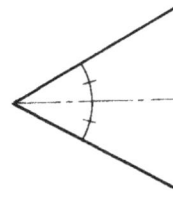

Fig. 11-4

PRINCIPLE 4:   *The locus of points equidistant from two given intersecting lines is the bisectors of the angles formed by the lines (Fig. 11-5).*

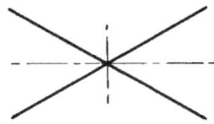

Fig. 11-5

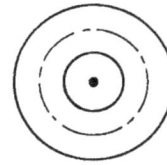

Fig. 11-6

PRINCIPLE 5:   *The locus of points equidistant from two concentric circles is the circle concentric with the given circles and midway between them (Fig. 11-6).*

PRINCIPLE 6:   *The locus of points at a given distance from a given point is a circle whose center is the given point and whose radius is the given distance (Fig. 11-7).*

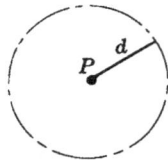

Fig. 11-7

Fig. 11-8

PRINCIPLE 7:   *The locus of points at a given distance from a given line is a pair of lines, parallel to the given line and at the given distance from the given line (Fig. 11-8).*

PRINCIPLE 8:   *The locus of points at a given distance from a given circle whose radius is greater than that distance is a pair of concentric circles, one on either side of the given circle and at the given distance from it (Fig. 11-9).*

PRINCIPLE 9:   *The locus of points at a given distance from a given circle whose radius is less than the distance is a circle, outside the given circle and concentric with it* (Fig. 11-10). (If $r = d$, the locus also includes the center of the given circle.)

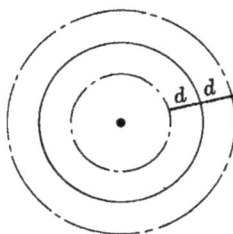

Fig. 11-9

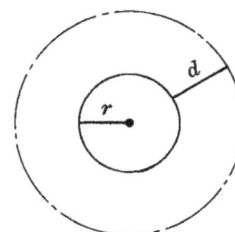

Fig. 11-10

SOLVED PROBLEMS

11.1   Determining loci

Determine the locus of (a) a runner moving equidistant from the sides of a straight track; (b) a plane flying equidistant from two separated aircraft batteries; (c) a satellite 100 mi above the earth; (d) the furthermost point reached by a gun with a range of 10 mi.

Solutions

See Fig. 11-11.

(a) The locus is a line parallel to the two given lines and midway between them.

(b) The locus is the perpendicular bisector of the line joining the two points.

(c) The locus is a circle concentric with the earth and of radius 100 mi greater than that of the earth.

(d) The locus is a circle of radius 10 mi with its center at the gun.

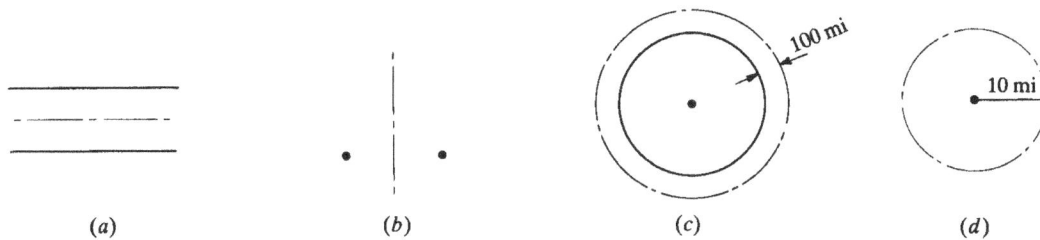

Fig. 11-11

11.2   Determining the locus of the center of a circle

Determine the locus of the center of a circular disk (a) moving so that it touches each of two parallel lines; (b) moving tangentially to two concentric circles; (c) moving so that its rim passes through a fixed point; (d) rolling along a large fixed circular hoop.

Solutions

See Fig. 11-12.

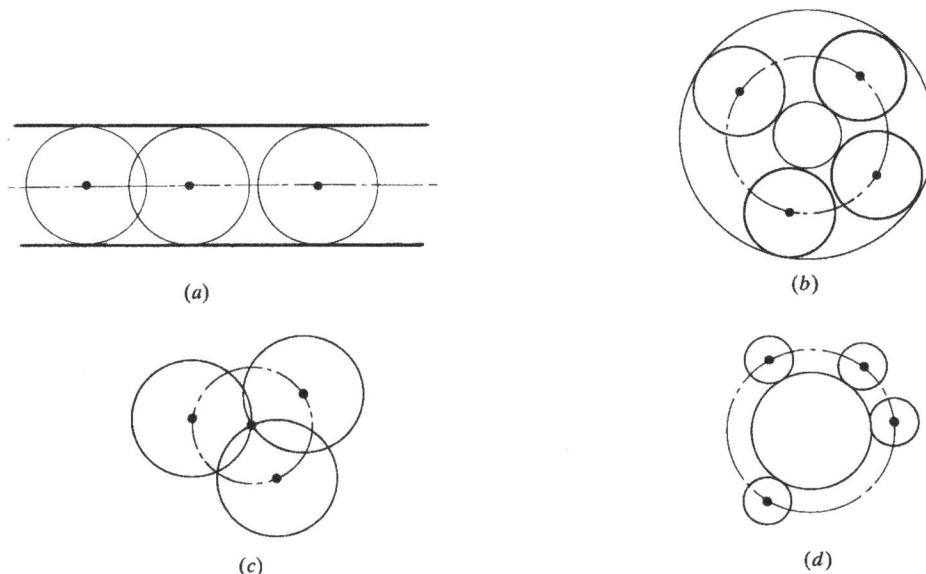

Fig. 11-12

(a) The locus is a line parallel to the two given lines and midway between them.

(b) The locus is a circle concentric with the given circles and midway between them.

(c) The locus is a circle whose center is the given point and whose radius is the radius of the circular disk.

(d) The locus is a circle outside the given circle and concentric to it.

11.3   Constructing loci

Construct (a) the locus of points equidistant from two given points; (b) the locus of points equidistant from two given parallel lines; (c) the locus of points at a given distance from a given circle whose radius is less than that distance.

Solutions

See Fig. 11-13.

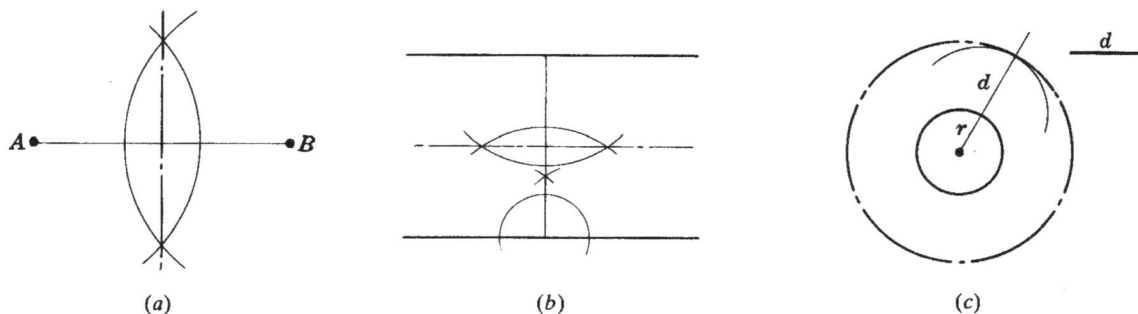

Fig. 11-13

11.2   Locating Points by Means of Intersecting Loci

A point or points which satisfy two conditions may be found by drawing the locus for each condition. The required points are the points of intersection of the two loci.

SOLVED PROBLEM

11.4   Locating points that satisfy two conditions

On a map locate buried treasure that is 3 ft from a tree (T) and equidistant from two points (A and B) in Fig. 11-14.

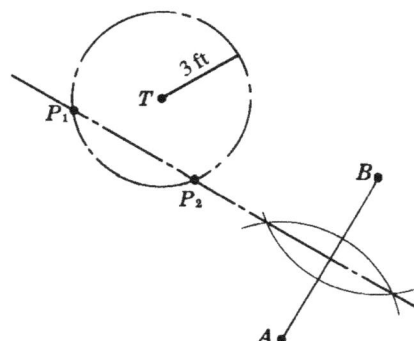

Fig. 11-14

Solution

The required loci are (1) the perpendicular bisector of $\overline{AB}$ and (2) a circle with its center at T and radius 3 ft. As shown, these meet in P_1 and P_2, which are the locations of the treasure.

Note: The diagram shows the two loci intersecting at P_1 and P_2. However, there are three possible kinds of solutions, depending on the location of T with respect to A and B:

1. The solution has two points if the loci intersect.
2. The solution has one point if the perpendicular bisector is tangent to the circle.
3. The solution has no points if the perpendicular bisector does not meet the circle.

11.3 Proving a Locus

To prove that a locus satisfies a given condition, it is necessary to prove the locus theorem *and* its converse or its inverse. Thus to prove that a circle A of radius 2 in is the locus of points 2 in from A, it is necessary to prove either that

1. Any point on circle A is 2 in from A.
2. Any point 2 in from A is on circle A (converse of statement 1).

or that

1. Any point on circle A is 2 in from A.
2. Any point not on circle A is not 2 in from A (inverse of statement 1).

These statements are easily proved using the principle that a point is outside, on, or inside a circle according as its distance from the center is greater than, equal to, or less than the radius of the circle.

SOLVED PROBLEM

11.5 Proving a locus theorem

Prove that the locus of points equidistant from two given points is the perpendicular bisector of the segment joining the two points.

Solution

First prove that any point on the locus satisfies the condition:

Given: Points A and B. $\overline{CD}$ is the $\perp$ bisector of $\overline{AB}$.
To Prove: Any point P on $\overline{CD}$ is equidistant from A and B; that is, $\overline{PA} \cong \overline{PB}$.
Plan: Prove $\triangle PEA \cong \triangle PEB$ to obtain $\overline{PA} \cong \overline{PB}$.

PROOF:

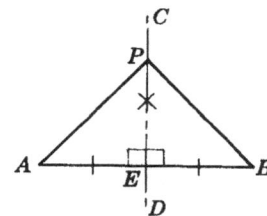

Statements	Reasons
1. $\overline{CD}$ is the $\perp$ bisector of $\overline{AB}$.	1. Given
2. $\angle PEA \cong \angle PEB$	2. Perpendiculars form right angles; all right angles are congruent.
3. $\overline{AE} \cong \overline{EB}$	3. To bisect is to divide into congruent parts.
4. $\overline{PE} \cong \overline{PE}$	4. Reflexive property
5. $\triangle PEA \cong \triangle PEB$	5. SAS
6. $\overline{PA} \cong \overline{PB}$	6. Corresponding parts of $\cong$ triangles are $\cong$.

Then prove that any point satisfying the condition is on the locus:

Given: Any point Q which is equidistant
from points A and B ($\overline{QA} \cong \overline{QB}$).

To Prove: Q is on the perpendicular bisector of $\overline{AB}$.

Plan: Draw $\overline{QG}$ perpendicular to $\overline{AB}$ and prove
by congruent triangles that $\overline{QG}$ bisects $\overline{AB}$.

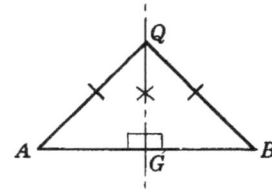

PROOF:

Statements	Reasons
1. Draw $\overline{QG} \perp \overline{AB}$.	1. Through an external point, a line can be drawn perpendicular to a given line.
2. $\overline{QA} \cong \overline{QB}$	2. Given
3. $\angle QGA$ and $\angle QGB$ are rt. $\angle$s; $\triangle QGA$ and $\triangle QGB$ are rt. $\triangle$s.	3. Perpendiculars form right angles; $\triangle$s with a rt.$\angle$ are rt. $\triangle$s.
4. $\overline{QG} \cong \overline{QG}$	4. Reflexive property
5. $\triangle QGA \cong \triangle QGB$	5. Hy-leg
6. $\overline{AG} \cong \overline{GB}$	6. Corresponding parts of $\cong$ triangles are $\cong$.
7. $\overline{QG}$ bisects $\overline{AB}$.	7. To bisect is to divide into two congruent parts
8. $\overline{QG}$ is $\perp$ bisector of $\overline{AB}$.	8. A line perpendicular to a segment and bisecting it is its perpendicular bisector.

SUPPLEMENTARY PROBLEMS

11.1. Determine the locus of   (11.1)

  (a) The midpoints of the radii of a given circle

  (b) The midpoints of chords of a given circle parallel to a given line

  (c) The midpoints of chords of fixed length in a given circle

  (d) The vertex of the right angle of a triangle having a given hypotenuse

  (e) The vertex of an isosceles triangle having a given base

  (f) The center of a circle which passes through two given points

  (g) The center of a circle tangent to a given line at a given point on that line

  (h) The center of a circle tangent to the sides of a given angle

11.2. Determine the locus of   (11.1)

  (a) A boat moving so that it is equidistant from the parallel banks of a stream

  (b) A swimmer maintaining the same distance from two floats

  (c) A police helicopter in pursuit of a car which has just passed the junction of two straight roads and which may be on either one of them

  (d) A treasure buried at the same distance from two intersecting straight roads

11.3. Determine the locus of (a) a planet moving at a fixed distance from its sun; (b) a boat moving at a fixed distance from the coast of a circular island; (c) plants laid at a distance of 20 ft from a straight row of other plants; (d) the outer extremity of a clock hand.   (11.1)

11.4. Excluding points lying outside rectangle *ABCD* in Fig. 11-15, find the locus of points which are   (11.1)

(a) Equidistant from $\overline{AD}$ and $\overline{BC}$

(b) Equidistant from $\overline{AB}$ and $\overline{CD}$

(c) Equidistant from *A* and *B*

(d) Equidistant from *B* and *C*

(e) 5 units from $\overline{BC}$

(f) 10 units from $\overline{AB}$

(g) 20 units from $\overline{CD}$

(h) 10 units from *B*

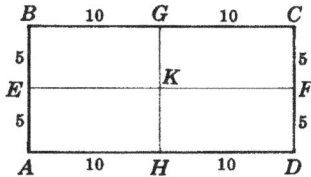

Fig. 11-15

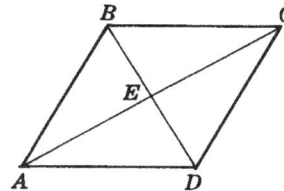

Fig. 11-16

11.5. Find the locus of points in rhombus *ABCD* in Fig. 11-16, which are equidistant from (a) $\overline{AB}$ and $\overline{AD}$; (b) $\overline{AB}$ and $\overline{BC}$; (c) *A* and *C*; (d) *B* and *D*; (e) each of the four sides.   (11.1)

11.6. In Fig. 11-17, find the locus of points which are on or inside circle *C* and   (11.1 and 11.2)

(a) 5 units from *O*

(b) 15 units from *O*

(c) Equidistant from circles *A* and *C*

(d) 10 units from circle *C*

(e) 10 units from circle *A*

(f) 5 units from circle *B*

(g) The center of a circle tangent to circles *A* and *C*

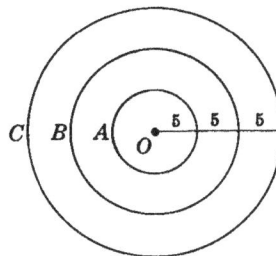

Fig. 11-17

11.7. Determine the locus of the center of (a) a coin rolling around and touching a smaller coin; (b) a coin rolling around and touching a larger coin; (c) a wheel moving between two parallel bars and touching both of them; (d) a wheel moving along a straight metal bar and touching it.   (11.2)

11.8. Find the locus of points that are in rectangle *ABCD* of Fig. 11-18 and the center of a circle   (11.2)

(a) Tangent to $\overline{AD}$ and $\overline{BC}$

(b) Tangent to $\overline{AB}$ and $\overline{CD}$

(c) Tangent to $\overline{AD}$ and $\overline{EF}$

(d) Of radius 10, tangent to $\overline{BC}$

(e) Of radius 20, tangent to $\overline{AD}$

(f) Tangent to $\overline{BC}$ at *G*

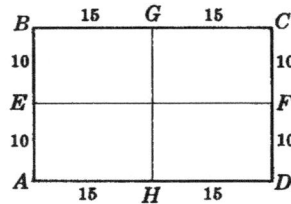

Fig. 11-18

11.9. Locate each of the following: (11.4)

(a) Treasure that is buried 5 ft from a straight fence and equidistant from two given points where the fence meets the ground

(b) Points that are 3 ft from a circle whose radius is 2 ft and are equidistant from two lines which are parallel to each other and tangent to the circle

(c) A point equidistant from the three vertices of a given triangle

(d) A point equidistant from two given points and equidistant from two given parallels

(e) Points equidistant from two given intersecting lines and 5 ft from their intersection

(f) A point that is equidistant from the sides of an angle and $\frac{1}{2}$ in from their intersection

11.10. Locate the point or points which satisfy the following conditions with respect to $\triangle ABC$ in Fig. 11-19: (11.4)

(a) Equidistant from its sides

(b) Equidistant from its vertices

(c) Equidistant from A and B and from $\overline{AB}$ and $\overline{BC}$

(d) Equidistant from $\overline{BC}$ and $\overline{AC}$ and 5 units from C

(e) 5 units from B and 10 units from A

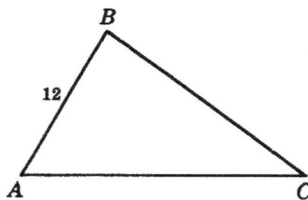

Fig. 11-19

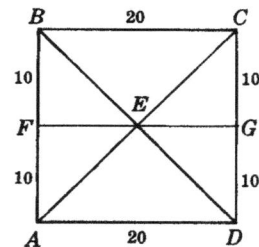

Fig. 11-20

11.11. Excluding points lying outside square $ABCD$ in Fig. 11-20, how many points are there that are (11.4)

(a) Equidistant from its vertices

(b) Equidistant from its sides

(c) 5 units from E and on one of the diagonals

(d) 5 units from E and equidistant from $\overline{AD}$ and $\overline{BC}$

(e) 5 units from $\overline{FG}$ and equidistant from $\overline{AB}$ and $\overline{CD}$

(f) 20 units from A and 10 units from B

11.12. Prove that the locus of points equidistant from the sides of an angle is the bisector of the angle. (11.5)

CHAPTER 12

Analytic Geometry

12.1 Graphs

A *number line* is a line on which distances from a point are marked off in equal units, positively in one direction and negatively in the other. The *origin* is the zero point from which distances are measured. Figure 12-1 shows a horizontal number line.

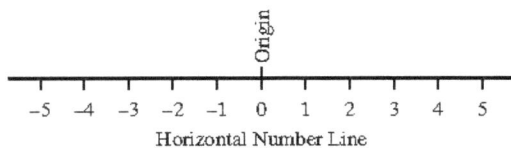

Fig. 12-1

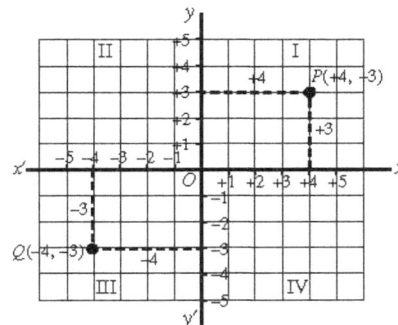

Fig. 12-2

The *graph* shown in Fig. 12-2 is formed by combining two number lines at right angles to each other so that their zero points coincide. The horizontal number line is called the *x-axis*, and the vertical number line is the *y-axis*. The point where the two lines cross each other is, again, called the *origin*.

A *point is located on a graph by its coordinates,* which are its distances from the axes. The *abscissa* or *x-coordinate* of a point is its distance from the *y*-axis. The *ordinate* or *y-coordinate* of a point is its distance from the *x*-axis.

When the coordinates of a point are stated, the *x*-coordinate precedes the *y*-coordinate. Thus, the coordinates of point *P* in Fig. 12-2 are written (4, 3); those for *Q* are $(-4, -3)$. Note the parentheses.

The *quadrants* of a graph are the four parts cut off by the axes. These are numbered **I**, **II**, **III**, and **IV** in a counterclockwise direction, as shown in Fig. 12-2.

12.1 Locating points on a graph

Give the coordinates of the following points in Fig. 12-3:

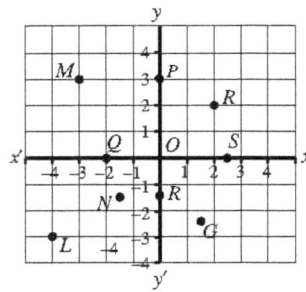

Fig. 12-3

(a) B	(c) O	(e) N	(g) P	(i) R
(b) M	(d) L	(f) G	(h) Q	(j) S

Solutions

(a) $(2,2)$	(c) $(0,0)$	(e) $(-1\frac{1}{2}, -1\frac{1}{2})$	(g) $(0,3)$	(i) $(0, -1\frac{1}{2})$
(b) $(-3,3)$	(d) $(-4,-3)$	(f) $(1\frac{1}{2}, -2\frac{1}{2})$	(h) $(-2,0)$	(j) $(2\frac{1}{2}, 0)$

12.2   Coordinates of points in the four quadrants

What are the signs of the coordinates of (a) a point in quadrant I; (b) a point in quadrant II; (c) a point in quadrant III; (d) a point in quadrant IV? Show which coordinate has a sign and which zero value for a point between quadrants (e) IV and I; (f) I and II; (g) II and III; (h) III and IV.

Solutions

(a) $(+,+)$	(c) $(-,-)$	(e) $(+,0)$	(g) $(-,0)$
(b) $(-,+)$	(d) $(+,-)$	(f) $(0,+)$	(h) $(0,-)$

12.3   Graphing a quadrilateral

If the vertices of a rectangle have the coordinates $A(3,1)$, $B(-5,1)$, $C(-5,-3)$, and $D(3-3)$, find its perimeter and area.

Solution

The base and height of the rectangle are 8 and 4 (see Fig. 12-4). Hence, the perimeter is $2b + 2h = 2(8) + 2(4) = 24$, and the area is $bh = (8)(4) = 32$.

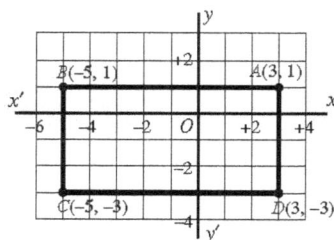

Fig. 12-4

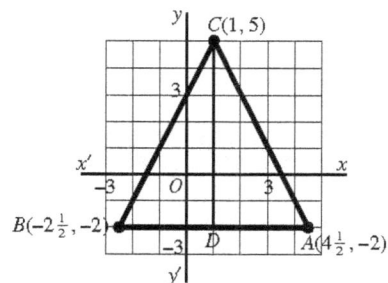

Fig. 12-5

12.4 Graphing a triangle

If the vertices of a triangle have the coordinates $A\left(4\tfrac{1}{2},-2\right)$, $B\left(-2\tfrac{1}{2},-2\right)$ and $C(1,5)$, find its area.

Solution

The length of the base is $BA = 7$ (see Fig. 12-5). The height is $CD = 7$. Then $A = \tfrac{1}{2}bh = \tfrac{1}{2}(7)(7) = 24\tfrac{1}{2}$.

12.2 Midpoint of a Segment

The coordinates (x_m, y_m) of the midpoint M of the line segment joining $P(x_1, y_1)$ to $Q(x_2, y_2)$ are

$$x_m = \tfrac{1}{2}(x_1 + x_2) \qquad \text{and} \qquad y_m = \tfrac{1}{2}(y_1 + y_2)$$

In Fig. 12-6, segment y_m is the median of trapezoid $CPQD$, whose bases are y_1 and y_2. Since the length of a median is one-half the sum of the bases, $y_m = \tfrac{1}{2}(y_1 + y_2)$. Similarly, segment x_m is the median of trapezoid $ABQP$, whose bases are x_1 and x_2; hence, $x_m = \tfrac{1}{2}(x_1 + x_2)$.

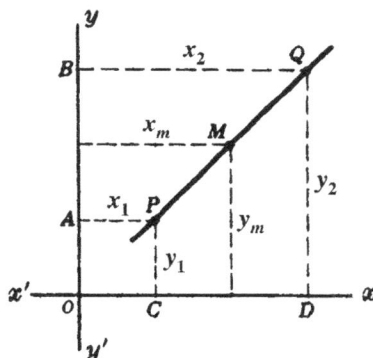

Fig. 12-6

SOLVED PROBLEMS

12.5 Applying the midpoint formula

If M is the midpoint of $\overline{PQ}$, find the coordinates of (a) M if the coordinates of P and Q are $P(3,4)$ and $Q(5,8)$; (b) Q if the coordinates of P and M are $P(1,5)$ and $M(3,4)$.

Solutions

(a) $x_m = \tfrac{1}{2}(x_1 + x_2) = \tfrac{1}{2}(3 + 5) = 4$; $y_m = \tfrac{1}{2}(y_1 + y_2) = \tfrac{1}{2}(4 + 8) = 6$.

(b) $x_m = \tfrac{1}{2}(x_1 + x_2)$, so $3 = \tfrac{1}{2}(1 + x_2)$ and $x_2 = 5$; $y_m = \tfrac{1}{2}(y_1 + y_2)$, so $4 = \tfrac{1}{2}(5 + y_2)$ and $y_2 = 3$.

12.6 Determining if segments bisect each other

The vertices of a quadrilateral are $A(0,0)$, $B(0,3)$, $C(4,3)$, and $D(4,0)$.

(a) Show that $ABCD$ is a rectangle.

(b) Show that the midpoint of $\overline{AC}$ is also the midpoint of $\overline{BD}$.

(c) Do the diagonals bisect each other? Why?

Solutions

(a) From Fig. 12-7, $AB = CD = 3$ and $BC = AD = 4$; hence, $ABCD$ is a parallelogram.
Since $\angle BAD$ is a right angle, $ABCD$ is a rectangle.

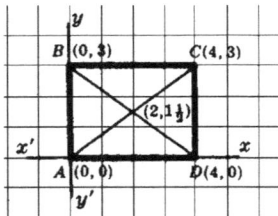

Fig. 12-7

(b) The coordinates of the midpoint of $\overline{AC}$ are $x = \frac{1}{2}(0 + 4) = 2$, $y = \frac{1}{2}(0 + 3) = 1\frac{1}{2}$.
The coordinates of the midpoint of $\overline{BD}$ are $x = \frac{1}{2}(0 + 4) = 2$, $y = \frac{1}{2}(3 + 0) = 1\frac{1}{2}$.
Hence, $(2, 1\frac{1}{2})$ is the midpoint of both $\overline{AC}$ and $\overline{BD}$.

(c) Yes, since the midpoints of both diagonals are the same point.

12.3 Distance Between Two Points

PRINCIPLE 1: *The distance between two points having the same ordinate (or y-value) is the absolute value of the difference of their abscissas.* (Hence, the distance between two points must be *positive*.)

Thus, the distance between the point $P(6, 1)$ and $Q(9, 1)$ is $9 - 6 = 3$.

PRINCIPLE 2: *The distance between two points having the same abscissa (or x-value) is the absolute value of the difference of their ordinates.*

Thus, the distance between the points $P(2, 1)$ and $Q(2, 4)$ is $4 - 1 = 3$.

PRINCIPLE 3: *The distance d between the points $P_1(x_1, y_1)$ and $P_2(x_2, y_2)$ is*

$$d = \sqrt{(x_2 - x_1)^2 + (y_2 - y_1)^2} \qquad \text{or} \qquad d = \sqrt{(\Delta x)^2 + (\Delta y)^2}$$

The difference $x_2 - x_1$ is denoted by the symbol Δx; the difference $y_2 - y_1$ is denoted by Δy. Delta (Δ) is the fourth letter of the Greek alphabet, corresponding to our d. The difference Δx and Δy may be positive or negative.

SOLVED PROBLEMS

12.7 Providing and using the distance formula

(a) Prove the distance formula (Principle 3) algebraically.

(b) Use it to find the distance between $A(2, 5)$ and $B(6, 8)$.

Solutions

(a) See Fig. 12-8. By Principle 1, $P_1S = x_2 - x_1 = \Delta x$. By Principle 2, $P_2S = y_2 - y_1 = \Delta y$. Also, in right triangle P_1SP_2,

$$(P_1P_2)^2 = (P_1S)^2 + (P_2S)^2$$

or

$$d^2 = (x_2 - x_1)^2 + (y_2 - y_1)^2$$

and

$$d = \sqrt{(x_2 - x_1)^2 + (y_2 - y_1)^2}$$

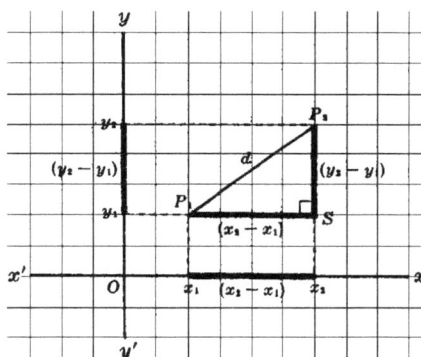

Fig. 12-8

(b) The distance from $A(2,5)$ to $B(6,8)$ is found as follows:

(x, y)
$B(6,8) \rightarrow x_2 = 6, y_2 = 8$
$A(2,5) \rightarrow x_1 = 2, y_1 = 5$

$d^2 = (x_2 - x_1)^2 + (y_2 - y_1)^2$
$d^2 = (6 - 2)^2 + (8 - 5)^2 = 4^2 + 3^2 = 25$ and $d = 5$

12.8 Finding the distance between two points

Find the distance between the points (a) $(-3, 5)$ and $(1, 5)$; (b) $(3, -2)$ and $(3, 4)$; (c) $(3, 4)$ and $(6, 8)$; $(-3, 2)$ and $(9, -3)$.

Solutions

(a) Since both points have the same ordinate (or y-value), $d = x_2 - x_1 = 1 - (-3) = 4$

(b) Since both points have the same abscissa (or x-value), $d = y_2 - y_1 = 4 - (-2) = 6$

(c) $d = \sqrt{(x_2 - x_1)^2 + (y_2 - y_1)^2} = \sqrt{(6 - 3)^2 + (8 - 4)^2} = \sqrt{3^2 + 4^2} = 5$

(d) $d = \sqrt{(x_2 - x_1)^2 + (y_2 - y_1)^2} = \sqrt{[9 - (-3)]^2 + (-3 - 2)^2} = \sqrt{12^2 + (-5)^2} = 13$

12.9 Applying the distance formula to a triangle

(a) Find the lengths of the sides of a triangle whose vertices are $A(1, 1), B(1, 4)$, and $C(5, 1)$.

(b) Show that the triangle whose vertices are $G(2, 10), H(3, 2)$, and $J(6, 4)$ is a right triangle.

Solutions

See Fig. 12-9.

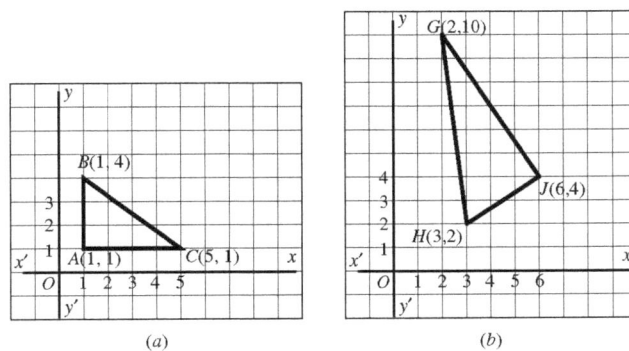

(a)

(b)

Fig. 12-9

(a) $AC = 5 - 1 = 4$ and $AB = 4 - 1 = 3$; $BC = \sqrt{(5-1)^2 + (1-4)^2} = \sqrt{4^2 + (-3)^2} = 5$.

(b) $(GJ)^2 = (6-2)^2 + (4-10)^2 = 52$; $(HJ)^2 = (6-3)^2 + (4-2)^2 = 13$; $(GH)^2 = (2-3)^2 + (10-2)^2 = 65$. Since $(GJ)^2 + (HJ)^2 = (GH)^2$, $\triangle GHJ$ is a right triangle.

12.10 Applying the distance formula to a parallelogram

The coordinates of the vertices of a quadrilateral are $A(2,2)$, $B(3,5)$, $C(6,7)$, and $D(5,4)$. Show that $ABCD$ is a parallelogram.

Solution

See Fig. 12-10, where we have

$$AB = \sqrt{(3-2)^2 + (5-2)^2} = \sqrt{1^2 + 3^2} = \sqrt{10}$$
$$CD = \sqrt{(6-5)^2 + (7-4)^2} = \sqrt{1^2 + 3^2} = \sqrt{10}$$
$$BC = \sqrt{(6-3)^2 + (7-5)^2} = \sqrt{3^2 + 2^2} = \sqrt{13}$$
$$AD = \sqrt{(5-2)^2 + (4-2)^2} = \sqrt{3^2 + 2^2} = \sqrt{13}$$

Thus, $\overline{AB} \cong \overline{CD}$ and $\overline{BC} \cong \overline{AD}$. Since opposite sides are congruent, $ABCD$ is a parallelogram.

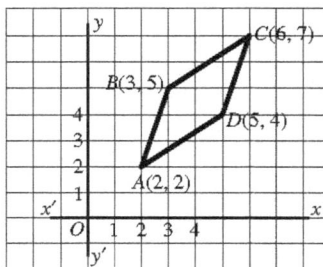

Fig. 12-10

12.11 Applying the distance formula to a circle

A circle is tangent to the x-axis and has its center at $(6,4)$. Where is the point $(9,7)$ with respect to the circle?

Solution

Since the circle is tangent to the x-axis, $\overline{AQ}$ in Fig. 12-11 is a radius. By Principle 2, $AQ = 4$.

By Principle 3, $BQ = \sqrt{(9-6)^2 + (7-4)^2} = \sqrt{3^2 + 3^2} = \sqrt{18}$. Since $\sqrt{18}$ is greater than 4, $\overline{BQ}$ is greater than a radius so B is outside the circle.

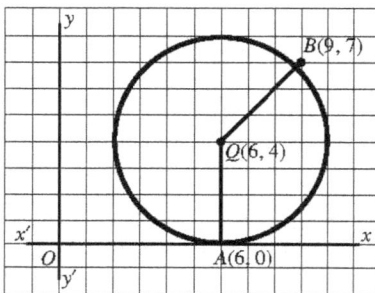

Fig. 12-11

12.4   Slope of a Line

PRINCIPLE 1:   *If a line passes through the points $P_1\,(x_1, y_1)$ and $P_2(x_2, y_2)$, then*

$$\text{Slope of } \overleftrightarrow{P_1P_2} = \frac{y_2 - y_1}{x_2 - x_1} = \frac{\Delta y}{\Delta x}$$

PRINCIPLE 2:   *The line whose equation is $y = mx + b$ has slope m.*

PRINCIPLE 3:   *The slope of a line equals the tangent of its inclination.*

The inclination i of a line is the angle above the x-axis that is included between the line and the positive direction of the x-axis (see Fig. 12-12). In the figure,

$$\text{Slope of } \overleftrightarrow{P_1P_2} = \frac{y_2 - y_1}{x_2 - x_1} = \frac{\Delta y}{\Delta x} = m = \tan i$$

The slope is independent of the order in which the end points are selected. Thus,

$$\text{Slope of } \overleftrightarrow{P_1P_2} = \frac{y_2 - y_1}{x_2 - x_1} = \frac{y_1 - y_2}{x_1 - x_2} = \text{slope of } \overleftrightarrow{P_2P_1}$$

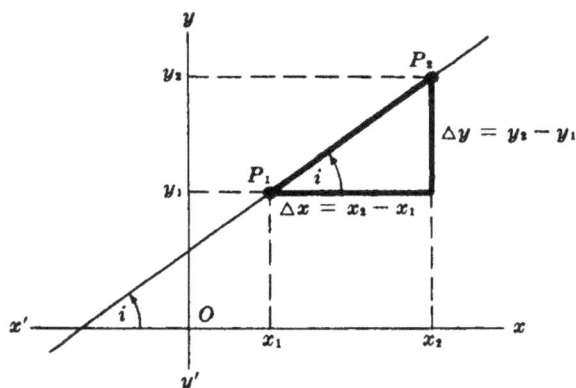

Fig. 12-12

12.4A   Positive and Negative Slopes

PRINCIPLE 4:   *If a line slants upward from left to right, its inclination i is an acute angle and its slope is positive (Fig. 12-13).*

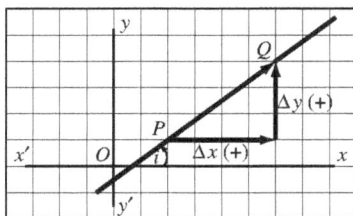

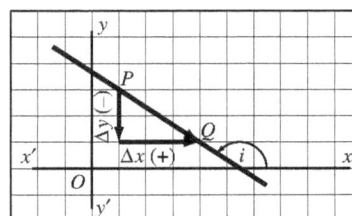

Slope of $PQ = \dfrac{\Delta y}{\Delta x} = \dfrac{+}{+} = +$

Fig. 12-13

Slope of $PQ = \dfrac{\Delta y}{\Delta x} = \dfrac{-}{+} = -$

Fig. 12-14

PRINCIPLE 5: *If a line slants downward from left to right, its inclination is an obtuse angle and its slope is negative* (Fig. 12-14).

PRINCIPLE 6: *If a line is parallel to the x-axis, its inclination is 0° and its slope is 0* (Fig. 12-15).

PRINCIPLE 7: *If a line is perpendicular to the x-axis, its inclination is 90° and it has no slope* (Fig. 12-16).

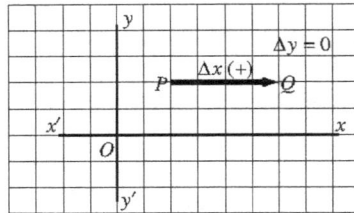

Slope of $PQ = \dfrac{\Delta y}{\Delta x} = \dfrac{0}{+} = 0$

Fig. 12-15

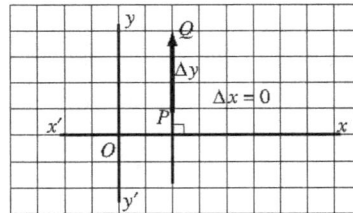

Slope of $PQ = \dfrac{\Delta y}{\Delta x} = \dfrac{+}{0}$ (meaningless)

Fig. 12-16

12-4B   Slopes of Parallel and Perpendicular Lines

PRINCIPLE 8: *Parallel lines have the same slope.*

In Fig. 12-17, $l \parallel l'$; hence, corresponding angles i and i' are equal, and $\tan i = \tan i'$ or $m = m'$, where m and m' are the slopes of l and l'.

PRINCIPLE 9: *Lines having the same slope are parallel to each other.* (This is the converse of Principle 8.)

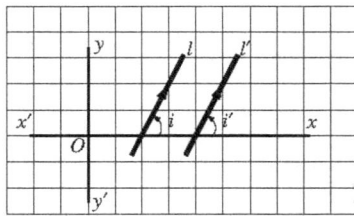

Fig. 12-17

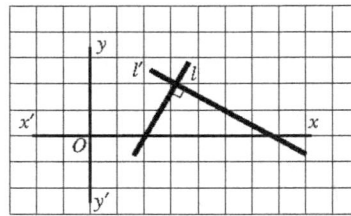

Fig. 12-18

PRINCIPLE 10: *Perpendicular lines have slopes that are negative reciprocals of each other.* (Negative reciprocals are numbers, such as $\frac{2}{5}$ and $-\frac{5}{2}$, whose product is -1.)

Thus in Fig. 12-18, if $l \perp l'$, then $m = -1/m'$ or $mm' = -1$, where m and m' are the slopes of l and l'.

PRINCIPLE 11: *Lines whose slopes are negative reciprocals of each other are perpendicular.* (This is the converse of Principle 10.)

12.4C   Collinear Points

Collinear points are points which lie on the same straight line. Thus, A, B, and C are collinear points here:

$$Q \quad C \quad B \quad A \quad P$$

PRINCIPLE 12: *The slope of a straight line is constant all along the line.*

Thus if $\overleftrightarrow{PQ}$ above is a straight line, the slope of the segment from A to B equals the slope of the segment from C to Q.

PRINCIPLE 13: *If the slope of a segment between a first point and a second equals the slope of the segment between either point and a third, then the points are collinear.*

SOLVED PROBLEMS

12.12 Slope and inclination of a line

(a) Find the slope of the line through $(-2, -1)$ and $(4, 3)$.

(b) Find the slope of the line whose equation is $3y - 4x = 15$.

(c) Find the inclination of the line whose equation is $y = x + 4$.

Solutions

(a) By principle 1, $m = \dfrac{y_2 - y_1}{x_2 - x_1} = \dfrac{3 - (-1)}{4 - (-2)} = \dfrac{4}{6} = \dfrac{2}{3}$

(b) We may rewrite $3y - 4x = 15$ as $y = \frac{4}{3}x + 5$, from which $m = \frac{4}{3}$.

(c) Since $y = x + 4$, we have $m = 1$; thus, $\tan i = 1$ and $i = 45°$.

12.13 Slopes of parallel or perpendicular lines

Find the slope of $\overleftrightarrow{CD}$ if (a) $\overleftrightarrow{AB} \parallel \overleftrightarrow{CD}$ and the slope of $\overleftrightarrow{AB}$ is $\frac{2}{3}$; (b) $\overleftrightarrow{AB} \perp \overleftrightarrow{CD}$ and the slope of $\overleftrightarrow{AB}$ is $\frac{3}{4}$.

Solutions

(a) By Principle 8, slope of $\overleftrightarrow{CD}$ = slope of $\overleftrightarrow{AB}$ = $\frac{2}{3}$.

(b) By Principle 10, slope of $\overleftrightarrow{CD} = -\dfrac{1}{\text{slope of } AB} = -\dfrac{1}{3/4} = -\dfrac{4}{3}$.

12.14 Applying principles 9 and 11 to triangles and quadrilaterals

Complete each of the following statements:

(a) In quadrilateral $ABCD$, if the slopes of $\overleftrightarrow{AB}, \overleftrightarrow{BC}, \overleftrightarrow{CD}$, and $\overleftrightarrow{DA}$ are $\frac{1}{2}, -2, \frac{1}{2}$, and -2, respectively, the quadrilateral is a $\underline{\ ?\ }$.

(b) In triangle LMP, if the slopes of $\overleftrightarrow{LM}$ and $\overleftrightarrow{MP}$ are 5 and $-\frac{1}{5}$, then LMP is a $\underline{\ ?\ }$ triangle.

Solutions

(a) Since the slopes of the opposite sides are equal, $ABCD$ is a parallelogram. In addition, the slopes of adjacent sides are negative reciprocals; hence, those sides are $\perp$ and $ABCD$ is a rectangle.

(b) Since the slopes of $\overleftrightarrow{LM}$ and $\overleftrightarrow{MP}$ are negative reciprocals, $\overline{LM} \perp \overline{MP}$ and the triangle is a right triangle.

12.15 Applying principle 12

(a) $\overleftrightarrow{AB}$ has a slope of 2 and points A, B, and C are collinear. What are the slopes of $\overline{AC}$ and $\overline{BC}$?

(b) Find y if $G(1, 4)$, $H(3, 2)$, and $J(9, y)$ are collinear.

Solutions

(a) By Principle 12, $\overline{AC}$ and $\overline{BC}$ have a slope of 2.

(b) By Principle 12, slope of $\overleftrightarrow{GJ}$ = slope of $\overleftrightarrow{GH}$. Hence $\dfrac{y-4}{9-1} = \dfrac{2-4}{3-1}$, so that $\dfrac{y-4}{8} = \dfrac{-2}{2} = -1$ and $y = -4$.

12.5   Locus in Analytic Geometry

A locus of points is the set of points, and only those points, satisfying a given condition. In geometry, a line or curve (or set of lines or curves) on a graph is the locus of analytic points that satisfy the equation of the line or curve.

Think of the locus as the path of a point moving according to a given condition or as the set of points satisfying a given condition.

PRINCIPLE 1:   *The locus of points whose abscissa is a constant k is a line parallel to the y-axis; its equation is x = k. (See Fig. 12-19.)*

PRINCIPLE 2:   *The locus of points whose ordinate is a constant k is a line parallel to the x-axis; its equation is y = k. (See Fig. 12-19.)*

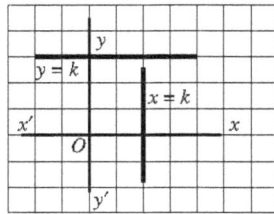

Fig. 12-19

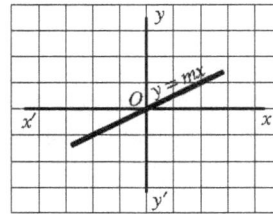

Fig. 12-20

PRINCIPLE 3:   *The locus of points whose ordinate equals the product of a constant m and its abscissa is a straight line passing through the origin; its equation is y = mx.*

The constant m is the slope of the line. (See Fig. 12-20.)

PRINCIPLE 4:   *The locus of points whose ordinate and abscissa are related by either of the equations*

$$y = mx + b \qquad or \qquad \frac{y - y_1}{x - x_1} = m$$

where m and b are constants, is a line (Fig. 12-21).

In the equation $y = mx + b$, m is the slope and b is the *y-intercept*. The equation $\dfrac{y - y_1}{x - x_1} = m$ tells us that the line passes through the fixed point (x_1, y_1) and has a slope of m.

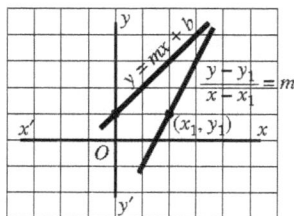

Fig. 12-21

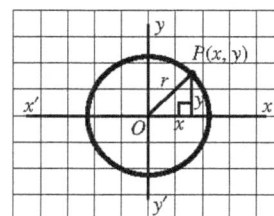

Fig. 12-22

PRINCIPLE 5: *The locus of points such that the sum of the squares of the coordinates is a constant is a circle whose center is the origin.*

The constant is the square of the radius, and the equation of the circle is

$$x^2 + y^2 = r^2$$

(see Fig. 12-22). Note that for any point $P(x, y)$ on the circle, $x^2 + y^2 = r^2$.

SOLVED PROBLEMS

12.16 Applying principles 1 and 2

Graph and give the equation of the locus of points (a) whose ordinate is -2; (b) that are 3 units from the y-axis; (c) that are equidistant from the points $(3, 0)$ and $(5, 0)$.

Solutions

(a) From Principle 2, the equation is $y = -2$; see Fig. 12-23(a).

(b) From Principle 1, the equation is $x = 3$ and $x = -3$; see Fig. 12-23(b).

(c) The equation is $x = 4$; see Fig. 12-23(c).

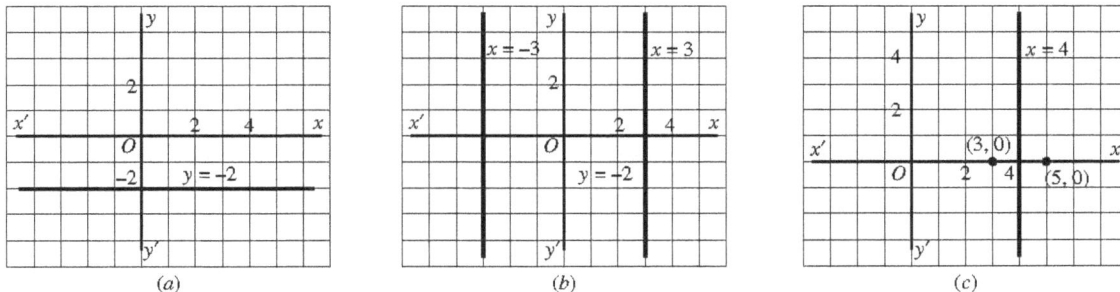

Fig. 12-23

12.17 Applying principles 3 and 4

Graph and describe the locus whose equation is (a) $y = \frac{1}{3}x + 1$; (b) $y = \frac{3}{2}x$; (c) $\dfrac{y - 1}{x - 1} = \dfrac{3}{4}$.

Solutions

(a) The locus is a line whose y-intercept is 1 and whose slope equals $\frac{1}{3}$. See Fig. 12-24(a).

(b) The locus is a line which passes through the origin and has slope $\frac{3}{2}$. See Fig. 12-24(b).

(c) The locus is a line which passes through the point $(1, 1)$ and has slope $\frac{3}{4}$. See Fig. 12-24(c).

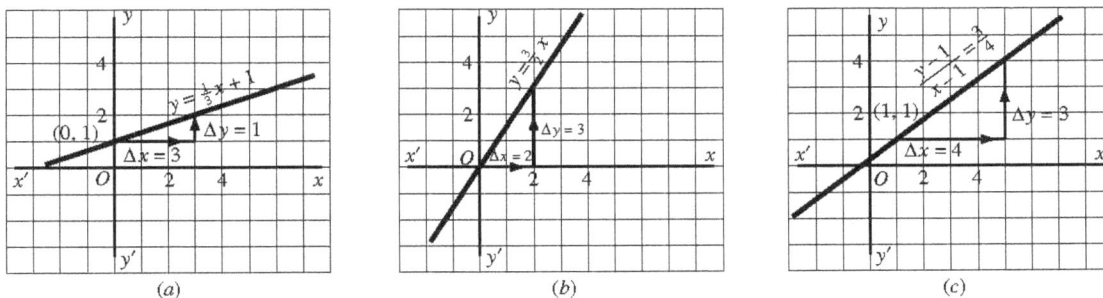

Fig. 12-24

12.18 Applying principle 5

Graph and give the equation of the locus of points (a) 2 units from the origin; (b) 2 units from the locus of $x^2 + y^2 = 9$.

Solutions

(a) The locus is a circle whose equation is $x^2 + y^2 = 4$. See Fig. 12-25(a).

(b) The locus is a pair of circles, each 2 units from the circle with center at O and radius 3. Their equations are $x^2 + y^2 = 25$ and $x^2 + y^2 = 1$. See Fig. 12-25(b).

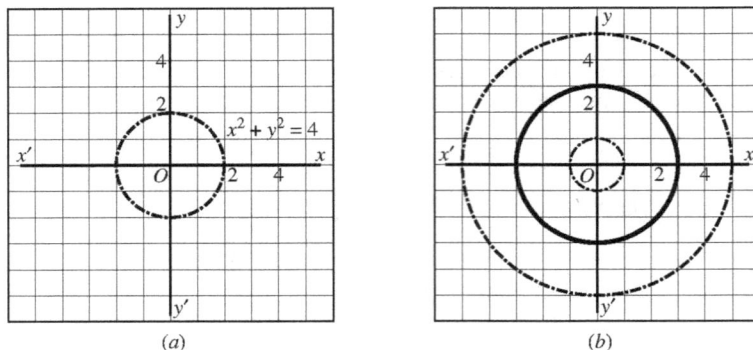

(a)   (b)

Fig. 12-25

12.6 Areas in Analytic Geometry

12.6A Area of a Triangle

If one side of a triangle is parallel to either coordinate axis, the length of that side and the length of the altitude to that side can be found readily. Then the formula $A = \frac{1}{2}bh$ can be used.

If no side of a triangle is parallel to either axis, then either

1. The triangle can be enclosed in a rectangle whose sides are parallel to the axes (Fig. 12-26), or
2. Trapezoids whose bases are parallel to the y-axis can be formed by dropping perpendiculars to the x-axis (Fig. 12-27).

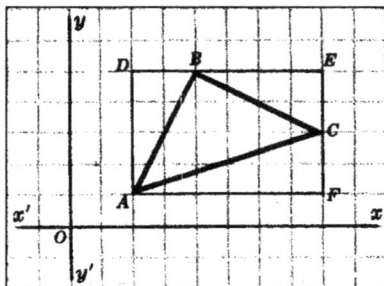

Fig. 12-26

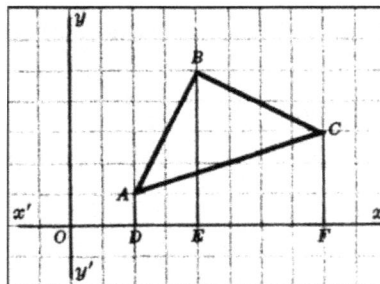

Fig. 12-27

The area of the triangle can then be found from the areas of the figures so formed:

1. In Fig. 12-26, area ($\triangle ABC$) = area(rectangle $ADEF$) − [area($\triangle ABD$) + area ($\triangle BCE$) + area ($\triangle ACF$)].
2. In Fig. 12-27, area ($\triangle ABC$) = area(trapezoid $ABED$) + area(trapezoid $BEFC$) − area (trapezoid $DFCA$).

12.6B Area of a Quadrilateral

The trapezoid method described above can be extended to finding the area of a quadrilateral if its vertices are given.

SOLVED PROBLEMS

12.19 Area of a triangle having a side parallel to an axis

Find the area of the triangle whose vertices are $A(1, 2)$, $B(7, 2)$, and $C(5, 4)$.

Solution

From the graph of the triangle (Fig. 12-28), we see that $b = 7 - 1 = 6$ and $h = 4 - 2 = 2$. Then $A = \frac{1}{2}bh = \frac{1}{2}(6)(2) = 6$.

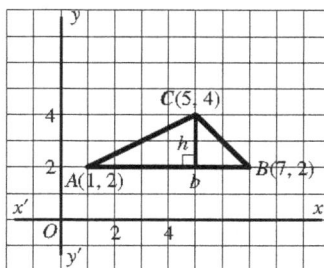

Fig. 12-28

12.20 Area of a triangle having no side parallel to an axis

Find the area of $\triangle ABC$ whose vertices are $A(2, 4)$, $B(5, 8)$ and $C(8, 2)$ (a) using the rectangle method; (b) using the trapezoid method.

Solutions

See Fig. 12-29.

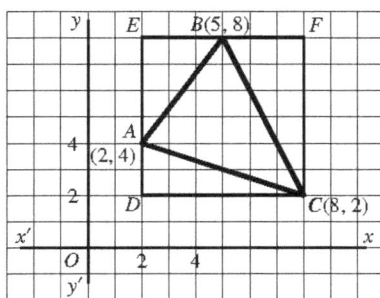 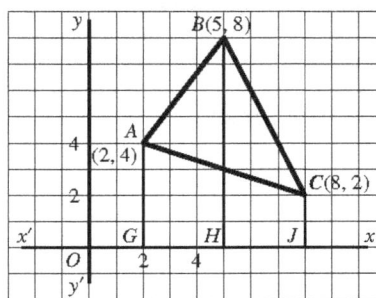

Fig. 12-29

(a) Area of rectangle $DEFC = bh = 6(6) = 36$. Then:

Area of $\triangle DAC = \frac{1}{2}bh = \frac{1}{2}(2)(6) = 6$.

Area of $\triangle ABE = \frac{1}{2}bh = \frac{1}{2}(3)(4) = 6$.

Area of $\triangle BCF = \frac{1}{2}bh = \frac{1}{2}(3)(6) = 9$.

So area($\triangle ABC$) = area($DEFC$) − area($\triangle DAC + \triangle ABE + \triangle BCF$) = 36 − (6 + 6 + 9) = 15.

(b) Area of trapezoid $ABHG = \frac{1}{2}h(b + b') = \frac{1}{2}(3)(4 + 8) = 18$.

Area of trapezoid $BCJH = \frac{1}{2}(3)(2 + 8) = 15$.

Area of trapezoid $ACJG = \frac{1}{2}(6)(2 + 4) = 18$.

Then area$(\triangle ABC) = $ area$(ABHG) + $ area$(BCJH) - $ area$(ACJG) = 18 + 15 - 18 = 15$.

12.7　Proving Theorems with Analytic Geometry

Many theorems of plane geometry can be proved with analytic geometry. The procedure for proving a theorem has two major steps, as follows:

1. *Place each figure in a convenient position on a graph.* For a triangle, rectangle, or parallelogram, place one vertex at the origin and one side of the figure on the x-axis. Indicate the coordinates of each vertex (Fig. 12-30).

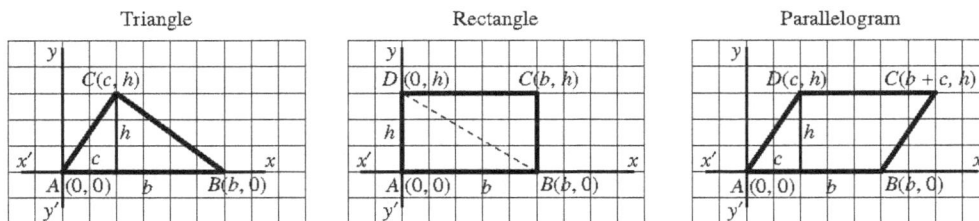

Fig. 12-30

2. *Apply the principles of analytic geometry.* For example, prove that lines are parallel by showing that their slopes are equal; or that lines are perpendicular by showing that their slopes are negative reciprocals of each other. Use the midpoint formula when the midpoint of a segment is involved, and use the distance formula to obtain the lengths of segments.

SOLVED PROBLEM

12.21　Proving a theorem with analytic geometry

Using analytic geometry, prove that the diagonals of a parallelogram bisect each other.

Solution

Given:　$\square ABCD$, diagonals $\overline{AC}$ and $\overline{BD}$.
To Prove:　$\overline{AC}$ and $\overline{BD}$ bisect each other.
Plan:　Use the midpoint formula to obtain the coordinates of the midpoints of the diagonals

Place $\square ABCD$ with vertex A at the origin and side $\overline{AD}$ along the x-axis (Fig. 12-31). Then the vertices have the coordinates $A(0,0)$, $B(a, b)$, $C(a + c, b)$, and $D(c, 0)$.

By the midpoint formula, the midpoint of $\overline{AC}$ has the coordinates $\left(\frac{a + c}{2}, \frac{b}{2}\right)$, and the midpoint of $\overline{BD}$ has the coordinates $\left(\frac{a + c}{2}, \frac{b}{2}\right)$. Then the diagonals bisect each other, since the midpoints of both diagonals are the same point.

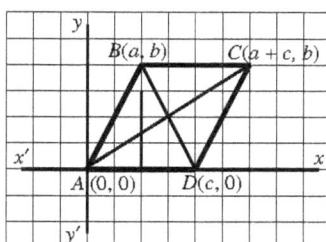

Fig. 12-31

SUPPLEMENTARY PROBLEMS

12.1. State the coordinates of each lettered point in Fig. 12-32. (12.1)

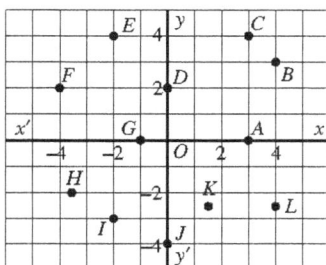

Fig. 12-32

12.2. Plot each of the following points: (12.2)

$A(-2,-3)$ $C(0,-1)$ $E(3,-4)$ $G(0,3)$
$B(-3,2)$ $D(-3,0)$ $F(1\frac{1}{2}, 2\frac{1}{2})$ $H(3\frac{1}{2}, 0)$

12.3. Plot the following points: $A(2,3), B(-3,3), C(-3,-2), D(2,-2)$. Then find the perimeter and area of square *ABCD*.

12.4. Plot the following points: $A(4,3), B(-1,3), C(-3,-3), D(2,-3)$. Then find the area of parallelogram *ABCD* and triangle *BCD*. (12.3, 12.4)

12.5. Find the midpoint of the segment joining (12.5)

(a) $(0,0)$ and $(8,6)$ (e) $(-20,-5)$ and $(0,0)$ (i) $(3,4)$ and $(7,6)$

(b) $(0,0)$ and $(5,7)$ (f) $(0,4)$ and $(0,16)$ (j) $(-2,-8)$ and $(-4,-12)$

(c) $(0,0)$ and $(-8,12)$ (g) $(8,0)$ and $(0,-2)$ (k) $(7,9)$ and $(3,3)$

(d) $(14,10)$ and $(0,0)$ (h) $(-10,0)$ and $(0,-5)$ (l) $(2,-1)$ and $(-2,-5)$

12.6. Find the midpoints of the sides of a triangle whose vertices are (12.5)

(a) $(0,0),(8,0),(0,6)$ (c) $(12,0),(0,-4),(0,0)$ (e) $(4,0),(0,-6),(-4,10)$

(b) $(-6,0),(0,0),(0,10)$ (d) $(3,5),(5,7),(3,11)$ (f) $(-1,-2),(0,2),(1,-1)$

12.7. Find the midpoints of the sides of the quadrilateral whose successive vertices are   (12.5)

    (a) $(0, 0), (0, 4), (2, 10), (6, 0)$ (c) $(-2, 0), (0, 4), (6, 2), (0, -10)$

    (b) $(-3, 5), (-1, 9), (7, 3), (5, -1)$ (d) $(-3, -7), (-1, 5), (9, 0), (5, -8)$

12.8. Find the midpoints of the diagonals of the quadrilateral whose successive vertices are   (12.5)

    (a) $(0, 0), (0, 5), (4, 12), (8, 1)$ (c) $(0, -5), (0, 1), (4, 9), (4, 3)$

    (b) $(-4, -1), (-2, 3), (6, 1), (2, -8)$

12.9. Find the center of a circle if the end points of a diameter are   (12.5)

    (a) $(0, 0)$ and $(-4, 6)$ (c) $(-3, 1)$ and $(0, -5)$ (e) (a, b) and $(3a, 5b)$

    (b) $(-1, 0)$ and $(-5, -12)$ (d) $(0, 0)$ and $(2a, 2b)$ (f) $(a, 2b)$ and $(a, 2c)$

12.10. If M is the midpoint of $\overline{AB}$, find the coordinates of   (12.5)

    (a) M if the coordinates of A and B are $A(2, 5)$ and $B(6, 11)$

    (b) A if the coordinates of M and B are $M(1, 3)$ and $B(3, 6)$

    (c) B if the coordinates of A and M are $A(-2, 1)$ and $M(2, -1)$

12.11. The trisection points of $\overline{AD}$ are B and C. Find the coordinates of   (12.5)

    (a) B if the coordinates of A and C are $A(1, 2)$ and $C(3, 5)$

    (b) D if the coordinates of B and C are $B(0, 5)$ and $C(1\frac{1}{2}, 4)$

    (c) A if the coordinates of B and C are $B(0, 6)$ and $C(2, 3)$

12.12. $A(0, 0), B(0, 5), C(6, 5)$, and $D(6, 0)$ are the vertices of quadrilateral $ABCD$.   (12.6)

    (a) Prove that $ABCD$ is a rectangle.

    (b) Show that the midpoints of $\overline{AC}$ and $\overline{BD}$ have the same coordinates.

    (c) Do the diagonals bisect each other? Why?

12.13. The vertices of $\triangle ABC$ are $A(0, 0), B(0, 4)$, and $C(6, 0)$.   (12.6)

    (a) If $\overline{AD}$ is the median to $\overline{BC}$, find the coordinates of D and the midpoint of $\overline{AD}$.

    (b) If $\overline{CE}$ is the median to $\overline{AB}$, find the coordinates of E and the midpoint of $\overline{CE}$.

    (c) Do the medians, $\overline{AD}$ and $\overline{CE}$, bisect each other? Why?

12.14. Find the distance between each of the following pairs of points: (12.8)

    (a) $(0, 0)$ and $(0, 5)$ (d) $(-6, -1)$ and $(-6, 11)$ (g) $(-3, -4\frac{1}{2})$ and $(-3, 4\frac{1}{2})$

    (b) $(4, 0)$ and $(-2, 0)$ (e) $(5, 3)$ and $(5, 8.4)$ (h) (a, b) and $(2a, b)$

    (c) $(0, -3)$ and $(0, 7)$ (f) $(-1.5, 7)$ and $(6, 7)$

12.15. Find the distances separating pairs of the following collinear points: (12.8)

(a) $(5, -2), (5, 1), (5, 4)$

(c) $(-4, 2), (-3, 2), (0, 2)$

(b) $(0, -6), (0, -2), (0, 12)$

(d) $(0, b), (a, b), (3a, b)$

12.16. Find the distance between each of the following pairs of points: (12.8)

(a) $(0, 0)$ and $(5, 12)$

(e) $(-3, -6)$, and $(3, 2)$

(i) $(3, 4)$ and $(4, 7)$

(b) $(-3, -4)$ and $(0, 0)$

(f) $(2, 3)$ and $(-10, 12)$

(j) $(-1, -1)$ and $(1, 3)$

(c) $(0, -6)$ and $(9, 6)$

(g) $(2, 2)$ and $(5, 5)$

(k) $(-3, 0)$ and $(0, \sqrt{7})$

(d) $(4, 1)$ and $(7, 5)$

(h) $(0, 5)$ and $(-5, 0)$

(l) $(a, 0)$ and $(0, a)$

12.17. Show that the triangles having the following vertices are isosceles triangles: (12.9)

(a) $A(3, 5), B(6, 9)$, and $C(2, 6)$

(c) $G(5, -5), H(-2, -2)$, and $J(8, 2)$

(b) $D(2, 0), E(6, 0)$, and $F(4, 4)$

(d) $K(7, 0), L(3, 4)$, and $M(2, -1)$

12.18. Which of the triangles having the following vertices are right triangles? (12.9)

(a) $A(7, 0), B(6, 3)$, and $C(12, 5)$

(c) $G(1, -1), H(5, 0)$, and $J(3, 8)$

(b) $D(2, 0), E(5, 2)$, and $F(1, 8)$

(d) $K(-4, 0), L(-2, 4)$, and $M(4, -1)$

12.19. The vertices of $\triangle ABC$ are $A(-2, 2), B(4, 4)$, and $C(8, 2)$. Find the length of the median to (a) $\overline{AB}$; (b) $\overline{AC}$; (c) $\overline{BC}$. (12.9)

12.20. (a) The vertices of quadrilateral $ABCD$ are $A(0, 0), B(3, 2), C(7, 7)$, and $D(4, 5)$. Show that $ABCD$ is a parallelogram. (12.10)

(b) The vertices of quadrilateral $DEFG$ are $D(3, 5), E(1, 1), F(5, 3)$, and $G(7, 7)$. Show that $DEFG$ is a rhombus.

(c) The vertices of quadrilateral $HJKL$ are $H(0, 0), J(4, 4), K(0, 8)$, and $L(-4, 4)$. Show that $HJKL$ is a square.

12.21. Find the radius of a circle that has its center at (12.11)

(a) $(0, 0)$ and passes through $(-6, 8)$ (d) $(2, 0)$ and passes through $(7, -12)$

(b) $(0, 0)$ and passes through $(3, -4)$ (e) $(4, 3)$ and is tangent to the y-axis

(c) $(0, 0)$ and passes through $(-5, 5)$ (f) $(-1, 7)$ and is tangent to the line $x = -4$

12.22. A circle has its center at the origin and a radius of 10. State whether each of the following points is on, inside, or outside of this circle: (a) $(6, 8)$; (b) $(-6, 8)$; (c) $(0, 11)$; (d) $(-10, 0)$; (e) $(7, 7)$; (f) $(-9, 4)$; (g) $(9, \sqrt{19})$. (12.11)

12.23. Find the slope of the line through each of the following pairs of points: (12.12)

(a) $(0, 0)$ and $(5, 9)$

(e) $(-2, -3)$ and $(7, 15)$

(i) $(3, -9)$ and $(0, 0)$

(b) $(0, 0)$ and $(9, 5)$

(f) $(-2, -3)$ and $(2, 1)$

(j) $(0, -2)$ and $(8, 10)$

(c) $(0, 0)$ and $(6, 15)$

(g) $(3, -4)$ and $(5, 6)$

(k) $(-1, -5)$ and $(1, -7)$

(d) $(2, 3)$ and $(6, 15)$

(h) $(0, 0)$ and $(-4, 8)$

(l) $(-3, -4)$ and $(-1, -2)$

12.24. Find the slope of the line whose equation is   (12.12)

    (a) $y = 3x - 4$ (e) $y = 5x$ (i) $3y = -12x + 6$ (m) $\frac{1}{5}y = x - 3$

    (b) $y = 4x - 3$ (f) $y = 5$ (j) $3y = 12 - 2x$ (n) $\frac{1}{3}y = 2x - 6$

    (c) $y = -\frac{1}{2}x + 5$ (g) $2y = 6x - 10$ (k) $y + x = 21$ (o) $\frac{1}{4}y = 7 - x$

    (d) $y = 8 - 7x$ (h) $2y = 10x - 6$ (l) $2x = 12 - y$ (p) $\frac{1}{4}y + 2x = 1$

12.25. Find the inclination, to the nearest degree, of each of the following lines: (12.12)

    (a) $y = 3x - 1$ (c) $2y = 5x + 10$ (e) $5y = 5x - 3$

    (b) $y = \frac{1}{3}x - 1$ (d) $y = \frac{2}{5}x + 5$ (f) $y = -3$

12.26. Find the slope of a line whose inclination is (a) 5°; (b) 17°; (c) 20°; (d) 35°; (e) 45°; (f) 73°; (g) 85°.   (12.12)

12.27. Find the inclination, to the nearest degree, of a line whose slope is (a) 0; (b) 0.4663; (c) 1, (d) 1.4281; (e) $\frac{1}{8}$; (f) $\frac{1}{2}$; (g) $\frac{3}{4}$; (h) $1\frac{1}{3}$; (i) $2\frac{1}{5}$.   (12.12)

12.28. In hexagon *ABCDEF* of Fig. 12-33, $\overline{CD} \parallel \overline{AF}$. Which sides or diagonals have (a) positive slope; (b) negative slope; (c) zero slope; (d) no slope?

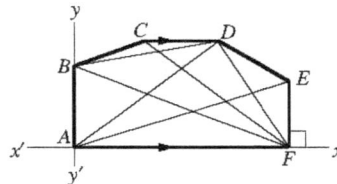

Fig. 12-33

12.29. Find the slope of a line that is parallel to a line whose slope is (a) 0; (b) has no slope; (c) 5; (d) −5; (e) 0.5; (f) −0.0005.   (12.13)

12.30. Find the slope of a line that is parallel to the line whose equation is   (12.13)

    (a) $y = 0$ (c) $x = 7$ (e) $y = 5x - 2$ (g) $3y - 6x = 12$

    (b) $x = 0$ (d) $y = 7$ (f) $x + y = 5$

12.31. Find the slope of a line that is parallel to a line which passes through (a) (0, 0) and (2, 3); (b) (2, −1) and (5, 6); (c) (3, 4) and (5, 2); (d) (1, 2) and (0, −4).   (12.13)

12.32. Find the slope of a line that is perpendicular to a line whose slope is   (12.13)

    (a) $\frac{1}{2}$ (c) 3   (e) 0.1   (g) $-\frac{4}{5}$ (i) 0

    (b) 1   (d) $2\frac{1}{2}$ (f) −1   (h) $-3\frac{1}{4}$ (j) has no slope

12.33. Find the slope of a line that is perpendicular to a line which passes through (a) (0, 0) and (0, 5); (b) (0, 0) and (2, 1); (c) (0, 0) and (3, −1); (d) (1, 1) and (3, 3).   (12.13)

12.34. In rectangle $DEFG$, the slope of $\overline{DE}$ is $\frac{2}{3}$. What is the slope of (a) $\overline{EF}$; (b) $\overline{FG}$; (c) $\overline{DG}$?   (12.14)

12.35. In $\square ABCD$ the slope of $\overline{AB}$ is 1 and the slope of $\overline{BC}$ is $-\frac{1}{2}$. What is the slope of (a) $\overline{AD}$; (b) $\overline{CD}$; (c) the altitude of $\overline{AD}$; (d) the altitude to $\overline{CD}$?   (12.14)

12.36. The vertices of $\triangle ABC$ are $A(0, 5)$, $B(3, 7)$, and $C(5, -1)$. What is the slope of the altitude to (a) $\overline{AB}$; (b) $\overline{BC}$; (c) $\overline{AC}$?   (12.14)

12.37. Which of the following sets of points are collinear: (a) $(2, 1), (4, 4), (8, 10)$; (b) $(-1, 1), (2, 4), (6, 8)$; (c) $(1, -1), (3, 4), (5, 8)$?   (12.15)

12.38. What values of k will make the following trios of points collinear (a) $A(0, 1), B(2, 7), C(6, k)$; (b) $D(-1, 5), E(3, k), F(5, 11)$; (c) $G(0, k), H(1, 1), I(3, -1)$?   (12.15)

12.39. State the equation of the line or pair of lines which is the locus of points   (12.16)

(a) Whose abscissa is -5

(f) 3 units from the line $x = 2$

(b) Whose ordinate is $3\frac{1}{2}$

(g) 6 units above the line $y = -2$

(c) 3 units from the x-axis

(h) 1 unit to the right of the y-axis

(d) 5 units below the x-axis

(i) Equidistant from the lines $x = 5$ and $x = 13$

(e) 4 units from the y-axis

12.40. State the equation of the locus of the center of a circle that   (12.18)

(a) Is tangent to the x-axis at $(6, 0)$

(d) Passes through the origin and $(10, 0)$

(b) Is tangent to the y-axis at $(0, 5)$

(e) Passes through $(3, 7)$ and $(9, 7)$

(c) Is tangent to the lines $x = 4$ and $x = 8$

(f) Passes through $(3, -2)$ and $(3, 8)$

12.41. State the equation of the line or pair of lines which is the locus of points   (12.16)

(a) Whose coordinates are equal

(e) The sum of whose coordinates is 12

(b) Whose ordinate is 5 more than the abscissa

(f) The difference of whose coordinates is 2

(c) Whose abscissa is 4 less than the ordinate

(g) Equidistant from the x-axis and y-axis

(d) Whose ordinate exceeds the abscissa by 10

(h) Equidistant from $x + y = 3$ and $x + y = 7$

12.42. Describe the locus of each of the following equations: (12.17)

(a) $y = 2x + 5$ (c) $\dfrac{y + 3}{x + 2} = \dfrac{5}{4}$ (e) $x + y = 7$

(b) $\dfrac{y - 3}{x - 2} = 4$ (d) $y = \frac{1}{2}x$ (f) $3y = x$

12.43. State the equation of a line which passes through the origin and has a slope of (a) 4; (b) -2; (c) $\frac{3}{2}$; (d) $-\frac{2}{5}$; (e) 0.   (12.17)

12.44. State the equation of a line which has a y- intercept of   (12.17)

(a) 5 and a slope of 4   (d) 8 and is parallel to $y = 3x - 2$

(b) 2 and a slope of -3 (e) -3 and is parallel to $y = 7 - 4x$

(c) -1 and a slope of $\frac{1}{3}$ (f) 0 and is parallel to $y - 2x = 8$

12.45. State the equation of a line which has a slope of 2 and passes through (a) $(1, 4)$; (b) $(-2, 3)$; (c) $(-4, 0)$;
(d) $(0, -7)$.   (12.17)

12.46. State the equation of a line   (12.17)

(a) Which passes through the origin and has a slope of 4

(b) Which passes through $(0, 3)$ and has a slope of $\frac{1}{2}$

(c) Which passes through $(1, 2)$ and has a slope of 3

(d) Which passes through $(-1, -2)$ and has a slope of $\frac{1}{3}$

(e) Which passes through the origin and is parallel to a line that has a slope of 2

12.47. (a) Describe the locus of the equation $x^2 + y^2 = 49$.   (12.18)

(b) State the equation of the locus of points 4 units from the origin.

(c) State the equations of the locus of points 3 units from the locus of $x^2 + y^2 = 25$.

12.48. State the equation of the locus of point 5 units from (a) the origin; (b) the circle $x^2 + y^2 = 16$; (c) the circle
$x^2 + y^2 = 49$.   (12.18)

12.49. What is the radius of the circle whose equation is (a) $x^2 + y^2 = 9$; (b) $x^2 + y^2 = \frac{16}{9}$; (c) $9x^2 + 9y^2 = 36$;
(d) $x^2 + y^2 = 3$?   (12.18)

12.50. What is the equation of a circle whose center is the origin and whose radius is (a) 4; (b) 11; (c) $\frac{2}{3}$; (d) $1\frac{1}{2}$; (e) $\sqrt{5}$;
(f) $\frac{1}{3}\sqrt{3}$?   (12.18)

12.51. Find the area of $\triangle ABC$, whose vertices are $A(0, 0)$ and   (12.19)

(a) $B(0, 5)$ and $C(4, 5)$ (c) $B(0, 8)$ and $C(-5, 8)$ (e) $B(6, 2)$ and $C(7, 0)$

(b) $B(0, 5)$ and $C(4, 2)$ (d) $B(0, 8)$ and $C(-5, 12)$ (f) $B(6, -5)$ and $C(10, 0)$

12.52. Find the area of a   (12.19)

(a) Triangle whose vertices are $A(0, 0), B(3, 4)$, and $C(8, 0)$

(b) Triangle whose vertices are $D(1, 1), E(5, 6)$, and $F(1, 7)$

(c) Rectangle three of whose vertices are $H(2, 2), J(2, 6)$, and $K(7, 2)$

(d) Parallelogram three of whose vertices are $L(3, 1), M(9, 1)$, and $P(5, 5)$

12.53. Find the area of $\triangle DEF$, whose vertices are $D(0, 0)$ and (a) $E(6, 4)$ and $F(8, 2)$; (b) $E(3, 2)$ and $F(6, -4)$;
(c) $E(-2, 3)$ and $F(10, 7)$.   (12.20)

12.54. Find the area of a triangle whose vertices are (a) $(0,0), (2,3)$, and $(4,1)$; (b) $(1,1), (7,3)$, and $(3,6)$; (c) $(-1,2)$, $(0,-2)$, and $(3,1)$. (12.20)

12.55. The vertices of $\triangle ABC$ are $A(2,1)$, $B(8,9)$, and $C(5,7)$. (a) Find the area of $\triangle ABC$. (b) Find the length of $\overline{AB}$. (c) Find the length of the altitude to $\overline{AB}$. (12.20)

12.56. Find the area of a quadrilateral whose vertices are (a) $(3,3), (10,4), (8,7)$, and $(5,6)$; (b) $(0,4), (5,8), (10,6)$, and $(14,0)$; (c) $(0,1), (2,4), (8,10)$, and $(12,2)$.

12.57. Find the area of the quadrilateral formed by the lines (12.19)

(a) $x = 0, x = 5, y = 0$, and $y = 6$

(d) $x = 0, x = 6, y = 0$, and $y = x + 1$

(b) $x = 0, x = 7, y = -2$, and $y = 5$

(e) $y = 0, y = 4, y = x$, and $y = x + 4$

(c) $x = -3, x = 5, y = 3$, and $y = -8$

(f) $y = 0, y = 6, y = 2x$, and $y = 2x + 6$

12.58. Prove each of the following with analytic geometry: (12.21)

(a) A line segment joining the midpoints of two sides of a triangle is parallel to the third side.

(b) The diagonals of a rhombus are perpendicular to each other.

(c) The median to the base of an isosceles triangle is perpendicular to the base.

(d) The length of a line segment joining the midpoints of two sides of a triangle equals one-half the length of the third side.

(e) If the midpoints of the sides of a rectangle taken in succession are joined, the quadrilateral formed is a rhombus.

(f) In a right triangle, the length of the median to the hypotenuse is one-half the length of the hypotenuse.

CHAPTER 13

Inequalities and Indirect Reasoning

13.1 Inequalities

An inequality is a statement that quantities are not equal. If two quantities are unequal, the first is either greater than or less than the other. The inequality symbols are: $\neq$, meaning unequal to; $>$, meaning greater than; and $<$, meaning less than. Thus, $4 \neq 3$ is read "four is unequal to three"; $7 > 2$ is read "seven is greater than two"; and $1 < 5$ is read "one is less than five."

Two inequalities may be of the same order or of opposite order. In inequalities of the same order, the same inequality symbol is used; in inequalities of the opposite order, opposite inequality symbols are used. Thus, $5 > 3$ and $10 > 7$ are inequalities of the same order; $5 > 3$ and $7 < 10$ are inequalities of opposite order.

Inequalities of the same order may be combined, as follows. The inequalities $x < y$ and $y < z$ may be combined into $x < y < z$, which states that y is greater than x and less than z. The inequalities $a > b$ and $b > c$ may be combined into $a > b > c$, which states b is less than a and greater than c.

13.1A Inequality Axioms

Axioms are statements that are accepted as true without proof and are used in the same way as theorems.

AXIOM 1: *A quantity may be substituted for its equal in any inequality.*

Thus if $x > y$ and $y = 10$, then $x > 10$.

AXIOM 2: *If the first of three quantities is greater than the second, and the second is greater than the third, then the first is greater than the third.*

Thus if $x > y$ and $y > z$, then $x > z$.

AXIOM 3: *The whole is greater than any of its parts.*

Thus, $AB > AM$ and $m\angle BAD > m\angle BAC$ in Fig. 13-1.

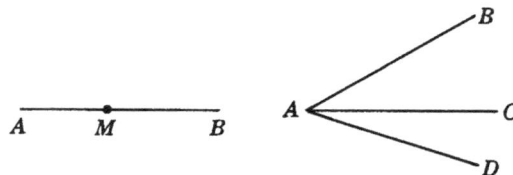

Fig. 13-1

13.1B Inequality Axioms of Operation

AXIOM 4: *If equals are added to unequals, the sums are unequal in the same order.*

Since $5 > 4$ and $4 = 4$, we know that $5 + 4 > 4 + 4$ (or $9 > 8$). If $x - 4 < 5$, then $x - 4 + 4 < 5 + 4$ or $x < 9$.

AXIOM 5: *If unequals are added to unequals of the same order, the sums are unequal in the same order.*

Since $5 > 3$ and $4 > 1$, we have $5 + 4 > 3 + 1$ (or $9 > 4$). If $2x - 4 < 5$ and $x + 4 < 8$, then $2x - 4 + x + 4 < 5 + 8$ or $3x < 13$.

AXIOM 6: *If equals are subtracted from unequals, the differences are unequal in the same order.*

Since $10 > 5$ and $3 = 3$, we have $10 - 3 > 5 - 3$ (or $7 > 2$). If $x + 6 < 9$ and $6 = 6$, then $x + 6 - 6 < 9 - 6$ or $x < 3$.

AXIOM 7: *If unequals are subtracted from equals, the differences are unequal in the opposite order.*

Since $10 = 10$ and $5 > 3$, we have $10 - 5 < 10 - 3$ (or $5 < 7$). If $x + y = 12$ and $y > 5$, then $x + y - y < 12 - 5$ or $x < 7$.

AXIOM 8: *If unequals are multiplied by the same positive number, the products are unequal in the same order.*

Thus if $\frac{1}{4}x < 5$, then $4(\frac{1}{4}x) < 4(5)$ or $x < 20$.

AXIOM 9: *If unequals are multiplied by the same negative number, the results are unequal in the opposite order.*

Thus if $\frac{1}{2}x < 5$, then $(-2)(\frac{1}{2}x) > (-2)(5)$ or $-x > -10$ or $x < 10$.

AXIOM 10: *If unequals are divided by the same positive number, the results are unequal in the same order.*

Thus if $4x > 20$, then $\frac{4x}{4} > \frac{20}{4}$ or $x > 5$.

AXIOM 11: *If unequals are divided by the same negative number, the results are unequal in the opposite order.*

Thus if $-7x < 42$, then $\frac{-7x}{-7} > \frac{42}{-7}$ or $x > -6$.

13.1C Inequality Postulate

POSTULATE 1: *The length of a line segment is the shortest distance between two points.*

13.1D Triangle Inequality Theorems

PRINCIPLE 1: *The sum of the lengths of two sides of a triangle is greater than the length of the third side. (Corollary: The length of the longest side of a triangle is less than the sum of the lengths of the other two sides and greater than their difference.)*

Thus in Fig. 13-2, $BC + CA > AB$ and $AB > BC - AC$.

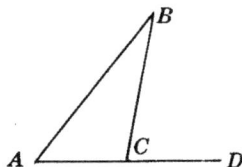

Fig. 13-2

PRINCIPLE 2: *In a triangle, the measure of an exterior angle is larger than the measure of either nonadjacent interior angle.*

Thus in Fig. 13-2, $m\angle BCD > m\angle BAC$ and $m\angle BCD > m\angle ABC$.

PRINCIPLE 3: *If the lengths of two sides of a triangle are unequal, the measures of the angles opposite these sides are unequal, the larger angle being opposite the longer side. (Corollary: The largest angle of a triangle is opposite the longest side.)*

Thus in Fig. 13-2, if $BC > AC$, then $m\angle A > m\angle B$.

PRINCIPLE 4:   *If the measures of two angles of a triangle are unequal, the lengths of the sides opposite these angles are unequal, the longer side being opposite the larger angle. (Corollary: The longest side of a triangle is opposite the largest angle.)*

Thus in Fig. 13-2, if $m\angle A > m\angle B$, then $BC > AC$.

PRINCIPLE 5:   *The perpendicular from a point to a line is the shortest segment from the point to the line.*

Thus in Fig. 13-3, if $\overline{PC} \perp \overline{AB}$ and $\overline{PD}$ is any other line from P to $\overline{AB}$, then $PC < PD$.

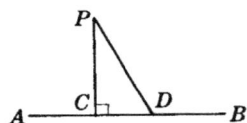   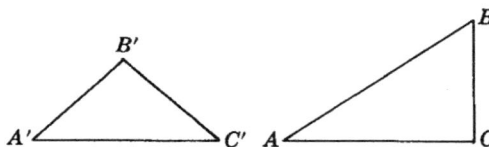

           Fig. 13-3   Fig. 13-4

PRINCIPLE 6:   *If two sides of a triangle are congruent to two sides of another triangle, the triangle having the greater included angle has the greater third side.*

Thus in Fig. 13-4, if $BC = B'C'$, $AC = A'C'$, and $m\angle C > m\angle C'$, then $AB > A'B'$.

PRINCIPLE 7:   *If two sides of a triangle are congruent to two sides of another triangle, the triangle having the greater third side has the greater angle opposite this side.*

Thus in Fig. 13-4, if $BC = B'C'$, $AC = A'C'$, and $AB > A'B'$, then $m\angle C > m\angle C'$.

13.1E   Circle Inequality Theorems

PRINCIPLE 8:   *In the same or equal circles, the greater central angle has the greater arc.*

Thus in Fig. 13-5, if $m\angle AOB > m\angle COD$, then $m\overset{\frown}{AB} > m\overset{\frown}{CD}$.

PRINCIPLE 9:   *In the same or equal circles, the greater arc has the greater central angle. (This is the converse of Principle 8.)*

Thus in Fig. 13-5, if $m\overset{\frown}{AB} > m\overset{\frown}{CD}$, then $m\angle AOB > m\angle COD$.

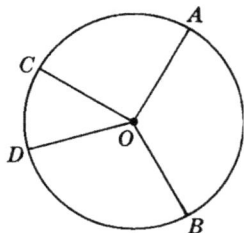   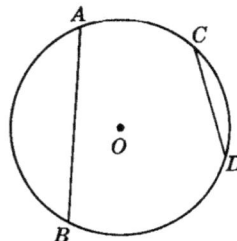

           Fig. 13-5   Fig. 13-6

PRINCIPLE 10:   *In the same or equal circles, the greater chord has the greater minor arc.*

Thus in Fig. 13-6, if $AB > CD$, then $m\overset{\frown}{AB} > m\overset{\frown}{CD}$.

PRINCIPLE 11:   *In the same or equal circles, the greater minor arc has the greater chord. (This is the converse of Principle 10.)*

Thus in Fig. 13-6, if $m\overset{\frown}{AB} > m\overset{\frown}{CD}$, then $AB > CD$.

PRINCIPLE 12:   *In the same or equal circles, the greater chord is at a smaller distance from the center.*

Thus in Fig. 13-7, if $AB > CD$, then $OE < OF$.

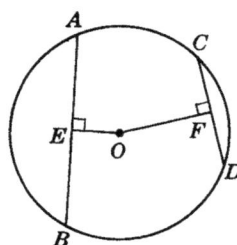

Fig. 13-7

PRINCIPLE 13: *In the same or equal circles, the chord at the smaller distance from the center is the greater chord.* (This is the converse of Principle 12.)

Thus in Fig. 13-7, if $OE < OF$, then $AB > CD$.

SOLVED PROBLEMS

13.1 Selecting inequality symbols

Determine which inequality symbol, $>$ or $<$, makes each of the following true:

(a) 5 __?__ 3   (c) -5 __?__ 3   (e) If $x = 3$, then x^2 __?__ x.

(b) 6 __?__ 9   (d) -5 __?__ -3 (f) If $x > 10$, then 10 __?__ x.

Solutions

(a) $>$ (b) $<$ (c) $<$ (d) $<$ (e) $>$ (f) $<$

13.2 Applying inequality axioms

Complete each of the following statements:

(a) If $a > b$ and $b > 8$, then a __?__ 8.

(b) If $x > y$ and $y = 15$, then x __?__ 15.

(c) If $c < 20$ and $d < 5$, then $c + d$ __?__ 25.

(d) If $x > y$ and $y > 6$, then x __?__ y __?__ 6.

(e) If $x > y$, then $\frac{1}{2}x$ __?__ $\frac{1}{2}y$.

(f) If $e < \frac{1}{4}f$, then $4e$ __?__ f.

(g) If $-y < z$ then y __?__ $-z$.

(h) If $-4x > p$, then x __?__ $-\frac{1}{4}p$.

(i) If Paul and Jack have equal amounts of money and Paul spends more than Jack, then Paul will have __?__ than Jack.

(j) If Anne is now older than Helen, then 10 years ago, Anne was __?__ than Helen.

Solutions

(a) $>$ (c) $<$ (e) $>$ (g) $>$ (i) less

(b) $>$ (d) $>,>$ (f) $<$ (h) $<$ (j) older

13.3 Applying triangle inequality theorems (Fig. 13-8)

(a) Determine the integer values that the length of side a of the triangle can have if the other two sides have lengths 3 and 7.

(b) Determine which is the longest side of the triangle if two angles have measures 59° and 60°.

(c) Determine which is the longest side of parallelogram $ABCD$ if E is the midpoint of the diagonals and $m\angle AEB > m\angle AED$.

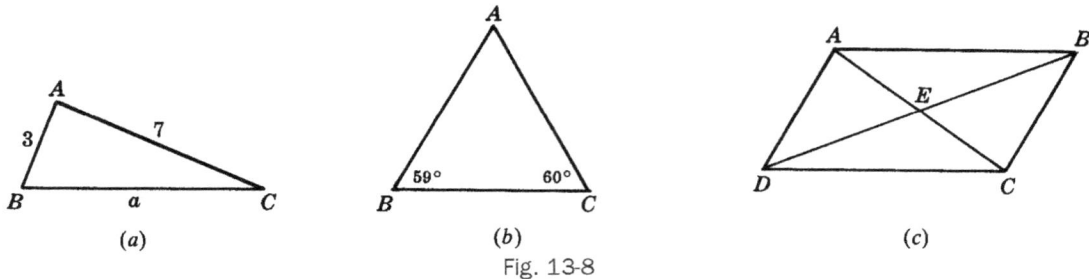

(a)   (b)   (c)

Fig. 13-8

Solutions

(a) Since a must be less than $3 + 7 = 10$ and greater than $7 - 3 = 4$, a can have the integer values of $5, 6, 7, 8, 9$.

(b) Since $m\angle B = 59°$ and $m\angle C = 60°$, $m\angle A = 180° - (59° + 60°) = 61°$. Then the longest side is opposite the largest angle, $\angle A$, so the longest side is BC.

(c) In $\square ABCD$, $AE = CE$ and $DE = EB$. Since $m\angle AEB > m\angle AED$, $AB > AD$ or $AB\,(= DC)$ is the longest side (Principle 6).

13.4 Applying circle inequality theorems

In Fig. 13-9, compare

(a) OD and OF if $\angle C$ is the largest angle of $\triangle ABC$

(b) AC and BC if $m\overset{\frown}{AC} > m\overset{\frown}{BC}$

(c) $m\overset{\frown}{BC}$ and $m\overset{\frown}{AC}$ if $OF > OE$

(d) $m\angle AOB$ and $m\angle BOC$ if $m\overset{\frown}{AB} > m\overset{\frown}{BC}$

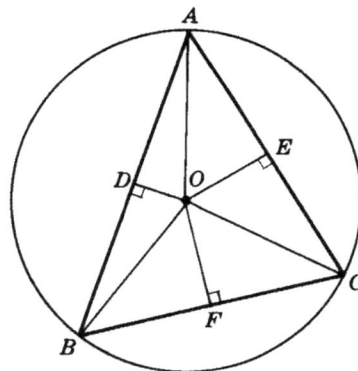

Fig. 13-9

Solutions

(a) Since $\angle C$ is the largest angle of the triangle, $\overline{AB}$ is the longest side, or $AB > BC$; hence, $OD < OF$ by Principle 12.

(b) Since $m\overset{\frown}{AC} > m\overset{\frown}{BC}$, $AC > BC$, the greater arc having the greater chord.

(c) Since $OF > OE$, $BC < AC$ by Principle 13; hence, $m\overset{\frown}{BC} < m\overset{\frown}{AC}$ by Principle 10.

(d) Since $m\overset{\frown}{AB} > m\overset{\frown}{BC}$, $m\angle AOB > m\angle BOC$, the greater arc having the greater central angle.

13.5 Proving an inequality problem

Prove that in $\triangle ABC$, if M is the midpoint of $\overline{AC}$ and $BM > AM$, then $m\angle A + m\angle C > m\angle B$.

Given: $\triangle ABC, M$ is midpoint of AC.
    $BM > AM$
To Prove: $m\angle A + m\angle C > m\angle B$
Plan: Prove $m\angle A > m\angle 1$ and $m\angle C > m\angle 2$
    and then add unequals.

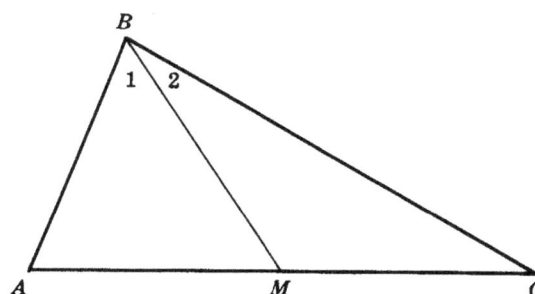

PROOF:

Statements	Reasons
1. M is the midpoint of $\overline{AC}$.	1. Given
2. $\overline{AM} \cong \overline{MC}$	2. A midpoint divides a line into two congruent parts.
3. $BM > AM$	3. Given
4. $BM > MC$	4. A quantity may be substituted for its equal in any inequality. Definition of congruent segments.
5. In $\triangle AMB, m\angle A > \angle 1$. In $\triangle BMC, m\angle C > \angle 2$.	5. In a triangle, the larger angle lies opposite the longer side.
6. $m\angle A + m\angle C > m\angle B$	6. If unequals are added to unequals, the sums are unequal in the same order.

13.2 Indirect Reasoning

We often arrive at a correct conclusion by *indirect* reasoning. In this form of reasoning, the correct conclusion is reached by eliminating all possible conclusions except one. The remaining possibility must be the correct one. Suppose we are given the years 1492, 1809, and 1960 and are assured that one of these years is the year in which a president of the United States was born. By eliminating 1492 and 1960 as impossibilities, we know by indirect reasoning that 1809 is the correct answer. (Had we known that 1809 was the year in which Lincoln was born, the reasoning would have been direct.)

In proving a theorem by indirect reasoning, a possible conclusion may be eliminated if we assume it is true and that assumption results in a contradiction of some given or known fact.

SOLVED PROBLEMS

13.6 Applying indirect reasoning in life situations

Explain how indirect reasoning is used in each of the following situations:

(a) A detective determines the murderer of a slain person.

(b) A librarian determines which volume of a set of books is in use.

Solutions

(a) The detective, using a list of all those who could have been a murderer in the case, eliminates all except one. He or she concludes that the remaining one is the murderer.

(b) The librarian finds all the books of the set except one by looking on the shelf and checking the records. He or she concludes that the missing one is the one in use.

13.7 Proving an inequality theorem by the indirect method

Prove that in the same or equal circles, unequal chords are unequally distant from the center.

Given: Circle O, $AB \neq CD$
$\overline{OE} \perp \overline{AB}$, $\overline{OF} \perp \overline{CD}$

To Prove: $OE \neq OF$

Plan: Assume the other possible conclusion, $OE = OF$, and arrive at a contradiction.

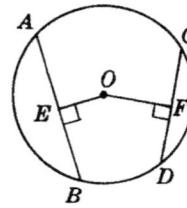

PROOF:

Statements	Reasons
1. Either $OE = OF$ or $OE \neq OF$.	1. Two quantities are either equal or unequal.
2. Assume $OE = OF$.	2. This is one of the possible conclusions.
3. If $OE = OF$, then $AB = CD$.	3. In the same or equal circles, chords equally distant from the center are equal.
4. But $AB \neq CD$.	4. Given
5. The assumption $OE = OF$ is not valid.	5. It leads to a contradiction.
6. Hence, $OE \neq OF$.	6. This is the only remaining possibility.

SUPPLEMENTARY PROBLEMS

13.1. Determine which inequality symbol, $>$ or $<$, makes each of the following true: (13.1)

(a) If $y > 15$, then $15 \underline{\ ?\ } y$.　　(d) If $a = 4$ and $b = \frac{1}{4}$, then $a/b \underline{\ ?\ } 15$.

(b) If $x = 2$, then $3x - 1 \underline{\ ?\ } 4$.　　(e) If $a = 5$, then $a^2 \underline{\ ?\ } 4a$.

(c) If $x = 2$ and $y = 3$, then $xy \underline{\ ?\ } 5$.　　(f) If $b = \frac{1}{2}$, then $b^2 \underline{\ ?\ } b$.

13.2. Complete each of the following statements: (13.2)

(a) If $y > x$ and $x = z$, then $y \underline{\ ?\ } z$.　　(c) If $a < b$ and $b < 15$, then $a \underline{\ ?\ } 15$.

(b) If $a + b > c$ and $b = d$, then $a + d \underline{\ ?\ } c$.　　(d) If $z > y$, $y > x$, and $x = 10$, then $z \underline{\ ?\ } 10$.

13.3. Complete each of the following statements about Fig. 13-10: (13.2)

(a) $BC \underline{\ ?\ } BD$　　(c) $\triangle ADC \underline{\ ?\ } \triangle ABC$

(b) $m\angle BAD \underline{\ ?\ } m\angle BAC$　　(d) If $m\angle A = m\angle C$, then $AB \underline{\ ?\ } BD$.

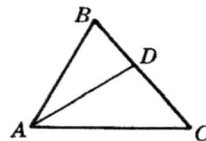

Fig. 13-10

13.4. Complete each of the following statements: (13.2)

(a) If Mary and Ann earn the same weekly wage and Mary is to receive a larger increase than Ann, then Mary will earn $\underline{\ ?\ }$ Ann earns.

(b) If Bernice, who is the same weight as Helen, loses more weight than Helen, then Bernice will weigh $\underline{\ ?\ }$ Helen weighs.

13.5. Complete each of the following statements: (13.2)

(a) If $a > 3$, then $4a \underline{\ ?\ } 12$.　　(d) If $f > 8$, then $f + 7 \underline{\ ?\ } 15$.

(b) If $x - 3 > 15$, then $x \underline{\ ?\ } 18$.　　(e) If $x = y$, then $x + 5 \underline{\ ?\ } y + 6$.

(c) If $3x < 18$, then $x \underline{\ ?\ } 6$.　　(f) If $g = h$, then $g - 10 \underline{\ ?\ } h - 9$.

13.6. Which of the following sets of numbers can be the lengths of the sides of a triangle? (13.3)

    (a) 3, 4, 8   (b) 5, 7, 12   (c) 3, 4, 6   (d) 2, 7, 8   (e) 50, 50, 5

13.7. What integer values can the length of the third side of a triangle have if the two sides have lengths (a) 2 and 6; (b) 3 and 8; (c) 4 and 7; (d) 4 and 6; (e) 4 and 5; (f) 7 and 7 ? (13.3)

13.8. In Fig. 13-11, arrange, in descending order of size, (a) the angles of $\triangle ABC$; (b) the sides of $\triangle DEF$; (c) the angles 1, 2, and 3. (13.3)

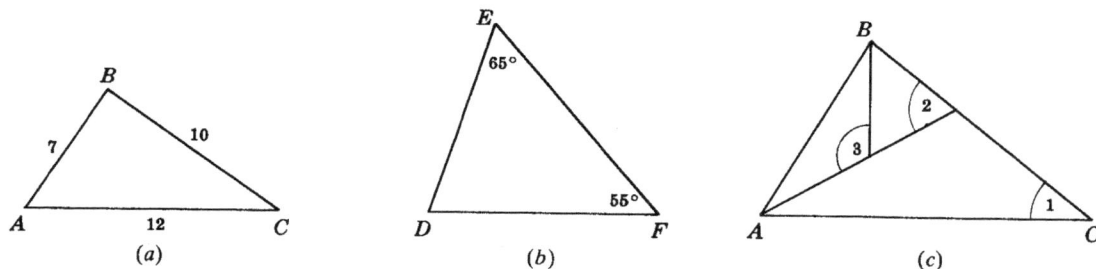

Fig. 13-11

13.9. (a) In quadrilateral $ABCD$ of Fig. 13-12, compare $m\angle BAC$ and $m\angle ACD$ if $AB = CD$ and $BC > AD$.

    (b) In $\triangle ABC$ of Fig. 13-13, compare AB and BC if $\overline{BM}$ is the median to $\overline{AC}$ and $m\angle AMB > m\angle BMC$. (13.3)

13.10. Arrange, in descending order of magnitude, (13.4)

    (a) The sides of $\triangle ABC$ in Fig. 13-14

    (b) The central angles AOB, BOC, and AOC in Fig. 13-14

    (c) The sides of trapezoid $ABCD$ in Fig. 13-15

    (d) The distances of the sides of $\triangle DEF$ from the center in Fig. 13-16

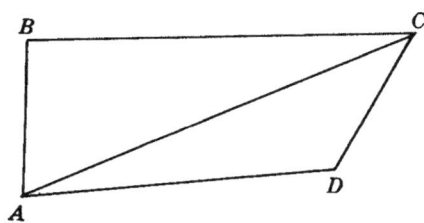

Fig. 13-12

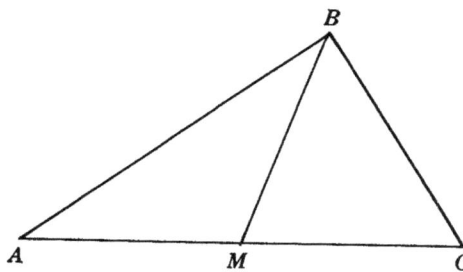

Fig. 13-13

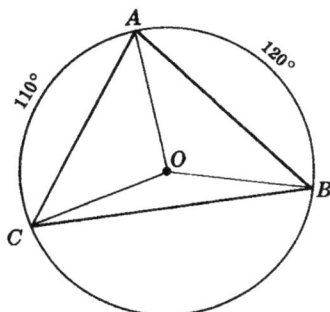

Fig. 13-14

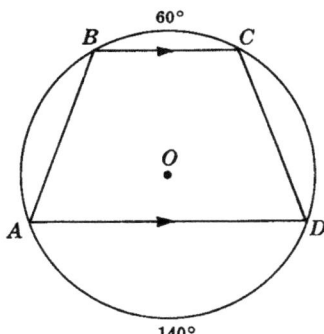

Fig. 13-15

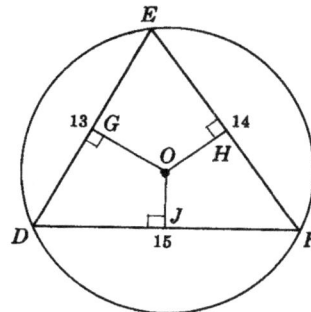

Fig. 13-16

13.11. Provide the proofs requested in Fig. 13-17. (13.5)

(*a*) **Given:** Parallelogram *ABCD*
$AC > BD$
To Prove: $m\angle BDA > m\angle CAD$

(*b*) **Given:** Rhombus *FGHJ*
$m\angle G > m\angle F$
To Prove: $FL > GL$

(*c*) **Given:** $\overline{AD}$ bisects $\angle A$
$\overline{CD}$ bisects $\angle C$, $AB > BC$
To Prove: $AD > CD$

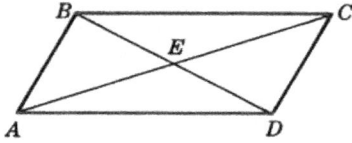 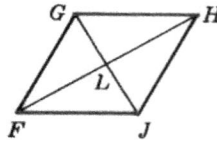 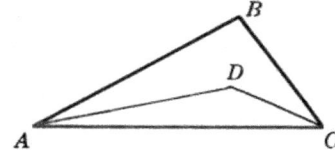

(*d*) **Given:** $\vec{AE} \parallel \vec{BC}$
$m\angle DAE > m\angle EAC$
To Prove: $AC > AB$

(*e*) **Given:** Quad. *ABCD*
$AB > BC$, $AD > CD$
To Prove: $m\angle C > m\angle A$

(*f*) **Given:** $AB = AC$
To Prove: $BD > CD$

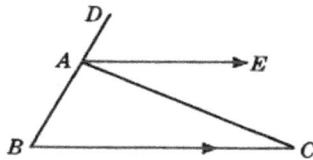 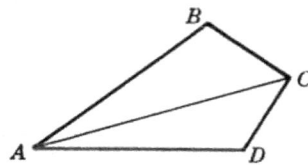 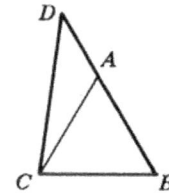

Fig. 13-17

13.12. Explain how indirect reasoning is used in each of the following situations: (13.6)

(a) A person determines which of his ties has been borrowed by his roommate.

(b) A girl determines that the electric motor in her train set is not defective even though her toy trains do not run.

(c) A teacher finds which of his students did not do their assigned homework.

(d) A mechanic finds the reason why the battery in a car does not work.

(e) A person accused of a crime proves her innocence by means of an alibi.

13.13. Prove each of the following: (13.7)

(a) The base angles of an isosceles triangle cannot be right angles.

(b) A scalene triangle cannot have two congruent angles.

(c) The median to the base of a scalene triangle cannot be perpendicular to the base.

(d) If the diagonals of a parallelogram are not congruent, then it is not a rectangle.

(e) If a diagonal of a parallelogram does not bisect a vertex angle, then the parallelogram is not a rhombus.

(f) If two angles of a triangle are unequal, the sides opposite are unequal, the longer side being opposite the larger angle.

Formulas for Reference

Angle Formulas

1. Complement of $a°$	1. $c = 90° - a°$
2. Supplement of $a°$	2. $s = 180° - a°$
3. Sum of measures of angles of a triangle	3. $S = 180°$
4. Sum of measures of angles of a quadrilateral	4. $S = 360°$
5. Sum of measures of exterior angles of an n-gon	5. $S = 360°$
6. Sum of measures of interior angles of an n-gon	6. $S = 180°(n - 2)$
7. Measure of each interior angle of an equiangular or regular n-gon	7. $S = \dfrac{180°(n - 2)}{n}$
8. Measure of each exterior angle of an equiangular or regular n-gon	8. $S = \dfrac{360°}{n}$
9. Measure of central $\angle O$ intercepting an arc of $a°$	9. $m\angle O = a°$
10. Measure of inscribed $\angle A$ intercepting an arc of $a°$	10. $m\angle A = \frac{1}{2}a°$
11. Measure of $\angle A$ formed by a tangent and a chord and intercepting an arc of $a°$	11. $m\angle A = \frac{1}{2}a°$
12. Measure of $\angle A$ formed by two intersecting chords and intercepting arcs of $a°$ and $b°$	12. $m\angle A = \frac{1}{2}(a° + b°)$
13. Measure of $\angle A$ formed by two intersecting tangents, two intersecting secants, or an intersecting tangent and secant and intercepting arcs of $a°$ and $b°$	13. $m\angle A = \frac{1}{2}(a° - b°)$
14. Measure of $\angle A$ inscribed in a semicircle	14. $m\angle A = 90°$
15. Opposite $\angle\!\!s A$ and B of an inscribed quadrilateral	15. $m\angle A = 180° - m\angle B$

Area Formulas

1. Area of a rectangle	1. $A = bh$
2. Area of a square	2. $A = s^2$, $\quad\quad\quad A = \frac{1}{2}d^2$
3. Area of a parallelogram	3. $A = bh$, $\quad\quad\quad A = ab\sin C$
4. Area of a triangle	4. $A = \frac{1}{2}bh$, $\quad\quad\quad A = \frac{1}{2}ab\sin C$
5. Area of a trapezoid	5. $A = \frac{1}{2}h(b + {}^nb)$, $\quad A = hm$
6. Area of an equilateral triangle	6. $A = \frac{1}{4}s^2\sqrt{3}$, $\quad\quad A = \frac{1}{3}h^2\sqrt{3}$
7. Area of a rhombus	7. $A = \frac{1}{2}dd$
8. Area of a regular polygon	8. $A = \frac{1}{2}pr$
9. Area of a circle	9. $A = \pi r^2$, $\quad\quad\quad A = \frac{1}{4}\pi d^2$
10. Area of a sector	10. $A = \dfrac{n}{360}(\pi r^2)$
11. Area of a minor segment	11. $A =$ area of sector $-$ area of triangle

Circle Intersection Formulas

1.	2.	3.
Intersecting Chords $ab = cd$	Intersecting Tangent and Secant $\frac{s}{t} = \frac{t}{e}, t^2 = se$	Intersecting Secants $se = s'e'$

Right-Triangle Formulas

1.	Pythagorean Theorem	1. $c^2 = a^2 + b^2$
2.	Leg opposite 30° angle Leg opposite 45° angle Leg opposite 60° angle	2. $b = \frac{1}{2}c$ $b = \frac{1}{2}c\sqrt{2}, b = a$ $a = \frac{1}{2}c\sqrt{3}, a = b\sqrt{3}$
3.	Altitude of equilateral triangle Side of equilateral triangle	3. $h = \frac{1}{2}s\sqrt{3}$ $s = \frac{2}{3}h\sqrt{3}$

(contd.)

Right-Triangle Formulas

4.	Side of square	4. $s = \frac{1}{2}d\sqrt{2}$
	Diagonal of square	$d = s\sqrt{2}$
5.	Altitude to hypotenuse	5. $\frac{p}{h} = \frac{h}{q}, h^2 = pq, h = \sqrt{pq}$
	Leg of right triangle	$\frac{c}{a} = \frac{a}{p}, a^2 = pc, a = \sqrt{pc}$
		$\frac{c}{b} = \frac{b}{q}, b^2 = qc, b = \sqrt{qc}$

Coordinate-Geometry Formulas

1.	Midpoint M	1. $x_M = \dfrac{x_1 + x_2}{2}, y_M = \dfrac{y_1 + y_2}{2}$
	Distance P_1P_2	$d = \sqrt{(x_2 - x_1)^2 + (y_2 - y_1)^2}$
	Slope of $\overleftrightarrow{P_1P_2}$	$m = \dfrac{y_2 - y_1}{x_2 - x_1}, m = \dfrac{\Delta y}{\Delta x}, m = \tan i$
2.	Slopes of parallels, L_1 and L_2	2. Same slope, m
	Slopes of perpendiculars, L_1 and L'	$mm' = -1$ $m' = -\dfrac{1}{m}, m = -\dfrac{1}{m'}$
3.	Equation of L_1, parallel to x-axis	3. $y = k'$
	Equation of L_2, parallel to y-axis	$x = k$
4.	Equation of L_1 with slope m and y-intercept b	4. $y = mx + b$
	Equation of L_2 with slope m passing through the origin	$y = mx$
	Equation of L_1 with x-intercept a and y-intercept b	$\dfrac{x}{a} + \dfrac{y}{b} = 1$
	Equation of L_3 with slope m and passing through (x_1, y_1)	$y - y_1 = m(x - x_1)$
5.	Equation of circle with center at origin and radius r	5. $x^2 + y^2 = r^2$

Table of Trigonometric Functions

Angle Measure	Sine	Cosine	Tangent	Angle Measure	Sine	Cosine	Tangent
1°	.0175	.9998	.0175	46°	.7193	.6947	1.0355
2°	.0349	.9994	.0349	47°	.7314	.6820	1.0724
3°	.0523	.9986	.0524	48°	.7431	.6691	1.1106
4°	.0698	.9976	.0699	49°	.7547	.6561	1.1504
5°	.0872	.9962	.0875	50°	.7660	.6428	1.1918
6°	.1045	.9945	.1051	51°	.7771	.6293	1.2349
7°	.1219	.9925	.1228	52°	.7880	.6157	1.2799
8°	.1392	.9903	.1405	53°	.7986	.6018	1.3270
9°	.1564	.9877	.1584	54°	.8090	.5878	1.3764
10°	.1736	.9848	.1763	55°	.8192	.5736	1.4281
11°	.1908	.9816	.1944	56°	.8290	.5592	1.4826
12°	.2097	.9781	.2126	57°	.8387	.5446	1.5399
13°	.2250	.9744	.2309	58°	.8480	.5299	1.6003
14°	.2419	.9703	.2493	59°	.8572	.5150	1.6643
15°	.2588	.9659	.2679	60°	.8660	.5000	1.7321
16°	.2756	.9613	.2867	61°	.8746	.4848	1.8040
17°	.2924	.9563	.3057	62°	.8829	.4695	1.8807
18°	.3090	.9511	.3249	63°	.8910	.4540	1.9626
19°	.3256	.9455	.3443	64°	.8988	.4384	2.0503
20°	.3420	.9397	.3640	65°	.9063	.4226	2.1445
21°	.3584	.9336	.3839	66°	.9135	.4067	2.2460
22°	.3746	.9272	.4040	67°	.9205	.3907	2.3559
23°	.3907	.9205	.4245	68°	.9272	.3746	2.4751
24°	.4067	.9135	.4452	69°	.9336	.3584	2.6051
25°	.4226	.9063	.4663	70°	.9397	.3420	2.7475
26°	.4384	.8988	.4877	71°	.9455	.3256	2.9042
27°	.4540	.8910	.5095	72°	.9511	.3090	3.0777
28°	.4695	.8829	.5317	73°	.9563	.2924	3.2709
29°	.4848	.8746	.5543	74°	.9613	.2756	3.4874
30°	.5000	.8660	.5774	75°	.9659	.2588	3.7321
31°	.5150	.8572	.6009	76°	.9703	.2419	4.0108
32°	.5299	.8480	.6249	77°	.9744	.2250	4.3315
33°	.5446	.8387	.6494	78°	.9781	.2079	4.7046
34°	.5592	.8290	.6745	79°	.9816	.1908	5.1446
35°	.5736	.8192	.7002	80°	.9848	.1736	5.6713
36°	.5878	.8090	.7265	81°	.9877	.1564	6.3138
37°	.6018	.7986	.7536	82°	.9903	.1392	7.1154
38°	.6157	.7880	.7813	83°	.9925	.1219	8.1443
39°	.6293	.7771	.8098	84°	.9945	.1045	9.5144
40°	.6428	.7660	.8391	85°	.9962	.0872	11.4301
41°	.6561	.7547	.8693	86°	.9976	.0698	14.3007
42°	.6691	.7431	.9004	87°	.9986	.0523	19.0811
43°	.6820	.7314	.9325	88°	.9994	.0349	28.6363
44°	.6947	.7193	.9657	89°	.9998	.0175	57.2900
45°	.7071	.7071	1.0000	90°	1.0000	.0000	

Answers to Supplementary Problems

Chapter 1

1. (a) point; (b) line; (c) plane; (d) plane; (e) line; (f) point

2. (a) $\overline{AE}, \overline{DE}$; (b) $\overline{ED}, \overline{CD}, \overline{BD}, \overline{FD}$; (c) $\overline{AD}, \overline{BE}, \overline{CE}, \overline{EF}$; (d) F

3. (a) $AB = 16$; (b) $AE = 10\frac{1}{2}$

4. (a) 18; (b) 90°; (c) 50°; (d) 130°; (e) 230°

5. (a) $\angle CBE$; (b) $\angle AEB$; (c) $\angle ABE$; (d) $ABC, \angle BCD, \angle BED$; (e) $\angle AED$

6. (a) 130°; (b) 120°; (c) 75°; (d) 132°

7. (a) 75°; (b) 40°; (c) $10\frac{1}{3}°$ or $10°20'$; (d) $9°11'$

8. (a) 90°; (b) 120°; (c) 135°; (d) 270°; (e) 180°

9. (a) 90°; (b) 60°; (c) 15°; (d) 165°

10. (a) $\overline{AB} \perp \overline{BC}$ and $\overline{AC} \perp \overline{CD}$; (b) 129°; (c) 102°; (d) 51°; (e) 129°

11. (a) $\triangle ABC$, hypotenuse $\overline{AB}$, legs $\overline{AC}$ and $\overline{BC}$
    $\triangle ACD$, hypotenuse $\overline{AC}$, legs $\overline{AD}$ and $\overline{CD}$
    $\triangle BCD$, hypotenuse $\overline{BC}$, legs $\overline{BD}$ and $\overline{CD}$
    (b) $\triangle DAB$ and $\triangle ABC$
    (c) $\triangle AEB$, legs $\overline{AE}$ and $\overline{BE}$, base $\overline{AB}$, vertex angle $\angle AEB$
    $\triangle CED$, legs $\overline{DE}$ and $\overline{CE}$, base $\overline{CD}$, vertex angle $\angle CED$

12. (a) $\overline{AR} \cong \overline{BR}$ and $\angle PRA \cong \angle PRB$; (b) $\angle ABF \cong \angle CBF$; (c) $\angle CGA \cong \angle CGD$; (d) $\overline{AM} \cong \overline{MD}$

13. (a) vert. ⦟; (b) comp. adj. ⦟; (c) adj. ⦟; (d) supp. adj. ⦟; (e) comp. ⦟; (f) vert. ⦟

14. (a) 25°, 65°; (b) 18°, 72°; (c) 60°, 120°; (d) 61°, 119°; (e) 50°, 130°; (f) 56°, 84°; (g) 90°, 90°

15. (a) $a + b = 75°, a - b = 21°, a = 48°$ and $b = 27°$

    (b) $a + b = 90°, a = 3b - 10°, a = 65°$ and $b = 25°$

    (c) $a + b = 180°, a = 4b + 20°, a = 148°$ and $b = 32°$

Chapter 2

1. (a) *A* is *H*; (b) *P* is *D*; (c) *R* is *S*; (d) *E* is *K*; (e) *A* is *G*; (f) triangles are geometric figures; (g) a rectangle is a quadrilateral

2. (a) $a = c = f$; (b) $g = 15$; (c) $f = a$; (d) $a = f, a = h, c = h$; (e) $b = g, b = e, d = e$

3. (a) 130; (b) 4; (c) yes; (d) $x = 8\frac{1}{2}°$; (e) $y = 15$; (f) $x = 6$; (g) $x = \pm6$

4. (a) $AC = 12, AE = 11, AF = 15, DF = 9$

    (b) $m\angle ADC = 92°, m\angle BAE = 68°, m\angle FAD = 86°, m\angle BAD = 128°$

5. (a) $AB = DF$; (b) $AB = AC$; (c) $\angle ECA \cong \angle DCB$; (d) $\angle BAD \cong \angle BCD$

6. (a) If equals are divided by equals, the quotients are equal.
    (b) Doubles of equals are equal.
    (c) If equals are multiplied by equals, the products are equal.
    (d) Halves of equals are equal.

7. (a) If equals are divided by equals, the quotients are equal.
    (b) If equals are multiplied by equals, the products are equal.
    (c) Doubles of equals are equal.
    (d) Halves of equals are equal.

8. (a) Their new rates of pay per hour will be the same. (Add. Post.)
    (b) Those stocks have the same value now. (Mult. Post.)
    (c) The classes have the same number of pupils now. (Subt. Post.)
    (d) $100°C = 212°F$ (Trans. Post.)
    (e) Their parts will be the same length. (Div. Post.)
    (f) He has a total of $10,000 in Banks A, B, and C. (Part. Post.)
    (g) Their values are the same. (Trans. Post.)

9. (a) Vertical angles are congruent.
    (b) All straight angles are congruent.
    (c) Supplements of congruent angles are congruent.
    (d) Perpendiculars form right angles and all right angles are congruent.
    (e) Complements of congruent angles are congruent.

10. In each answer, (H) indicates the hypothesis and (C) indicates the conclusion.
    (a) (H) Stars, (C) twinkle.
    (b) (H) Jet planes, (C) are the speediest.
    (c) (H) Water, (C) boils at 212° Fahrenheit.
    (d) (H) If it is the American flag, (C) its colors are red, white, and blue.
    (e) (H) If you fail to do homework in the subject, (C) you cannot learn geometry.
    (f) (H) If the umpire calls a fourth ball, (C) a batter goes to first base.
    (g) (H) If A is B's brother and C is B's daughter, (C) then A is C's uncle.
    (h) (H) An angle bisector, (C) divides the angle into two equal parts.
    (i) (H) If it is divided into three equal parts, (C) a segment is trisected.
    (j) (H) A pentagon (C) has five sides and five angles.
    (k) (H) Some rectangles (C) are squares.
    (l) (H) If their sides are made longer, (C) angles do not become larger.
    (m) (H) If they are congruent and supplementary, (C) angles are right angles.
    (n) (H) If one of its sides is not a straight line segment, (C) the figure cannot be a polygon.

11. (a) An acute angle is half a right angle. Not necessarily true.
    (b) A triangle having one obtuse angle is an obtuse triangle. True.
    (c) If the batter is out, then the umpire called a third strike. Not necessarily true.
    (d) If you are shorter than I, then I am taller than you. True.
    (e) If our weights are unequal, then I am heavier than you. Not necessarily true.

Chapter 3

1. (a) $\triangle I \cong \triangle II \cong \triangle III$, SAS; (b) $\triangle I \cong \triangle III$, ASA; (c) $\triangle I \cong \triangle II \cong \triangle III$, SSS.

2. (a) ASA; (b) SAS; (c) SSS; (d) SAS; (e) ASA; (f) SAS; (g) SAS; (h) ASA.

3. (a) $\overline{AD} \cong \overline{DC}$; (b) $\angle ABD = \angle DBC$; (c) $\angle 1 \cong \angle 4$; (d) $\overline{BE} \cong \overline{ED}$; (e) $\overline{BD} \cong \overline{AC}$; (f) $\angle BAD \cong \angle CDA$

4. (a) $\angle 1 \cong \angle 3, \angle 2 \cong \angle 4, \overline{BD} \cong \overline{BE}$; (b) $\overline{AB} \cong \overline{AC}, \overline{BD} \cong \overline{DC}, \angle B \cong \angle C$;
   (c) $\angle E \cong \angle C, \angle A \cong \angle F, \angle EDF \cong \angle ABC$

5. (a) $x = 19, y = 8$; (b) $x = 4, y = 12$; (c) $x = 48, y = 12$

8. (a) $\angle b \cong \angle d, \angle E \cong \angle G$; (b) $\angle A \cong \angle 1 \cong \angle 4, \angle 2 \cong \angle C$; (c) $\angle 1 \cong \angle 5, \angle 4 \cong \angle 6, \angle EAD \cong \angle EDA$

9. (a) $\overline{BE} \cong \overline{EC}$; (b) $\overline{AB} \cong \overline{BD} \cong \overline{AD}, \overline{BC} \cong \overline{CD}$; (c) $\overline{BD} \cong \overline{DE}, \overline{EF} \cong \overline{FC}, \overline{AB} \cong \overline{AC}$

Chapter 4

1. (a) $x = 105°, y = 75°$; (b) $x = 60°, y = 40°$; (c) $x = 85°, y = 95°$; (d) $x = 50°, y = 50°$;
   (e) $x = 65°, y = 65°$; (f) $x = 40°, y = 30°$; (g) $x = 60°, y = 120°$; (h) $x = 90°, y = 35°$;
   (i) $x = 30°, y = 40°$; (j) $x = 80°, y = 10°$; (k) $x = 30°, y = 150°$; (l) $x = 85°, y = 95°$

2. (a) $x = 22°, y = 102°$; (b) $x = 40°, y = 100°$; (c) $x = 80°, y = 40°$

3. (a) Each angle measures $105°$. (b) Each angle measures $70°$. (c) Angles measure $72°$ and $108°$.

7. (a) 25; (b) 9; (c) 20; (d) 8

8. (a) 8; (b) 10; (c) 2; (d) 14

10. (a) P is equidistant from B and C. P is on $\perp$ bisector of $\overline{BC}$.
    Q is equidistant from A and B. Q is on $\perp$ bisector of $\overline{AB}$.
    R is equidistant from A, C, and D. R is on $\perp$ bisectors of $\overline{AD}$ and $\overline{CD}$.
    (b) P is equidistant from $\overline{AB}$ and $\overline{AD}$. P is on bisector of $\angle A$.
    Q is equidistant from $\overline{AB}$ and $\overline{BC}$. Q is on bisector of $\angle B$.
    R is equidistant from $\overline{BC}, \overline{CD}$, and $\overline{AD}$. R is on the bisectors of $\angle C$ and $\angle D$.

11. (a) P is equidistant from $\overline{AD}, \overline{AB}$, and $\overline{BC}$. Q is equidistant from $\overline{AD}$ and $\overline{AB}$ and equidistant from A and D. R is equidistant from $\overline{AB}$ and $\overline{BC}$ and equidistant from A and D.
    (b) P is equidistant from $\overline{AD}$ and $\overline{CD}$ and equidistant from B and C. Q is equidistant from A, B, and C. R is equidistant from $\overline{AD}$ and $\overline{CD}$ and equidistant from A and B.

12. (a) $x = 50°, y = 110°$; (b) $x = 65°, y = 65°$; (c) $x = 30°, y = 100°$; (d) $x = 51°, y = 112°$;
    (e) $x = 52°, y = 40°$; (f) $x = 120°, y = 90°$

13. (a) $x = 55°, y = 125°$; (b) $x = 80°, y = 90°$; (c) $x = 56°, y = 68°$; (d) $x = 100°, y = 30°$;
    (e) $x = 30°, y = 120°$; (f) $x = 90°, y = 30°$

14. (a) $18°, 54°, 108°$; (b) $40°, 50°, 90°$; (c) $36°, 36°, 108°$; (d) $36°, 72°, 108°, 144°$; (e) $50°, 75°$;
    (f) $100°, 60°$, and $20°$

16. (a) Since $x = 45$, each angle measures $60°$.
    (b) Since $x = 25, x + 15 = 40$ and $3x - 35 = 40$; that is, two angles each measure $40°$.
    (c) If $2x, 3x$, and $5x$ represent the angles, $x = 18$ and $5x = 90$; that is, one of the angles measures $95°$.
    (d) If x and $5x - 10$ represent the unknown angles, $x = 21$ and $5x - 10 = 95$; that is, one of the angles measures $95°$.

17. (a) 7 st. $\angle s$, 30 st. $\angle s$; (b) $1620°, 5400°, 180,000°$; (c) $30, 12, 27, 202$

18. (a) $20°, 18°, 9°$; (b) $160°, 162°, 171°$; (c) $3, 9, 20, 180$; (d) $3, 12, 36, 72, 360$

19. (a) $65°, 90°, 95°, 110°$; (b) $140°, 100°, 60°, 60°$

20. (a) $\triangle I \cong \triangle III$ by hy-leg; (b) $\triangle I \cong \triangle III$ by SAA.

Chapter 5

1. (a) $x = 15, y = 25$; (b) $x = 20, y = 130$; (c) $x = 20, y = 140$

4. (a) $\square EFGH$; (b) $\square ABCD$ and $EBFD$; (c) $\square GHKJ, HILK, GILJ$; (d) $\square ACHB, CEFH$

5. (a) Two sides are congruent and $\parallel$. (b) Opposite sides are congruent.
   (c) Opposite angles are congruent. (d) $\overline{AD}$ and $\overline{BC}$ are congruent and parallel $(\overline{AD} \cong \overline{EF} \cong BC)$.

6. (a) $x = 6, y = 12$; (b) $x = 5, y = 9$; (c) $x = 120, y = 30$; (d) $x = 15, y = 45$

7. (a) $x = 14, y = 6$; (b) $x = 18, y = 4\frac{1}{2}$; (c) $x = 8, y = 5$; (d) $x = 3, y = 9$

10. (a) $x = 5, y = 7$; (b) $x = 10, y = 35$; (c) $x = 2\frac{1}{2}, y = 17\frac{1}{2}$; (d) $x = 8, y = 4$; (e) $x = 25, y = 25$;
    (f) $x = 11, y = 118$

13. (a) $x = 6, y = 40$; (b) $x = 3, y = 5\frac{1}{2}$; (c) $x = 8\frac{1}{3}, y = 22$

14. (a) $x = 28, y = 25\frac{1}{2}$; (b) $x = 12$ (since y does not join midpoints, Pr. 3 does not apply);
    (c) $x = 19, \ y = 23\frac{1}{2}$

15. (a) $m = 19$; (b) $b' = 36$; (c) $b = 73$

16. (a) $x = 11, y = 33$; (b) $x = 32, y = 26$; (c) $x = 12, y = 36$

17. (a) $22\frac{1}{2}$; (b) 70

18. (a) 21; (b) 30; (c) 14; (d) 26

Chapter 6

5. (a) square; (b) isosceles triangle; (c) trapezoid; (d) right triangle

6. (a) $140°$; (b) $60°$; (c) $90°$; (d) $(180 - x)°$; (e) $x°$; (f) $(90 + x)°$

7. (a) $100°$; (b) $50°, 80°$; (c) $54°, 27°$; (d) $45°$; (e) $35°$; (f) $45°$

8. (a) $x = 22$; (b) $y = 6$; (c) $AB + CD = 22$; (d) perimeter = 44; (e) $x = 21$; (f) $r = 14$

9. (a) 0; (b) 40; (c) 33; (d) 7

10. (a) tangent externally; (b) tangent internally; (c) the circles are 5 units apart; (d) overlapping

11. (a) concentric; (b) tangent internally; (c) tangent externally; (d) outside each other; (e) the smaller entirely inside the larger; (f) overlapping

13. (a) 40; (b) 90; (c) 170; (d) 180; (e) $2x$; (f) $180 - x$; (g) $2x - 2y$

14. (a) 20; (b) 45; (c) 85; (d) 90; (e) 130; (f) 174; (g) x; (h) $90 - \frac{1}{2}x$; (i) $x - y$

15. (a) 85; (b) 170; (c) c; (d) $2i$; (e) 60; (f) 30

16. (a) 60, 120, 180; (b) 80, 120, 160; (c) 100, 120, 140; (d) 36, 144, 180

17. (a) $m\angle x = 136°$; (b) $m\widehat{y} = 11°$; (c) $m\angle x = 130°$; (d) $m\angle y = 126°$; (e) $m\angle x = 110°$; $m\widehat{y} = 77°$

18. (a) $135°$; (b) $90°$; (c) $(180 - x)°$; (d) $(90 + x)°$; (e) $100°$; (f) $80°$; (g) $55°$; (h) $72°$

19. (a) $85°$; (b) $y°$; (c) $110°$; (d) $95°$; (e) $72°$; (f) $50°$; (g) $145°$; (h) $87°$

20. (a) 50; (b) 60

21. (a) $m\widehat{x} = 65°, m\widehat{y} = 65°$; (b) $m\angle x = 90°, m\angle y = 55°$; (c) $m\angle x = 37°, m\angle y = 50°$

22. (a) 19; (b) 45; (c) 69; (d) 90; (e) 125; (f) 167; (g) $\frac{1}{2}x$; (h) $180 - \frac{1}{2}x$; (i) $x + y$

23. (a) 110; (b) 135; (c) 180; (d) 270; (e) $180 - 2x$; (f) $360 - 2x$; (g) $2x - 2y$; (h) $7x$

24. (a) $45°$; (b) $60°$; (c) $30°$; (d) $18°$

25. (a) $m\widehat{x} = 120°, m\angle y = 60°$; (b) $m\angle x = 62°, m\angle y = 28°$; (c) $m\angle x = 46°, m\angle y = 58°$

26. (a) 75°; (b) 75°; (c) 115°; (d) 100°; (e) 140°; (f) 230°; (g) 80°; (h) 48°

27. (a) 85°; (b) 103°; (c) 80°; (d) 72°; (e) 90°; (f) 110°; (g) 130°; (h) 110°

28. (a) $m\widehat{x} = 68°, m\angle y = 95°$; (b) $m\angle x = 90°, m\angle y = 120°$; (c) $m\widehat{x} = 34°, m\angle \widehat{y} = 68°$

29. (a) 30°; (b) 37°; (c) 20°; (d) 36°; (e) 120°; (f) 130°; (g) 94°; (h) 25°

30. (a) 45°; (b) 75°; (c) 50°; (d) $36\frac{1}{2}°$; (e) 90°; (f) 140°; (g) 115°; (h) 45°; (i) 80°

31. (a) 20°; (b) 85°; (c) $(180 - x)°$; (d) $(90 + x)°$; (e) 90°; (f) 25°; (g) 42°; (h) 120°; (i) 72°; (j) 110°; (k) 145°; (l) $(180 - y)°$; (m) 240°; (n) $(180 + x)°$; (o) 270°

32. (a) $m\widehat{x} = 43°, m\angle y = 43°$, (b) $m\widehat{x} = 190°, m\angle y = 55°$; (c) $m\widehat{x} = 140°, m\angle y = 40°$

33. (a) 120°; (b) 150°; (c) 180°; (d) 50°; (e) $22\frac{1}{2}°$; (f) 45°

34. (a) $m\widehat{x} = 150°, m\widehat{y} = 40°$; (b) $m\widehat{x} = 190°, m\widehat{y} = 70°$; (c) $m\widehat{x} = 252°, m\widehat{y} = 108°$

35. (a) 25°; (b) 39°; (c) 50°; (d) 30°; (e) 40°; (f) 76°; (g) 45°; (h) 95°; (i) 75°; (j) 120°

36. (a) 74°; (b) 90°; (c) 55°; (d) 60°; (e) 40°; (f) 37°; (g) 84°; (h) 110°; (i) 66°; (j) 98°; (k) 75°; (l) 79°

37. (a) $m\angle x = 120°, m\angle y = 60°$; (b) $m\angle x = 45°, m\angle y = 22\frac{1}{2}°$; (c) $m\angle x = 36°, m\angle y = 72°$

38. (a) $m\widehat{x} = 40°, m\angle y = 80°$; (b) $m\widehat{x} = 45°, m\angle y = 67\frac{1}{2}°$; (c) $m\angle x = 78°, m\angle y = 103°$

Chapter 7

1. (a) 4; (b) $\frac{1}{3}$; (c) $\frac{6}{5}$; (d) $\frac{10}{7}$; (e) $\frac{9}{7}$; (f) 2; (g) $\frac{1}{5}$; (h) $\frac{3}{7}$; (i) $\frac{1}{3}$; (j) $\frac{7}{8}$; (k) 2; (l) $\frac{5}{7}$; (m) 20; (n) $\frac{1}{3}$; (o) 3

2. (a) 6; (b) $\frac{14}{5}$; (c) $\frac{1}{7}$; (d) $\frac{3}{2}$; (e) 3; (f) $\frac{7}{2}$; (g) 2; (h) 250; (i) $\frac{1}{20}$; (j) 8; (k) $\frac{5}{3}$; (l) $\frac{9}{2}$

3. (a) $2:3:10$; (b) $12:6:1$; (c) $5:2:1$; (d) $1:4:7$; (e) $4:3:1$; (f) $8:2:1$; (g) $50:5:1$; (h) $6:2:1$; (i) $8:2:1$

4. (a) $\frac{6}{7}$; (b) 12; (c) $\frac{13}{3}$; (d) $\frac{1}{4}$; (e) 6; (f) $\frac{16}{9}$; (g) $\frac{1}{3}$; (h) $\frac{3}{2}$; (i) $\frac{2}{7}$; (j) 11; (k) $\frac{4}{5}$; (l) 60; (m) 3; (n) $\frac{3}{20}$; (o) $\frac{1}{2}$; (p) 14

5. (a) $\frac{1}{3}$; (b) $3c$; (c) $\frac{d}{2}$; (d) $\frac{2r}{D}$; (e) $\frac{b}{a}$; (f) $\frac{4}{S}$; (g) $\frac{S}{6}$; (h) $\frac{3r}{2t}$; (i) $1:4:10$; (j) $3:2:1$; (k) $x^2:x:1$; (l) $6:5:4:1$

6. (a) $5x$ and $4x$, sum $= 9x$; (b) $9x$ and x, sum $= 10x$; (c) $2x, 5x,$ and $11x$, sum $= 18x$; (d) $x, 2x, 2x, 3x,$ and $7x$, sum $= 15x$

7. (a) $5x + 4x = 45, x = 5, 25°$ and $20°$; (b) $5x + 4x = 90, x = 10, 50°$ and $40°$; (c) $5x + 4x = 180, x = 20, 100°$ and $80°$; (d) $5x + 4x + x = 180, x = 18, 90°$ and $72°$

8. (a) $7x + 6x = 91, x = 7, 49°, 42°$ and $35°$; (b) $7x + 5x = 180, x = 15, 105°, 90°$ and $75°$; (c) $7x + 3x = 90, x = 9, 63°, 54°$ and $45°$; (d) $7x + 6x + 5x = 180, x = 10, 70°, 60°$ and $50°$

9. (a) 16; (b) 16; (c) ± 6; (d) $\pm 2\sqrt{5}$; (e) ± 5; (f) 2; (g) $\frac{bc}{a}$; (h) $\pm 6y$

10. (a) 21; (b) $4\frac{2}{3}$; (c) ± 6; (d) $\pm \sqrt{5}$; (e) 8; (f) ± 4; (g) 3; (h) $\pm \sqrt{ab}$

11. (a) 15; (b) 3; (c) 6; (d) $2\frac{2}{3}$; (e) $3\frac{1}{3}$; (f) 30; (g) 32; (h) $6a$

12. (a) 6; (b) 6; (c) 3; (d) $4b$; (e) $\sqrt{10}$; (f) $\sqrt{27}$ or $3\sqrt{3}$; (g) $\sqrt{pq}$; (h) $a\sqrt{b}$

13. (a) $\frac{c}{b} = \frac{d}{x}$; (b) $\frac{a}{p} = \frac{q}{x}$; (c) $\frac{h}{a} = \frac{a}{x}$; (d) $\frac{3}{7} = \frac{1}{x}$; (e) $\frac{c}{a} = \frac{b}{x}$

14. (a) $\frac{x}{y} = \frac{1}{2}$; (b) $\frac{x}{y} = \frac{3}{4}$; (c) $\frac{x}{y} = \frac{1}{2}$; (d) $\frac{x}{y} = \frac{h}{a}$; (e) $\frac{x}{y} = b$

15. Only (b) is not a proportion since $3(12) \neq 5(7)$; that is, $36 \neq 35$.

16. (a) $\frac{x}{2} = \frac{9}{3}$, $x = 6$; (b) $\frac{x}{1} = \frac{4}{5}$, $x = \frac{4}{5}$; (c) $\frac{x}{a} = \frac{b}{2}$, $x = \frac{ab}{2}$; (d) $\frac{x}{5} = \frac{1}{10}$, $x = \frac{1}{2}$; (e) $\frac{x}{20} = \frac{5}{4}$, $x = 25$

17. (a) d; (b) 35; (c) 5 (d) 4

18. (a) 21; (b) $\frac{3}{2}$; (c) 5

19. (a) 16; (b) $6\frac{2}{3}$; (c) 10

20. (a) yes, since $\frac{15}{10} = \frac{18}{12}$; (b) no, since $\frac{10}{13} \neq \frac{7}{9}$; (c) yes, since $\frac{3x}{5x} = \frac{36}{60}$

21. (a) 12; (b) 8; (c) 60

22. (a) 15; (b) 15; (c) $6\frac{1}{2}$

24. (a) $35°$; (b) $53°$

25. (a) $a = 16$; (b) $b = 15$; (c) $c = 126$

27. (a) $\angle ABE \cong \angle EDC$, $\angle BAE \cong \angle DCE$ (also vert. $\angle$s at E)
    (b) $\angle BAF \cong \angle FEC$, $\angle B \cong \angle D$ (also $\angle EAD \cong \angle BFA$)
    (c) $\angle A \cong \angle EDF$, $\angle F \cong \angle BCA$
    (d) $\angle A \cong \angle A$, $\angle B \cong \angle C$
    (e) $\angle C \cong \angle D$, $\angle CAB \cong \angle CAD$
    (f) $\angle A \cong \angle A$, $\angle C \cong \angle DBA$

28. (a) $\angle D \cong \angle B$, $\angle AED \cong \angle FGB$; (b) $\angle ADB \cong \angle ABC$, $\angle A \cong \angle A$; (c) $\angle ABC \cong \angle AED$, $\angle BAE \cong \angle EDA$

29. (a) $\angle C \cong \angle F$, $\frac{14}{20} = \frac{21}{30}$; (b) $\angle A \cong \angle A$, $\frac{10}{25} = \frac{6}{15}$; (c) $\angle B \cong \angle B$, $\frac{16}{28} = \frac{20}{35}$

30. (a) $\frac{6}{18} = \frac{8}{24} = \frac{10}{30}$; (b) $\frac{24}{36} = \frac{28}{42} = \frac{30}{45}$; (c) $\frac{12}{18} = \frac{16}{24} = \frac{18}{27}$

32. (a) $q = 20$; (b) $p = 8$; (c) $b = 7$; (d) $a = 12$; (e) $AB = 35$; (f) $d = 2\frac{1}{4}$

33. (a) 8; (b) 6; (c) $26\frac{2}{3}$

34. (a) 42 ft; (b) 66ft

37. (a) 8:5; (b) 3:5; (c) halved (in each case)

38. (a) 15; (b) 60; (c) 25, 35, 40; (d) 4; (e) 6, 3

39. (a) 3:7; (b) 7:2; (c) quadrupled; (d) 7

43. (a) 5; (b) 14; (c) 6; (d) 5; (e) 12; (f) 13; (g) 48; (h) 2

44. 30, 18

45. (a) 8; (b) 6; (c) 12; (d) 5; (e) 7; (f) 12; (g) 30; (h) $7\frac{1}{2}$; (i) 5; (j) 8

46. (a) 8; (b) 13; (c) 21; (d) 6; (e) 9; (f) 14; (g) 3; (h) 8

47. (a) $a = 4$, $h = \sqrt{12}$ or $2\sqrt{3}$; (b) $c = 9$, $h = \sqrt{20}$ or $2\sqrt{5}$; (c) $q = 4$ and $b = \sqrt{80}$ or $4\sqrt{5}$;
    (d) $p = 18$, $h = \sqrt{108} = 6\sqrt{3}$

48. (a) 25; (b) 39; (c) $\sqrt{41}$; (d) 10; (e) $7\sqrt{2}$

49. (a) $b = 16$; (b) $a = 2\sqrt{7}$; (c) $a = 8$; (d) $b = 2\sqrt{3}$; (e) $b = 5\sqrt{2}$; $b = \sqrt{3}$

50. (a) 9, 12; (b) 10, 24; (c) 80, 150; (d) $2\sqrt{5}$, $4\sqrt{5}$

51. (a) 41; (b) $5\sqrt{5}$

52. (a) 12; (b) $10\sqrt{2}$; (c) $5\sqrt{5}$

53. All except (h)

54. (a) yes; (b) no, since $(2x)^2 + (3x)^2 \neq (4x)^2$

55. (a) 8; (b) 6; (c) $\sqrt{19}$; (d) $5\sqrt{3}$

56. (a) 15; (b) $2\sqrt{5}$; (c) 6

57. (a) 16; (b) 30; (c) $4\sqrt{3}$; (d) 10

58. (a) 10; (b) 12; (c) 28; (d) 15

59. (a) 5; (b) 20; (c) 15; (d) 25

60. (a) 12; (b) 24

61. 12

62. 30

63. (a) 10 and $10\sqrt{3}$; (b) $7\sqrt{3}$ and 14; (c) 5 and 10

64. (a) $11\sqrt{3}$; (b) $a\sqrt{3}$; (c) 48; (d) $16\sqrt{3}$

65. (a) 25 and $25\sqrt{3}$; (b) 35 and $35\sqrt{3}$

66. (a) 28, $8\sqrt{3}$; (b) 17, $14\sqrt{3}$

67. (a) $17\sqrt{2}$; (b) $a\sqrt{2}$; (c) $34\sqrt{2}$; (d) 30

68. (a) $20\sqrt{2}$; (b) $40\sqrt{2}$

69. (a) 45, $13\sqrt{2}$; (b) 11, $27\sqrt{2}$; (c) $15\sqrt{2}$, 55

70. $6\sqrt{2}, 5\sqrt{2}$

Chapter 8

1. (a) 0.4226, 0.7431, 0.8572, 0.9998; (b) 0.9659, 0.6157, 0.2756, 0.0349;
(c) 0.0699, 0.6745, 1.4281, 19.0811; (d) sine and tangent; (e) cosine; (f) tangent

2. (a) $x = 20°$; (b) $A = 29°$; (c) $B = 71°$; (d) $A' = 21°$; (e) $y = 45°$; (f) $Q = 69°$; (g) $W = 19°$;
(h) $B' = 67°$

3. (a) 26°; (b) 47°; (c) 69°; (d) 8°; (e) 40°; (f) 74°; (g) 7°; (h) 27°; (i) 80°; (j) 13° since $\sin x = 0.2200$;
(k) 45° since $\sin x = 0.707$; (l) 59° since $\cos x = 0.5200$; (m) 68° since $\cos x = 0.3750$;
(n) 30° since $\cos x = 0.866$; (o) 16° since $\tan x = 0.2857$; (p) 10° since $\tan x = 0.1732$

4. (a) $\sin A = \frac{4}{5}, \cos A = \frac{3}{4}, \tan A = \frac{4}{5}$; (b) $\sin A = \frac{3}{5}, \cos A = \frac{4}{5}, \tan A = \frac{3}{4}$;
(c) $\sin A = \dfrac{\sqrt{7}}{4}, \cos A = \frac{3}{4}, \tan A = \dfrac{\sqrt{7}}{3}$

5. (a) $m\angle A = 27°$ since $\cos A = 0.8900$; (b) $m\angle A = 58°$ since $\sin A = 0.8500$;
(c) $m\angle A = 52°$ since $\tan A = 1.2800$

6. (a) $m\angle A = 42°$ since $\sin B = 0.6700$; (b) $m\angle B = 74°$ since $\cos B = 0.2800$;
(c) $m\angle B = 68°$ since $\tan B = 2.500$; (d) $m\angle B = 30°$ since $\tan B = 0.577$

8. (a) 23°, 67°; (b) 28°, 62°; (c) 16°, 74°; (d) 10°, 80°

9. (a) $x = 188, y = 313$; (b) $x = 174, y = 250$; (c) $x = 123, y = 182$

10. (a) 82 ft; (b) 88 ft

11. 156 ft

12. (a) 2530 ft; (b) 2560 ft

13. (a) 21 in; (b) 79 in

14. 14

15. 16 and 18 in

16. 31 ft

17. 15 yd

18. (a) 1050 ft; (b) 9950 ft

19. 7°

20. 282 ft

21. (a) 81°; (b) 45°

22. (a) 22 ft; (b) 104 ft

23. 754 ft

24. 404 ft

25. (a) 295 ft; (b) 245 ft; (c) 960 ft

26. (a) 234 ft; (b) 343 ft

27. (a) 96 ft; (b) 166 ft

28. 9.1

Chapter 9

1. (a) 99 in^2; (b) 3 ft^2 or 432 in^2; (c) 500; (d) 120; (e) $36\sqrt{3}$; (f) $100\sqrt{3}$; (g) 300; (h) 150

2. (a) 48; (b) 432; (c) $25\sqrt{3}$; (d) 240

3. (a) 7 and 4; (b) 12 and 6; (c) 9 and 6; (d) 6 and 2; (e) 10 and 7; (f) 20 and 8

4. (a) 1296 in^2; (b) 100 square decimeters (100 dm^2)

5. (a) 225; (b) $12\frac{1}{4}$; (c) 3.24; (d) $64a^2$; (e) 121; (f) $6\frac{1}{4}$; (g) $9b^2$; (h) 32; (i) $40\frac{1}{2}$; (j) 64

6. (a) 128; (b) 72; (c) 100; (d) 49; (e) 400

7. (a) 1600; (b) 400; (c) 100

8. (a) 9; (b) 36; (c) $9\sqrt{2}$; (d) $4\frac{1}{2}$; (e) $\frac{9}{2}\sqrt{2}$

9. (a) $2\frac{1}{2}$; (b) 52; (c) 10; (d) $5\sqrt{2}$; (e) 6; (f) 4

10. (a) 16 ft^2; (b) 6 ft^2 or 864 in^2; (c) 70; (d) 1.62 m^2

11. (a) $3x^2$; (b) $x^2 + 3x$; (c) $x^2 - 25$; (d) $12x^2 + 11x + 2$

12. (a) 36; (b) 15; (c) 16

13. (a) $2\frac{2}{3}$; (b) 20; (c) 9; (d) 3; (e) 15; (f) 12; (g) 8; (h) 7

14. (a) 11 in^2; (b) 3 ft^2 (c) $4x - 28$; (d) $10x^2$; (e) $2x^2 + 18x$; (f) $\frac{1}{2}(x^2 - 16)$; (g) $x^2 - 9$

15. (a) 84; (b) 48; (c) 30; (d) 120; (e) 148; (f) 423; (g) $8\sqrt{3}$; (h) 9

16. (a) 24; (b) 2; (c) 4

17. (a) 8; (b) 10; (c) 8; (d) 18; (e) $9\frac{3}{5}$; (f) $12\frac{1}{2}$; (g) 12; (h) 18

18. (a) $25\sqrt{3}$; (b) $36\sqrt{3}$; (c) $12\sqrt{3}$; (d) $25\sqrt{3}$; (e) $b^2\sqrt{3}$; (f) $4x^2\sqrt{3}$; (g) $3r^2\sqrt{3}$

19. (a) $2\sqrt{3}$; (b) $\frac{49}{2}\sqrt{3}$; (c) $24\sqrt{3}$ (d) $18\sqrt{3}$

20. (a) $24\sqrt{3}$; (b) $54\sqrt{3}$; (c) $150\sqrt{3}$

21. (a) 15; (b) 8; (c) 12; (d) 5

22. (a) 140; (b) 69; (c) 225; (d) $60\sqrt{2}$; (e) 94

23. (a) 150; (b) 204; (c) 39; (d) $64\sqrt{2}$; (e) 160

24. (a) 4; (b) 7; (c) 18 and 9; (d) 9 and 6; (e) 10 and 5

25. (a) 17 and 9; (b) 23 and 13; (c) 17 and 11; (d) 5; (e) 13

26. (a) 36; (b) $38\frac{1}{2}$; (c) $12\sqrt{3}$; (d) $12x^2$; (e) 120; (f) 96; (g) 18; (h) $\frac{49}{2}\sqrt{2}$; (i) $32\sqrt{3}$; $98\sqrt{3}$

27. (a) 737; (b) 14; (c) 77

28. (a) 10; (b) 12 and 9; (c) 20 and 10; (d) 5; (e) $\sqrt{10}$

29. 12

34. (a) 1:49; (b) 49:4; (c) 1:3; (d) 1:25; (e) 81: x^2; (f) 9:x; (g) 1:2

35. (a) 49:100; (b) 4:9; (c) 25:36; (d) 1:9; (e) 9:4; (f) 1:2

36. (a) 10:1; (b) 1:7; (c) 20:9; (d) 5:11; (e) 2:y; (f) $3x$:1; (g) $\sqrt{3}$:2; (h) 1:$\sqrt{2}$; (i) x:$\sqrt{5}$; (j) $\sqrt{x}$:4

37. (a) 6:5; (b) 3:7; (c) $\sqrt{3}$:1; (d) $\sqrt{5}$:2; (e) $\sqrt{3}$:3 or 1:$\sqrt{3}$

38. (a) 100; (b) $12\frac{1}{2}$; (c) 12; (d) 100; (e) 105; (f) 18; (g) $20\sqrt{3}$

39. (a) 12; (b) 63; (c) 48; (d) $2\frac{1}{2}$; (e) 45

Chapter 10

1. (a) 200; (b) 24.5; (c) 112; (d) 13; (e) 9; (f) $3\frac{1}{3}$; (g) 4.5

2. (a) $12\frac{1}{2}$; (b) 23.47; (c) $7\sqrt{3}$; (d) 18.5; (e) $3\sqrt{2}$

3. (a) 24°; (b) 24°; (c) 156°

4. (a) 40°; (b) 9; (c) 140°

5. (a) 15°; (b) 15°; (c) 24

6. (a) 5°; (b) 72°; (c) 175°

7. (a) regular octagon; (b) regular hexagon; (c) equilateral triangle; (d) regular decagon; (e) square; (f) regular dodecagon (12 sides)

9. (a) 9; (b) 30; (c) $6\sqrt{3}$; (d) 6; (e) $13\sqrt{3}$; (f) 6; (g) $20\sqrt{3}$; (h) 60

10. (a) $18\sqrt{2}$; (b) $7\sqrt{2}$; (c) 40; (d) $8\sqrt{2}$; (e) 3.4; (f) 28; (g) $5\sqrt{2}$; (h) $2\sqrt{2}$

11. (a) $30\sqrt{3}$; (b) 14; (c) 27; (d) 18; (e) $8\sqrt{3}$; (f) $4\sqrt{3}$; (g) $48\sqrt{3}$; (h) 42; (i) 6; (j) 10; (k) $\frac{5}{2}\sqrt{3}$; (l) $3\sqrt{3}$

12. (a) 817; (b) 3078

13. (a) $54\sqrt{3}$; (b) $96\sqrt{3}$; (c) $600\sqrt{3}$

14. (a) 576; (b) 324; (c) 100

15. (a) $36\sqrt{3}$; (b) $27\sqrt{3}$; (c) $\frac{16}{3}\sqrt{3}$; (d) $144\sqrt{3}$; (e) $3\sqrt{3}$; (f) $48\sqrt{3}$

16. (a) 10; (b) 10; (c) $5\sqrt{3}$

17. (a) 18; (b) $9\sqrt{3}$; (c) $6\sqrt{3}$; (d) $3\sqrt{3}$

18. (a) 1:8; (b) 4:9; (c) 9:10; (d) 8:11; (e) 3:1; (f) 2:5; (g) $4\sqrt{2}$:3; (h) 5:2

19. (a) 5:2; (b) 1:5; (c) 1:3; (d) 3:4; (e) 5:1

20. (a) 5:1; (b) 4:7; (c) x:2; (d) $\sqrt{2}$:1; (e) $\sqrt{3}$:y; (f) $\sqrt{x}$:$3\sqrt{2}$ or $\sqrt{2x}$:6

21. (a) 1:4; (b) 1:25; (c) 36:1; (d) 9:100; (e) 49:25

22. (a) 12π; (b) 14π; (c) 10π; (d) $2\pi\sqrt{3}$

23. (a) 9π; (b) 25π; (c) 64π; (d) $\frac{1}{4}\pi$, (e) 18π

24. (a) $C = 10\pi, A = 25\pi$; (b) $r = 8, A = 64\pi$; (c) $r = 4, C = 8\pi$

25. (a) 12π; (b) 4π; (c) 7π; (d) 26π; (e) $8\sqrt{3}\pi$ (f) 3π

26. (a) 98π; (b) 18π; (c) 32π; (d) 25π; (e) 72π; (f) 100π

27. (a) (1) $C = 8\pi, A = 16\pi$; (2) $C = 4\sqrt{3}\pi, A = 12\pi$

(b) (1) $C = 16\pi, A = 64\pi$; (2) $C = 8\sqrt{3}\pi, A = 48\pi$

(c) (1) $C = 12\pi, A = 36\pi$; (2) $C = 6\pi, A = 9\pi$

(d) (1) $C = 16\pi, A = 64\pi$; (2) $C = 8\pi, A = 16\pi$

(e) (1) $C = 20\sqrt{2}\pi, A = 200\pi$; (2) $C = 20\pi, A = 100\pi$

(f) (1) $C = 6\sqrt{2}\pi, A = 18\pi$; (2) $C = 6\pi, A = 9\pi$

28. (a) 10 ft; (b) 17 ft; (c) $3\sqrt{5}$ ft or 6.7 ft

29. (a) 2π; (b) 10π; (c) 8; (d) 11π; (e) 6π; (f) 10π

30. (a) 3π; (b) $12\frac{1}{2}$; (c) 5π; (d) 2π; (e) π; (f) 4π

31. (a) 6π; (b) $\pi/6$; (c) $25\pi/6$; (d) 25π; (e) $4\frac{1}{2}$; (f) 13; (g) 24π; (h) $8\pi/3$

32. (a) 6π; (b) 20; (c) 3π; (d) 16π

33. (a) $120°$; (b) $240°$; (c) $36°$; (d) $180°$; (e) $135°$; (f) $(180/\pi)°$ or $57.3°$ to nearest tenth

34. (a) $72°$; (b) $270°$; (c) $40°$; (d) $150°$; (e) $320°$

35. (a) $90°$; (b) $270°$; (c) $45°$; (d) $36°$

36. (a) 12; (b) 9; (c) 10; (d) 6; (e) 5; (f) $3\sqrt{2}$

37. (a) 4; (b) 10; (c) 10 cm; (d) 9

38. (a) $6\pi - 9\sqrt{3}$; (b) $24\pi - 36\sqrt{3}$; (c) $\frac{3}{2}\pi - \frac{9}{4}\sqrt{3}$; (d) $\frac{\pi r^2}{6} - \frac{r^2\sqrt{3}}{4}$; (e) $\frac{2\pi r^2}{3} - r^2\sqrt{3}$

39. (a) $4\pi - 8$; (b) $150\pi - 225\sqrt{3}$; (c) $24\pi - 36\sqrt{3}$; (d) $16\pi - 32$; (e) $50\pi - 100$

40. (a) $\frac{64\pi}{3} - 16\sqrt{3}$; (b) $24\pi - 16\sqrt{2}$; (c) $\frac{80\pi}{3} - 16$

41. (a) $\frac{16\pi}{3} - 4\sqrt{3}$; (b) $\frac{8\pi}{3} - 4\sqrt{3}$; (c) $4\pi - 8$

42. (a) $12\pi - 9\sqrt{3}$; (b) $\frac{3}{2}\pi - \frac{9}{4}\sqrt{3}$; (c) $9\pi - 18$

43. (a) $200 - 25\pi/2$; (b) $48 + 26\pi$; (c) $25\sqrt{3} - 25\pi/2$; (d) $100\pi - 96$; (e) $128 - 32\pi$; (f) $300\pi + 400$; (g) 39π; (h) 100

44. (a) 36π; (b) $36\sqrt{3} + 18\pi$, (c) 14π

Chapter 11

1. The description of each locus is left for the reader.

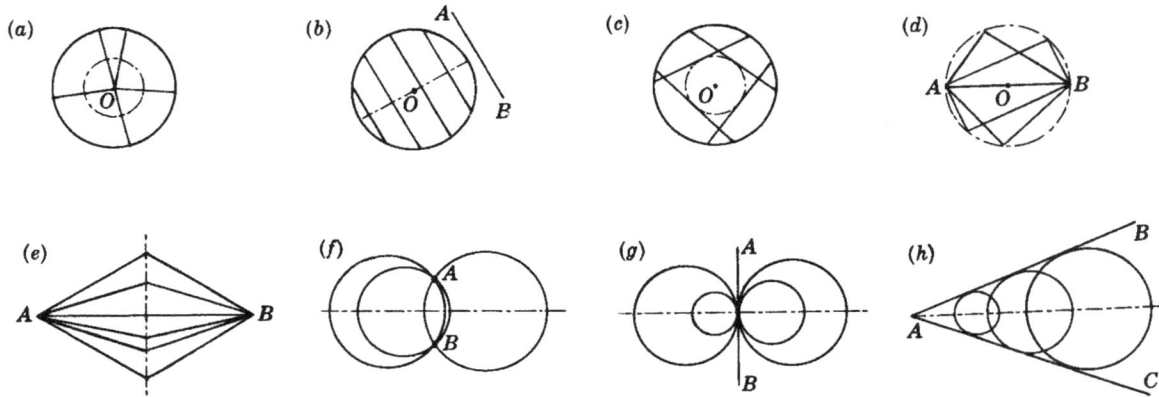

2. The diagrams are left for the reader.
    (a) The line parallel to the banks and midway between them
    (b) The perpendicular bisector of the segment joining the two floats
    (c) The bisector of the angle between the roads
    (d) The pair of bisectors of the angles between the roads

3. The diagrams are left for the reader.
    (a) A circle having the sun as its center and the fixed distance as its radius
    (b) A circle concentric to the coast, outside it, and at the fixed distance from it
    (c) A pair of parallel lines on either side of the row and 20 ft from it
    (d) A circle having the center of the clock as its center and the length of the clock hand as its radius.

4. (a) $\overline{EF}$; (b) $\overline{GF}$; (c) $\overline{EF}$; (d) $\overline{GH}$; (e) $\overline{EF}$; (f) $\overline{GH}$; (g) $\overline{AB}$; (h) a 90° arc from A to G with B as center

5. (a) $\overline{AC}$; (b) $\overline{BD}$; (c) $\overline{BD}$; (d) $\overline{AC}$; (e) E

6. In each case, the letter refers to the circumference of the circle. (a) A; (b) C; (c) B; (d) A; (e) C; (f) A and C; (g) B

7. The description of each locus is left for the reader.

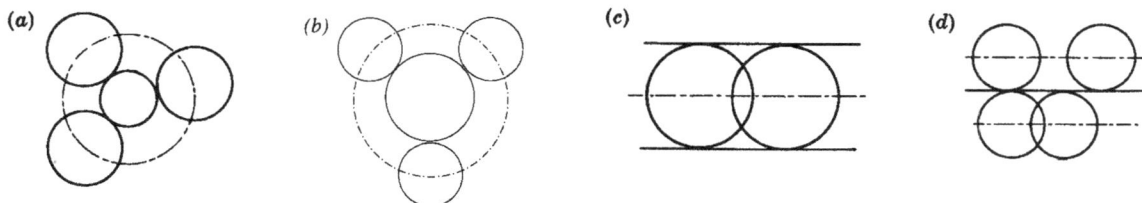

8. (a) $\overline{EF}$; (b) $\overline{GH}$; (c) line parallel to $\overline{AD}$ and $\overline{EF}$ midway between them; (d) $\overline{EF}$; (e) $\overline{BC}$; (f) $\overline{GH}$

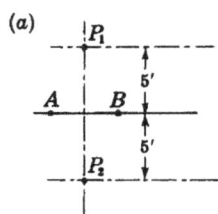

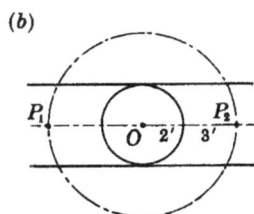

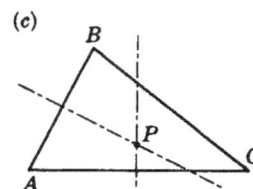

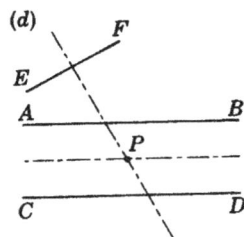

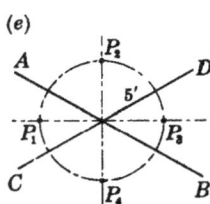

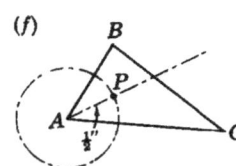

9. The explanation is left for the reader.

10. (a) The intersection of two of the angle bisectors
    (b) The intersection of two of the $\perp$ bisectors of the sides
    (c) The intersection of the $\perp$ bisector of $\overline{AB}$ and the bisector of $\angle B$
    (d) The intersection of the bisector of $\angle C$ and a circle with C as center and 5 as radius
    (e) The intersections of two circles, one with B as center and 5 as radius and the other with A as center and 10 as radius

11. (a) 1; (b) 1; (c) 4; (d) 2; (e) 2; (f) 1

Chapter 12

1. $A(3,0)$; $B(4,3)$; $C(3,4)$; $D(0,2)$; $E(-2,4)$; $F(-4,2)$; $G(-1,0)$; $H(-3\frac{1}{2},-2)$; $I(-2,-3)$; $J(0,-4)$; $K(1\frac{1}{2},-2\frac{1}{2})$; $L(4,-2\frac{1}{2})$

3. Perimeter of square formed is 20 units; its area is 25 square units.

4. Area of parallelogram $= 30$ square units.

    Area of $\triangle BCD = 15$ square units.

5. (a) $(4,3)$; (b) $(2\frac{1}{2},3\frac{1}{2})$; (c) $(-4,6)$; (d) $(7,-5)$; (e) $(-10,-2\frac{1}{2})$; (f) $(0,10)$; (g) $(4,-1)$;

    (h) $(-5,-2\frac{1}{2})$; (i) $(5,5)$; (j) $(-3,-10)$; (k) $(5,6)$; (l) $(0,-3)$

6. (a) $(4,0),(0,3),(4,3)$; (b) $(-3,0),(0,5),(-3,5)$; (c) $(6,-2),(0,-2),(6,0)$;

    (d) $(4,6),(4,9),(3,8)$; (e) $(2,-3),(-2,2),(0,5)$; (f) $(-\frac{1}{2},0),(\frac{1}{2},\frac{1}{2}),(0,-1\frac{1}{2})$

7. (a) $(0,2),(1,7),(4,5),(3,0)$; (b) $(-2,7),(3,6),(6,1),(1,2)$;

    (c) $(-1,2),(3,3),(3,-4),(-1,-5)$; (d) $(-2,1),(4,2\frac{1}{2}),(7,-4),(1,-7\frac{1}{2})$

8. (a) $(2,6),(4,3)$; (b) $(1,0),(0,-2\frac{1}{2})$; (c) common midpoint, $(2,2)$

9. (a) $(-2,3)$; (b) $(-3,-6)$; (c) $(-1\frac{1}{2},-2)$; (d) (a,b); (e) $(2a,3b)$; (f) $(a,b+c)$

10. (a) $M(4,8)$; (b) $A(-1,0)$; (c) $B(6,-3)$

11. (a) $B(2,3\frac{1}{2})$; (b) $D(3,3)$; (c) $A(-2,9)$

12. (a) Prove that *ABCD* is a parallelogram (since opposite sides are congruent) and has a rt. $\angle$.
(b) The point $(3, 2\frac{1}{2})$ is the midpoint of each diagonal.
(c) Yes, since the midpoint of each diagonal is their common point.

13. (a) $D(3, 2)$, $(1\frac{1}{2}, 1)$; (b) $E(0, 2)$, $(3, 1)$; (c) no, since the midpoint of each median is not a common point

14. (a) 5; (b) 6; (c) 10; (d) 12; (e) 5.4; (f) 7.5; (g) 9; (h) *a*

15. (a) 3, 3, 6; (b) 4, 14, 18; (c) 1, 3, 4; (d) $a, 2a, 3a$

16. (a) 13; (b) 5; (c) 15; (d) 5; (e) 10; (f) 15; (g) $3\sqrt{2}$; (h) $5\sqrt{2}$; (i) $\sqrt{10}$; (j) $2\sqrt{5}$; (k) 4; (l) $a\sqrt{2}$

18. (a) $\triangle ABC$; (b) $\triangle DEF$; (c) $\triangle GHJ$; (d) $\triangle KLM$ is *not* a rt. $\triangle$

19. (a) $5\sqrt{2}$; (b) $\sqrt{5}$; (c) $\sqrt{65}$

21. (a) 10; (b) 5; (c) $5\sqrt{2}$; (d) 13; (e) 4; (f) 3

22. (a) on; (b) on; (c) outside; (d) on; (e) inside; (f) inside; (g) on

23. (a) $\frac{9}{5}$; (b) $\frac{5}{9}$; (c) $\frac{5}{2}$; (d) 3; (e) 2; (f) 1; (g) 5; (h) –2; (i) –3; (j) $\frac{3}{2}$; (k) –1; (l) 1

24. (a) 3; (b) 4; (c) $-\frac{1}{2}$; (d) –7; (e) 5; (f) 0; (g) 3; (h) 5; (i) –4; (j) $-\frac{2}{3}$; (k) –1; (l) –2; (m) 5; (n) 6;
(o) –4; (p) –8

25. (a) 72°; (b) 18°; (c) 68°; (d) 22°; (e) 45°; (f) 0°

26. (a) 0.0875; (b) 0.3057; (c) 0.3640; (d) 0.7002; (e) 1; (f) 3.2709; (g) 11.430

27. (a) 0°; (b) 25°; (c) 45°; (d) 55°; (e) 7°; (f) 27°; (g) 37°; (h) 53°; (i) 66°

28. (a) $\overline{BC}, \overline{BD}, \overline{AD}, \overline{AE}$; (b) $\overline{BF}, \overline{CF}, \overline{DE}$; (c) $\overline{AF}, \overline{CD}$; (d) $\overline{AB}, \overline{EF}$

29. (a) 0; (b) no slope; (c) 5; (d) –5; (e) 0.5; (f) –0.0005

30. (a) 0; (b) no slope; (c) no slope; (d) 0; (e) 5; (f) –1; (g) 2

31. (a) $\frac{3}{2}$; (b) $\frac{7}{3}$; (c) –1; (d) 6

32. (a) –2; (b) –1; (c) $-\frac{1}{3}$; (d) $-\frac{2}{5}$; (e) –10; (f) 1; (g) $\frac{5}{4}$; (h) $\frac{4}{13}$; (i) no slope; (j) 0

33. (a) 0; (b) –2; (c) 3; (d) –1

34. (a) $-\frac{3}{2}$; (b) $\frac{2}{3}$; (c) $-\frac{3}{2}$

35. (a) $-\frac{1}{2}$; (b) 1; (c) 2; (d) –1

36. (a) $-\frac{3}{2}$; (b) $\frac{1}{4}$; (c) $\frac{5}{6}$

37. (a) and (b)

38. (a) 19; (b) 9; (c) 2

39. (a) $x = -5$; (b) $y = 3\frac{1}{2}$; (c) $y = 3$ and $y = -3$; (d) $y = -5$; (e) $x = 4$ and $x = -4$; (f) $x = 5$ and $x = -1$;
(g) $y = 4$; (h) $x = 1$; (i) $x = 9$

40. (a) $x = 6$; (b) $y = 5$; (c) $x = 6$; (d) $x = 5$; (e) $x = 6$; (f) $y = 3$

41. (a) $x = y$; (b) $y = x + 5$; (c) $x = y - 4$; (d) $y - x = 10$; (e) $x + y = 12$; (f) $x - y = 2$ or $y - x = 2$;
(g) $x = y$ and $x = -y$; (h) $x + y = 5$

42. (a) line having *y*-intercept 5, slope 2; (b) line passing through (2, 3), slope 4;
(c) line passing through (–2, –3), slope $\frac{5}{4}$; (d) line passing through origin, slope $\frac{1}{2}$;
(e) line having *y*-intercept 7, slope –1; (f) line passing through origin, slope $\frac{1}{3}$

43. (a) $y = 4x$; (b) $y = -2x$; (c) $y = \frac{3}{2}x$ or $2y = 3x$; (d) $y = -\frac{2}{5}x$ or $5y = -2x$; (e) $y = 0$

44. (a) $y = 4x + 5$; (b) $y = -3x + 2$; (c) $y = \frac{1}{3}x - 1$ or $3y = x - 3$; (d) $y = 3x + 8$; (e) $y = -4x - 3$;
(f) $y = 2x$ or $y - 2x = 0$

45. (a) $\dfrac{y-4}{x-1} = 2$ or $y = 2x + 2$; (b) $\dfrac{y-3}{x+2} = 2$ or $y = 2x + 7$; (c) $\dfrac{y}{x+4} = 2$ or $y = 2x + 8$;

(d) $\dfrac{y+7}{x} = 2$ or $y = 2x - 7$

46. (a) $y = 4x$; (b) $y = \frac{1}{2}x + 3$; (c) $\dfrac{y-2}{x-1} = 3$; (d) $\dfrac{y+2}{x+1} = \frac{1}{3}$; (e) $y = 2x$

47. (a) circle with center at origin and radius 7; (b) $x^2 + y^2 = 16$; (c) $x^2 + y^2 = 64$ and $x^2 + y^2 = 4$

48. (a) $x^2 + y^2 = 25$; (b) $x^2 + y^2 = 81$; (c) $x^2 + y^2 = 4$ or $x^2 + y^2 = 144$

49. (a) 3; (b) $\frac{4}{3}$; (c) 2; (d) $\sqrt{3}$

50. (a) $x^2 + y^2 = 16$; (b) $x^2 + y^2 = 121$; (c) $x^2 + y^2 = \frac{4}{9}$ or $9x^2 + 9y^2 = 4$; (d) $x^2 + y^2 = \frac{9}{4}$ or $4x^2 + 4y^2 = 9$;

(e) $x^2 + y^2 = 5$; (f) $x^2 + y^2 = \frac{3}{4}$ or $4x^2 + 4y^2 = 3$

51. (a) 10; (b) 10; (c) 20; (d) 20; (e) 7; (f) 25

52. (a) 16; (b) 12; (c) 20; (d) 24

53. (a) 10; (b) 12; (c) 22

54. (a) 5; (b) 13; (c) $7\frac{1}{2}$

55. (a) 6; (b) 10; (c) 1.2

56. (a) 15; (b) 49; (c) 53

57. (a) 30; (b) 49; (c) 88; (d) 24; (e) 16; (f) 18

Chapter 13

1. (a) $<$; (b) $>$; (c) $>$; (d) $>$; (e) $>$; (f) $<$

2. (a) $>$; (b) $>$; (c) $<$; (d) $>$

3. (a) $>$; (b) $<$; (c) $<$; (d) $>$

4. (a) more; (b) less

5. (a) $>$; (b) $>$; (c) $<$; (d) $>$; (e) $<$; (f) $<$

6. (c), (d), and (e)

7. (a) 5 to 7; (b) 6 to 10; (c) 4 to 10; (d) 3 to 9; (e) 2 to 8; (f) 1 to 13

8. (a) $\angle B, \angle A, \angle C$; (b) $\overline{DF}, \overline{EF}, \overline{DE}$; (c) $\angle 3, \angle 2, \angle 1$

9. (a) $m\angle BAC > m\angle ACD$; (b) $AB > BC$

10. (a) $\overline{BC}, \overline{AB},\ \overline{AC}$; (b) $\angle BOC, \angle AOB, \angle AOC$; (c) $\overline{AD}, \overline{AB} \cong \overline{CD}, \overline{BC}$; (d) $\overline{OG}, \overline{OH}, \overline{OJ}$

www.ingramcontent.com/pod-product-compliance
Lightning Source LLC
Chambersburg PA
CBHW081239220326
41597CB00023BA/4131